Y0-AAD-740

ELECTRONIC TESTING AND TROUBLESHOOTING

ELECTRONIC TESTING and TROUBLESHOOTING

George Loveday

Lecturer in Electronic Engineering
Havering Technical College

Adapted by

Richard S. Sandige, Ph. D.

R & D Engineer
Desktop Computer Division
Hewlett Packard Co.

John Wiley & Sons

New York • Chichester • Brisbane • Toronto • Singapore

Library of Congress Cataloging in Publication Data:

Loveday, George
  Electronic testing and troubleshooting.

  Originally published as: Electronic testing and
fault diagnosis. 1980.
  Includes index.
  1. Electronic apparatus and appliances–Maintenance
and repair. 2. Electronic apparatus and appliances–
Testing. I. Sandige, Richard S. II. Title.
TK7870.2.L68 1982   621.381'028'7   81-12948
ISBN 0-471-08718-1   AACR2

Printed in the United States of America

10 9 8 7 6 5 4 3 2 1

Preface to the American Edition

Electronic Testing and Troubleshooting is a practical textbook for electronic technicians and engineers whether in school or in industry. It is intended primarily for testing and troubleshooting courses in both 2- and 4-year technical programs. The book may also be used in conjunction with a basic course in electronic circuits, since it provides many practical aspects of the subject that are often not covered in other texts. Practicing technicians and engineers may use the book as a training or reference manual, since it includes some essential information on modern components, in particular on digital and linear integrated circuits.

Thirty-three laboratory exercises are provided to allow the reader to build, test, and evaluate practical circuits under certain failure modes and thus gain valuable experience. Answers to all of the laboratory exercises are provided at the end of the book.

Topics included in the book are specifications, reliability, electronic components, digital logic circuits, circuits using linear integrated circuits, power supply and power control circuits, and system maintenance and troubleshooting. The term *troubleshooting* in this text means either the process of debugging new designs or the process of fixing old designs as the situation requires.

I hope that this text will provide valuable knowledge and experience to students and practitioners alike.

Richard S. Sandige

Contents

CHAPTER 1

Specifications

1.1 INTRODUCTION

Before it can be stated that an instrument is capable of making some measurement, or that a component is suitable for a particular job, details about the important characteristics of the instrument or component must be known. What is required is called a specification, a list of data that tells the user what performance to expect from the instrument or component and under what environmental conditions the performance is guaranteed. A full definition of the word SPECIFICATION is: "a detailed description of the required characteristics of a device, equipment, system, product, or process."

As an example, suppose that a signal generator is wanted for general laboratory use. The characteristics required might be as follows:

(a) Frequency range: 0.1 Hz to 1 MHz continuously variable.

(b) Output waveform: sine or square.

(c) Sine wave distortion: less than 0.5% total harmonic distortion.

(d) Accuracy: better than ±2.5% of dial reading.

(e) Frequency stability: less than ±0.2% per 24 hours at 23°C.

(f) Frequency temperature coefficient: 0.01% per °C.

(g) Maximum output voltage: 10 V peak to peak.

(h) Output impedance: 60 Ω.

(i) Output square wave rise and fall time: better than 100 nanoseconds.

(j) Power required: 120 V, 60 Hz.

These are the basic figures that should be contained in the PERFORMANCE SPECIFICATION section of manufacturer's sales leaflets and instruction manuals.

For a particular application, the user has to decide on the performance required and then has to check that the manufacturer's specification can meet those requirements. Since most manufacturers wish to enjoy a good reputation, they will not make unjustified claims for their products, and they will try hard to ensure that all instruments or components from their production meet their specification. To do this they must test each instrument to a standard that is at least as good as or, in most cases, much better than, their published performance figures. Therefore, inside the factory a TEST SPECIFICATION is issued and used to guide the test department in checking all aspects of the instrument's or component's performance.

The performance specification can be used by the potential purchaser of the equipment to compare and contrast the various instruments or systems offered for sale by several manufacturers. In this way the user can select the best available on both technical and economic considerations.

One of the main purposes of testing, therefore, is to ensure that manufactured items and instruments conform to an agreed specification.

1.2 HOW SPECIFICATIONS ARISE

Specifications are an essential part of all manufacturing processes. Within the electronics industry, and the Original Equipment Manufacturers (O.E.M.s) in particular, there are various ways in which a

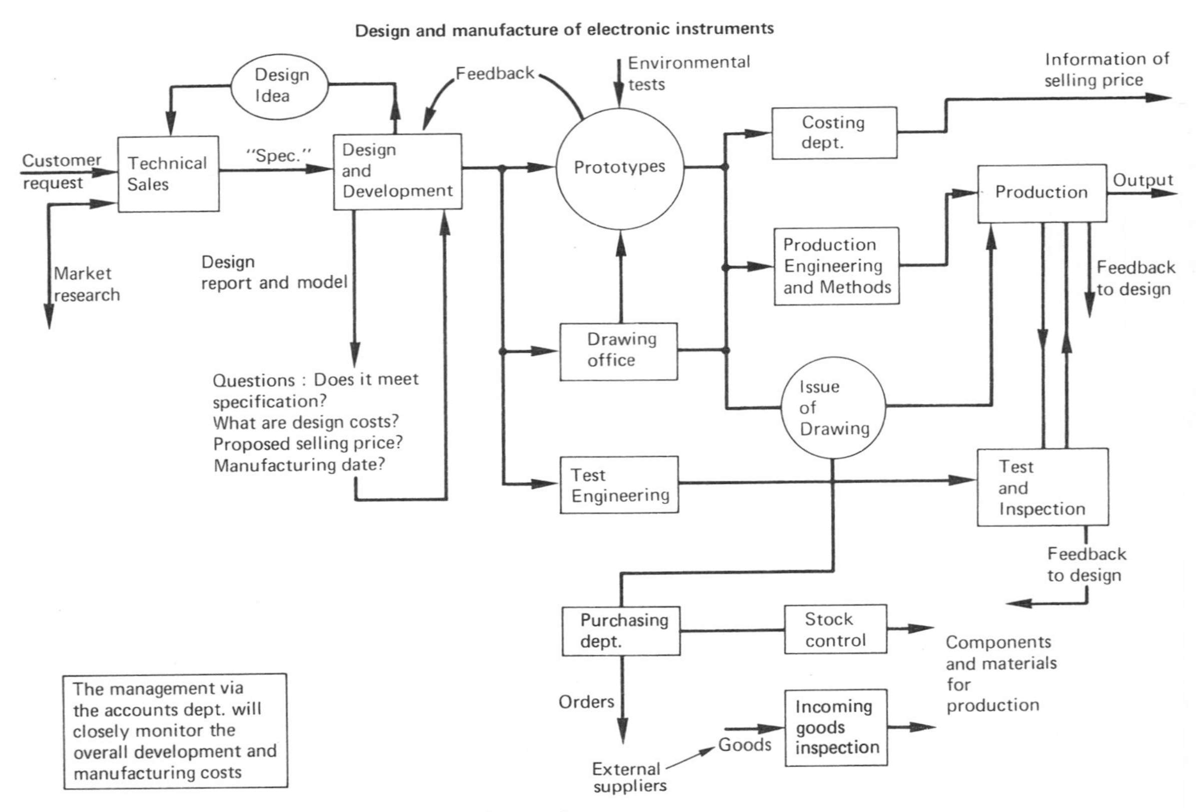

FIGURE 1.1 Typical structure of an electronic instrument company.

performance specification can be written and agreed. A block diagram showing the possible structure of an electronic instrument company is given in Figure 1.1. This can be used to illustrate how the specifications arise.

The requirement for a particular instrument may come about by:

A customer request to the technical sales department.

Market research initiated by sales/marketing department.

An innovating design idea from the design department.

If the prospects are promising, in other words a good potential market exists, the management will then decide to call for a design appraisal, allocating funds to the design department. This department will then investigate the feasibility of producing an instrument to meet the proposed specification. At the end of the appraisal a report and possibly a prototype model will be produced. The report must answer the basic questions:

(a) Can the proposed specification be met economically?

(b) What will the overall design costs be?

(c) How much will the total manufacturing costs be (and therefore the likely selling price)?

(d) How soon can production start?

Assuming that the answers to these questions are satisfactory, the management will then give the go ahead for a full design. The design department can proceed; finalizing circuits and layout; producing engineering drawings; issuing parts lists; and evaluating the prototype models produced. Complete environmental tests will be made on the prototype, checking the performance in the expected environmental conditions of temperature, humidity, and pressure. Prototypes may also be vibration-tested and given a mechanical shock test. Results from these tests will indicate any modifications that should be made to improve design and reliability, well before the instrument goes into production.

It is also the job of the design engineers to produce the TEST SPECIFICATION. This will be issued, well before production is scheduled to begin, to the test engineering and test departments. In some factories the last two departments are amalgamated. Using this specification, a TEST INSTRUCTION document will be produced for use by test engineers. Alternatively the test engineer will be issued with a copy of the full test specification with any additional necessary notes.

It is worth noting here the importance of feedback loops, whereby information from a particular department, say test engineering or production, can be rapidly fed back to the design department, so that modifications to improve performance and/or reliability can be made earlier rather than later. Making changes to equipment while it is in production may sometimes be necessary and unavoidable but such action is always costly. Even higher costs are incurred if postdesign of part of the equipment becomes necessary.

1.3 LIMITS

In the preparation of any specification, the writer always has to bear in mind the maximum and minimum values of any parameter or characteristic. Suppose, for example, that an oscillator frequency in an electronic circuit has to be measured and recorded. The required frequency is 400 Hz and the limits are plus and minus 10% of this value. There are several ways in which these limits can be written:

400 Hz ± 10%.

400 Hz ± 40 Hz.

Minimum 360 Hz; typical 400 Hz; maximum 440 Hz.

Not less than 360 Hz and not greater than 440 Hz.

Between 360 Hz and 440 Hz.

Less than 440 Hz and greater than 360 Hz.

Note that all of these are stating the same limits to the frequency of 400 Hz but some are less ambiguous and easier to follow than others. A limit written as a percentage requires that the person making the measurement converts the percentage to actual figures, in this case to ±40 Hz. This is a relatively easy matter when the percentage is 10, but suppose the required limits were +12% and - 2.5%. Errors caused by miscalculation could then result from writing the measurement as

400 Hz +12% - 2.5%

An improvement would be

400 Hz +48 Hz - 10 Hz

or better still

not less than 390 Hz and not greater than 448 Hz.

This last method is the clearest indication of the new limits. The desire to shorten this statement by using mathematical symbols should be resisted, for example, by using $\ngtr$ instead of "not greater than" because using such symbols may cause confusion in the minds of the reader of the specification.

Within any particular organization there usually exists a standard form for writing specifications and limits. It is designed to minimize errors arising from misinterpretation. An example of the use of limits is given later in this chapter.

1.4 STANDARD SPECIFICATIONS

It is naturally important that customers and manufacturers use the same terminology with regard to instrument and component performance. It is also important that they can refer to some external standard of measurement and quality. In the United States the American National Standards Institute (ANSI) serves this purpose. The following ANSI standards apply to electrical and electronic diagrams: (1) Graphic Symbols for Electrical and Electronic Diagrams (Including Reference Designation Class Designation Letters), which is referred to as document number ANSI Y32.2-1975, or IEEE (Institute of Electrical and Electronics Engineers) Std 315-1975; and (2) Graphic Symbols for Logic

Diagrams (two-state devices), which is referred to as document number ANSI Y32.14-1973, or IEEE Std 91-1973. Both of these standards have been adopted by the U.S. Department of Defense. Both also include approved International Electrotechnical Commission (IEC) recommendations. The first document (Graphical Symbols for Electrical and Electronic Diagrams) has also been approved by the Canadian Standards Association (CSA) as document number CSA Z99-1975.

An American National Standard is only intended as a guide to aid manufacturers, consumers, and the general public. Information concerning American National Standards may be obtained by calling or writing the American National Standards Institute, 1430 Broadway, New York, N.Y. 10018.

Throughout the following chapters the graphic symbols and designation letters that are used generally are those approved by ANSI. The Electronic Industries Association (EIA), Engineering Department, 2001 Eye Street, N.W., Washington, D.C. 20006 also publishes a document entitled JEDEC Recommendations for Letter Symbols, Abbreviations, Terms, and Definitions for Semiconductor Device Data Sheets and Specifications, publication no. 77B. This publication provides information to facilitate the use of Joint Electron Device Engineering Council (JEDEC) registration data formats. Semiconductor devices with a prefix of 1N, 2N, 3N, or 4N are devices with JEDEC registration. In general, devices with JEDEC registration will be used throughout the text, and EIA publication no. 77B will be followed when referring to semiconductor letter symbols.

1.5 COMPONENT SPECIFICATIONS

Sound engineering practice requires the designer to use the most effective and lowest-cost component for a given application. The temptation to overspecify, for example to use a 10 A rated rectifier when a 3 A rating is required, should be avoided, since this may not necessarily improve reliability or quality but will result in increased costs.

The importance of adequate component data to the user is self-evident, since it enables comparison of performance with other similar devices, it shows how the device can be used, and gives details of device characteristics under defined conditions. Before looking at actual specifications, we should first consider the various components used in the electronic industry. These can be grouped under the general headings of

1. **Mechanical parts**: metal chassis and brackets; wires, printed circuit boards; connectors; plugs and sockets.
2. **Passive devices**: fixed and variable resistors; fixed and variable capacitors; inductors.
3. **Active devices**: transistors; diodes; thyristors; field effect transistors; and integrated circuits.

The construction of some of these types is dealt with in Chapter 3. What concerns us here is which parameters and characteristics are important to the user, and within what limits do these characteristics need to be specified? The designer has to use the specification in order to choose the most suitable device for the particular application, dependent also on price, availability, and any standardization requirements. The latter point is worth considering since if, in a large organization, engineers were given a completely free hand in the choice of components, the result would probably be a proliferation within the factory of several different types of resistors, capacitors, and semiconductors. This would increase costs in purchasing, storage, and production. Apart from a few special items, most manufacturing organizations insist on common types of components for general applications: in other words, one range of fixed resistors, one of capacitors, and a selected range of general purpose semiconductor devices.

Within each class of component there are usually several different types. For example, in the class of fixed resistors the designer can use any of the following:

Carbon composition

Carbon film

Metal oxide

Metal film

Metal glaze

Wire-wound

whereas for fixed capacitors some of the available types are:

Silver mica } relatively low capacitance values
Ceramic }

Paper and metallized paper

Polyester

Polystyrene

Tantalum foil } high capacitance values
Tantalum solid }
Aluminum electrolytic }

From such an array of types, the designer has to select the one most suitable for the immediate application, dependent on the constraints of

(a) Price.

(b) Availability.

(c) Any standardization within his organization.

The data sheet must be written in a clear concise manner. A typical format for a COMPONENT SPECIFICATION could be divided into the following sections:

1. Device, type and family.
2. Brief description of the device and intended application to assist in this choice.
3. Outline drawing showing mechanical dimensions and connections.
4. Brief details of the most important electrical characteristics and the absolute maximum rated values of voltage, current, and power.
5. Complete electrical data including figures, necessary graphs, and characteristic curves.
6. Details of any manufacturing inspection methods.
7. Figures of reliability or failure rate.

single turn sealed wirewound potentiometer

A small general purpose sealed wirewound panel control potentiometer, with humidity proof construction. Phenolic body moulding, beryllium copper contact, nickel plated brass spindle and bush.

CLR 1232

½ W at 70°C Tolerance ± 10%. Range 100, 10K, 100, 500, 1K, 2.5K, 5K, 10K.

TECHNICAL DATA

Linearity	1% typical, 2% max.
Terminal resistance	0.2 Ω max. or 0.01% of nominal resistance whichever is the greater.
Limiting element voltage	250V d.c. or a.c. r.m.s.
Isolation voltage	360V a.c. peak.
Insulation resistance	1000M Ω min. at 100V d.c.
Rotational noise	100 E.N.R. max.
Angle of rotation:-	mechanical 290° ± 10°
	electrical 265° ± 15°

Spindle length code	
CLR 1232/268	5/8″ long Slotted
CLR 1232/11S	¾″ long Slotted

FIGURE 1.2 Manufacturer's data sheet. *(Courtesy Colvern Ltd.)*

Two examples of manufacturer's data sheets are shown in Figures 1.2 and 1.3. Note that the purpose of the specification is to clearly show the intended use, the absolute maximum ratings, and the limits of other important electrical data. The two examples have been chosen purely to illustrate the usual format. To fully appreciate component specifications, it is always best to consult manufacturers' data books. Within these there is always a wealth of useful information.

A question that naturally arises is one of how to interpret the data so that the correct device is chosen: To do this it is important to understand the way in which the device works and to have a good idea of the most important characteristics for the application. For example, one of the most commonly used components is a fixed

COMPONENTS

LIGHT EMITTING DIODE FOR HIGH-RELIABILITY APPLICATIONS

5082-4420

FEATURES

- DESIGNED FOR HIGH-RELIABILITY ENVIRONMENTS
- HERMETICALLY SEALED
- LONG LIFE
- HIGH BRIGHTNESS OVER A WIDE VIEWING ANGLE
- IC COMPATIBLE/LOW POWER CONSUMPTION

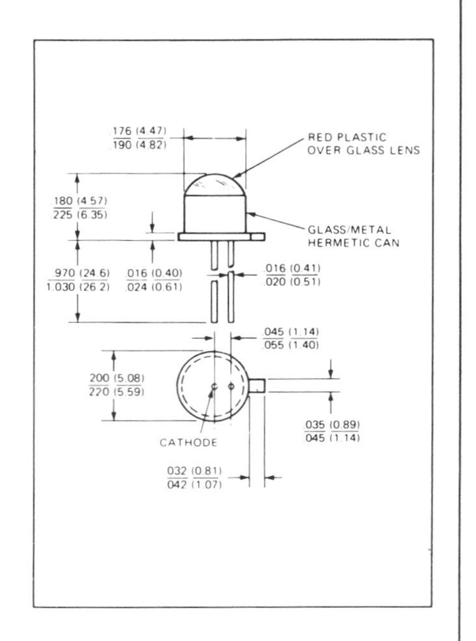

DESCRIPTION

5082-4420 – High performance Light Emitting Diode designed for high-reliability applications.

MAXIMUM RATINGS (25°C)

DC Power Dissipation . . . . 85 mW

DC Forward Current . . . . 50 mA

Peak Forward Current . . . . 1 Amp
(1 μsec pulse width, 300 pps)

Isolation Voltage (between lead and case) . . . . 500 V

Operating and Storage
Temperature Range . . . . −55°C to +100°C

Lead Soldering Temperature . . . . 230°C for 7 sec

ELECTRICAL CHARACTERISTICS (25°C)

Symbol	Parameters	5082-4420			Units	Test Conditions
		Min.	Typ.	Max.		
I	Luminous Intensity	500		3000	μcd	I_F = 20 mA
λ_{pk}	Wavelength		655		nm	Measurement at Peak
τ_s	Speed of Response		10		ns	
c	Capacitance		200	300	pF	V = 0 f = 1MHz
V_F	Forward Voltage		1.6	2.0	V	I_F = 20 mA
BV_R	Reverse Breakdown Voltage	4	5		V	I_R = 10 μA

FIGURE 1.3 Manufacturer's data sheet. *(Courtesy Hewlett Packard.)*

resistor. Apart from price, we must consider all the following aspects (which are not presented in order of importance):

1. **Physical dimensions** Length, diameter, lead shape and size.
2. **Resistance range** Minimum and maximum ohmic value.
3. **Selection tolerance** The maximum and minimum selection value of the resistor, i.e., ±2%, ±5%, ±10%, or ±20% (preferred values Chapter 3).
4. **Power rating** The maximum power in watts that can be dissipated, usually quoted at a temperature of 70°C (commercial), 125°C (military).
5. **Temperature coefficient** The change of resistance, with temperature quoted in parts per million (ppm) per °C. Since "coefficient" implies that a linear function occurs, the term temperature characteristic is now preferred.
6. **Voltage coefficient** The change of resistance, usually negative, with applied voltage, quoted in ppm per volt.
7. **Maximum working voltage** The maximum voltage that can be applied across the resistor.
8. **Breakdown voltage** The maximum voltage that can be applied between the resistor body and a touching external conductor, i.e., it is the breakdown voltage of the insulating coating of the resistor.
9. **Insulation resistance** The resistance of the insulating coating.
10. **Load life stability** The change of resistance after a stated operating time at full load and 70°C. The operating time is usually taken as 1000 hours.
11. **Shelf stability** The change of resistance during storage, usually quoted for 1 year.
12. **Working temperature range** The minimum and maximum allowable values of ambient temperature.
13. **Maximum surface temperature** The maximum permissible value of temperature of the resistor body, sometimes referred to as hot spot temperature.
14. **Noise** The electrical noise caused by applied voltage stressing the resistor. Quoted in μV/V.
15. **Humidity classification** The change of resistance following a standard high temperature and humidity cycling test. The change must be within defined limits.
16. **Soldering effect** Any change of resistance following a standard soldering test.

Having seen the various parameters that have to be considered, it is then very useful to compare the various resistor types that are of roughly the same physical size. This is shown in Table 1.1. The figures given are typical in most cases, apart from some maximum values. By studying the table it

TABLE 1.1 Comparison of Common Types of General Purpose Resistors

Resistor Type	Carbon Composition	Carbon Film	Metal Oxide	Metal Glaze	General Purpose Wire-Wound
Range	10 Ω to 22 MΩ	10 Ω to 2 MΩ	10 Ω to 1 MΩ	10 Ω to 1 MΩ[a]	0.25 Ω to 10 kΩ
Selection tolerance	±10%	±5%	±2%	±2%	±5%
Power rating	250 mW	250 mW	500 mW	500 mW	2.5 W
Load stability	10%	2%	1%	0.5%	1%
Max. voltage	150 V	200 V	350 V	250 V	200 V
Insulation resistance	10^9 Ω	10^{10} Ω	10^{10} Ω	10^{10} Ω	10^{10} Ω
Breakdown voltage	500 V	500 V	1 kV	500 V	500 V
Voltage coeff.	2000 ppm/V	100 ppm/V[b]	10 ppm/V	10 ppm/V	1 ppm/V
Ambient temp. range	-40°C to +105°C	-40°C to +125°C	-55°C to +150°C	-55°C to +150°C	-55°C to 185°C
Temperature coeff.	±1200 ppm/°C	-1200 ppm/°C	±250 ppm/°C	±100 ppm/°C	±200 ppm/°C
Noise	1 kΩ 2 μV/V 10 MΩ 6 μV/V	1 μV/V	0.1 μV/V	0.1 μV/V	0.01 μV/V
Soldering effect	2%	0.5%	0.15%	0.15%	0.05%
Shelf life 1 year	5% max.	2% max.	0.1% max.	0.1% max.	0.1% max.
Damp heat 95% RH	15% max.	4% max.	1%	1%	0.1%

[a] Some high values up to 68 M are now available (Mullard VR37 and VR68 range).
[b] 100 ppm = 100×10^{-6} ohms per ohm per volt, i.e., $100 \times 10^{-6} \times 100\% = 0.01\%$.

becomes obvious that resistors for professional applications should be metal oxide or metal glaze (*cermet*), because these have a wide range, good stability, and low temperature coefficient.

For special applications the designer has to look more closely at individual specifications of the particular type of component, but it is clearly an advantage to have short form data as in the table because this assists in narrowing down the search for the right component. For example, suitable transistors, of which there are several thousand types, are more readily found by using short form data. The common parameters for transistors are given in Table 1.2.

Consider how the data for three similar npn transistors can be presented in quick reference form, so that any small differences in performance can be quickly seen (Table 1.3).

The power ratings at 25°C ambient (free air), the maximum average current, and maximum $V_{(BR)EBO}$ are identical. However, for an application requiring the highest collector/base breakdown voltage, the 2N2243 would seem to be the correct

TABLE 1.2 The Common Parameters of Transistors

	Parameter	Given Conditions
h_{FE}	Large-signal common emitter current gain I_C/I_B	I_C, V_{CE}
h_{fe}	Small-signal common emitter current gain	
$V_{(BR)CBO}$	Collector/base breakdown voltage (emitter open circuit)	I_C
$V_{(BR)CEO}$	Collector/emitter breakdown voltage (base open circuit)	I_E
I_{CM}	Peak value of collector current	
$I_{C(av)}$	Maximum average value of collector current	
I_{CBO}	Collector/base leakage current	$V_{CB}(I_{E=0})$
$V_{CE(sat)}$	Common-emitter/collector-emitter saturation voltage	I_C:I_B in the ratio of 10:1
$V_{BE(sat)}$	Common-emitter/base-emitter saturation voltage	I_C:I_B in the ratio of 10:1
f_T	Gain bandwidth product. Frequency at which h_{FE} is unity.	I_C, V_{CE}
P_T	The maximum total power dissipation.	Temp. (ambient) or temp. (case).

TABLE 1.3

Device	Outline	P_T at 25°C	$V_{(BR)CBO}$	$V_{(BR)CEO}$	$V_{(BR)EBO}$	$I_{C(av)}$	h_{FE} at 10 mA	f_T
2N2243	TO–5	800 mW	120 V	80 V	7 V	1 A	30	50 MHz
2N2869	TO–5	800 mW	60 V	40 V	7 V	1 A	30	50 MHz
2N2939	TO–5	800 mW	75 V	60 V	7 V	1 A	240 at 150 mA	150 MHz

choice. The designer would then look up the full data to fully check this component's suitability.

It would not be worthwhile detailing here the possible specifications for all types of components. Excellent data and information are given by most component manufacturers in their data books. What is useful is to consider the steps required in choosing the correct component. These are as follows:

(a) **Define the application**
For example: a preset potentiometer for voltage level adjustment in a regulated power supply.

(b) **List the requirements**
Mechanical dimensions and method of adjustment (single turn).
Mounting method (p.c.b. or panel mounting).

Value (500 Ω).
Tolerance (±20%).
Maximum power dissipated (100 mW).
Temperature characteristic (better than ±250 ppm/°C).
Environmental conditions (temperature range −10°C to +40°C; mechanical shock and vibration levels likely to be encountered; humidity levels).
Reliability (a figure of predicted failure rate of the type of component used).
The figures given in parenthesis are intended to illustrate possible values required in this application.

(c) **Check short form data for suitable types**
Cermet or wire-wound; not carbon because the temperature coefficient of this type would not meet requirements.

(d) **Note other possible limitations**
Price.
Availability (preferably second sourced).
Standardization on a particular make within the organization.
Second or multiple sourced components improve the likelihood of continued availability over a long period.

(e) **Check complete component specification**
The full data must be used so that all aspects of the component's use, including the effect of derating, are considered. *Derating* is a technique of improving reliability by operating devices well below maximum rated values of voltage, current, and power.

(f) **Evaluate**
Obtain samples for inspection and then test them in the circuit.

Although this procedure may seem lengthy, it should ensure that the correct component is chosen and therefore should avoid costly modification in production or reduced operating reliability.

1.6 EQUIPMENT PERFORMANCE SPECIFICATIONS

For a complete electronic instrument or system, the specification must be clearly written so that all the necessary information on important characteristics is shown without other confusing details. Before we look at actual specifications, however, we should consider the various types of electronic equipment in common use. These may be broadly classified into the following categories:

Electronic Measuring Instruments
Analog multimeters
Digital multimeters
Frequency meters, counters, timers
Oscilloscopes
Chart recorders
XY plotters
Noise meters
Spectrum analyzers
Bridges

Signal Generating Instruments
A.F. (1 Hz to 1 MHz) sine/square generators

R.F. generators
Pulse generators
Function generators
Voltage and current calibrators

Power Sources
Low voltage and high voltage regulated d.c. supplies
Constant voltage supplies (linear and switching mode types)
Converters
Inverters
Battery chargers
Regulated a.c. supplies

Communication Equipment
Radio, television and telephone systems
Telemetry
Radar
Ultrasonics (sonar)

Data Processing Instruments
Computers
Printers
Calculators
Minicomputers
Storage systems

Consumer Electronics
Audio equipment
Electronic musical instruments
Tape and cassette recorders
TV games
Sound-to-light units

Control Systems
Machine tool controllers
Automatic process controllers

This list is not complete and there is obviously some overlap between the categories. It is intended to show the diversity of equipment being manufactured. All of them must have a performance specification which usually follows a standard format of:

1. **Description and type number**
   A concise note stating clearly what the instrument is designed to do and its intended applications.
2. **Electrical data**
   (a) Principal characteristics, e.g.,
   Outputs
   Voltage level
   Frequency
   Impedance
   Ranges
   Accuracy
   Distortion
   Temperature characteristics
   (b) **Power requirements**
   A.C. power line or mains voltage 120 V or 240 V a.c. single phase
   Frequency 50 Hz to 60 Hz
   Power 200 VA
3. **Environmental data**
   Working temperature range
   Humidity classification
   Vibration tests
   Figure for MTBF
   } Not usually supplied for general purpose instruments.
   (Mean Time Between Failures)
4. **Mechanical data**
   Dimensions
   Weight

An example of a performance specification is shown in Figure 1.4. This illustrates the usual layout and also shows the important parameters for a low frequency signal generator. The user then checks the specification data against requirements. This

AMF VENNER

TSA 625 Wide Range Oscillator

The TSA 625 is a versatile signal source providing either a low distortion sinusoidal output or a square wave with a fast rise time.

WIDE FREQUENCY RANGE

Output frequencies over the range 10 Hz to 1 MHz are selected by means of five range push buttons and a clearly calibrated dial. The effective scale length for the whole frequency range exceeds 3 feet. Output signal amplitude is continuously variable (in four switched ranges) from 0 to 2·5 V.

SINE OR SQUARE WAVE OUTPUT

The TSA 625 employs a Wien-bridge variable frequency oscillator as signal source; this is followed by an inverter/amplifier when sine wave output is selected, but by a Schmitt trigger circuit when square wave output is selected. The signals of selected waveform and frequency are applied to a complementary emitter follower output stage and hence pass via an attenuator to the output terminals.

RECHARGEABLE POWER SUPPLY

A nickel cadmium rechargeable battery and mains charging unit gives all the advantages of batteries including portability and freedom from output hum, together with the convenience of mains operation when used for long periods as a bench oscillator. A push button is provided for battery test, and the unit may be used on mains while the battery is being recharged.

SPECIFICATION

Frequency Range
10 Hz–1 MHz in 5 decade ranges with calibrated fine control 10–100 Hz, 100 Hz–1 kHz, 1kHz–10 kHz, 10 kHz–100 kHz, 100 kHz–1 MHz.

Calibration Accuracy
±3% of full-scale range selected.

Output Amplitude
2·5 V r.m.s. maximum sine and square wave.

Output Ranges
0–2·5 mV, 0–25 mV, 0–0·25 V, 0–2·5 V.

Sine Wave Distortion
Better than 2% up to 100 kHz.
Typically <0·5% over range 50 Hz–10 kHz.

Square Wave Rise Time
<100 ns (typically 50 ns).

Output Impedance
600 Ω on 2·5 mV, 25 mV and 0·25 V ranges.
Typically 100 Ω at maximum output.

Operating Temperature
0 °C to 45 °C.

Power Input
Rechargeable NiCd power supply with mains charger (230 V 50 Hz).

Dimensions and weight			
	Height	*130 mm*	*5 in.*
	Width	*210 mm*	*8 in.*
	Depth	*130 mm*	*5 in.*
	Weight	*2·3 kg*	*5 lb*

FIGURE 1.4 Performance Specification. *(Courtesy AMF Venner).*

naturally presupposes that the user knows exactly what is required, and what is meant by the various terms. Square Wave Rise Time, for example, means that the time taken for the output to change from 10 to 90% of full square wave amplitude is less than 100 ns (Figure 1.5).

Note also that the Calibration Accuracy in this specification is quoted as ±3% of full-scale range selected. This means that the accuracy of the dial setting for frequency could be rather poor at the low end

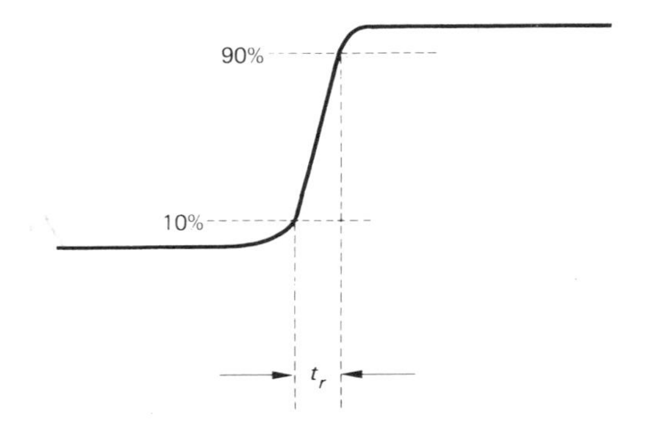

FIGURE 1.5 Square wave rise time.

of each range. For example, at the 30 Hz setting the calibration error could be ±3% of 100 Hz, i.e., ±3 Hz, which is ±10% of the dial setting. This, of course, may not be important for most applications but could be very important for some. It is obviously wise practice to make a list of all the requirements before considering the purchase of any new instrument or system.

Apart from a low frequency signal generator, the other most commonly used electronic instruments are regulated d.c. power supplies, digital multimeters, and medium bandwidth oscilloscopes. It is useful to list the important characteristics and parameters together with some typical values for these instruments. Before looking at the data in Table 1.4, make your own list of the requirements for each instrument and then compare it with the table.

The figures given are intended to be typical of general purpose instruments and not necessarily the best figures that can be obtained. For example, a very high performance power supply might be required to have an output ripple of less than 100 μV pk-pk, a load regulation of 0.005%, and a temperature coefficient of ± 100 ppm per °C. The procedure is to check the data for a number of similar instruments and compare their performance figures.

It is also important to have a clear understanding of what is meant by the various terms and expressions used in a specification. This is especially true when purchasing an unfamiliar new instrument. If there is any doubt over the exact meaning of some part of the specification, always clarify the point with the manufacturer or consult a standard specification or reference book. There is nothing to be gained by ignoring that part of the specification or in pretending to understand. No one can be an expert in every field. The engineer who wrote the specification should be the person who can explain any difficult points and can show how they will affect the application.

1.7 PREPARATION OF A TEST SPECIFICATION

A Test Specification is the information required by the test, service, and installation engineers for them to be able to check that the instrument or system meets the required figures of performance. The Test Specification must, then, be a comprehensive document that covers all aspects of the instrument's characteristics: those that must be checked, adjusted, measured, and recorded. The instructions need to be explicit and unambiguous and must be accompanied by limits for all the measured values.

Most large organizations have their own standard sheets on which their Test Specifications are written. These, showing a logical sequence of tests and adjustments, take the following form:

(a) Title, Instrument Type number. Specification serial number. Date and issue.

(b) List of test equipment necessary to perform tests.

(c) Continuity, insulation, and resistance checks (with power off).

(d) Voltage and signal level adjustments, measurements, and recordings on individual subassemblies.

TABLE 1.4 Typical Requirements for Common Test Instruments

(A) Regulated D.C. Power Supply	Bench Type	(B) Digital Multimeter	3½ Digits	(C) General Purpose Oscilloscope	(Single Beam)
Parameter or Function	**Typical Value**	**Parameter or Function**	**Typical Value**	**Parameter or Function**	**Typical Value**
Output voltage	0 to 50 V	Ranges:		Display c.r.t.	8 cm × 10 cm
Output current	1 A max.	d.c. and a.c. voltages	200 mV to 1 kV	phosphor	P31
Load regulation	0.02% zero to full load	d.c. and a.c. current	200 μA to 1 A	Vertical deflection (Y)	
Line regulation	0.01% for 10% mains change	Resistance	200 Ω to 20 MΩ	Bandwidth	d.c. to 10 MHz
Ripple and noise	5 mV pk-pk	Accuracy (90 days)		Sensitivity	5 mV/div. to 20 V/div.
Output impedance	10 mΩ at 1 kHz	d.c.	±0.3% of reading ±1 digit	Accuracy	±5%
Temperature coefficient	±0.01% per °C	a.c.	±0.5% of reading ±1 digit	Input impedance	1 MΩ in parallel with 25 pF.
Current limit	110% of full load	Resistance	±0.2% of reading ±1 digit	Max. input	500 V peak
		Response time d.c.	0.5 sec	Horizontal deflection (X)	
		a.c.	3 sec	Time	0.1 sec/div. to 0.1 μ sec/div.
		Temperature coeff.	±300 ppm/°C	Accuracy	±3%
		Input impedance (V)	10 MΩ in parallel with 100 pF	Trigger sensitivity	Internal 1 div. External 1 V
		Common mode rejection d.c.	120 dB		
		a.c.	60 dB		
		Reading rate	3 per sec		
		Maximum input voltage	1.2 kV		

Some of these tests may be made before final test. (Power on.)

(e) System or instrument performance tests.

(f) Burn-in test (sometimes called soak test).

One of the questions that naturally arises is what exactly should be tested, adjusted, measured, and recorded for any particular instrument or system? The degree of testing required falls between the two extremes of (a) switch on power and test for output; and (b) measure and record every possible voltage and signal level. The whole purpose of testing is to ensure a satisfactory level of quality. This implies that each instrument or system conforms well to an agreed specification. A simple basic test just to check operation is never sufficient for this. The output may be present even though a partial failure exists in another portion of the instrument (for example, high a.c. ripple levels from the power supply); the safety aspects such as insulation and chassis ground must be tested and proved; and detailed records of important characteristics or parameters must be taken.

On the other hand there is no virtue in overtesting; this can be costly in test engineer's time and also unproductive. It is the *essential* voltages and signals that must be measured and recorded. The recording of data is very important as the results can be used to monitor quality; to give the essential figures for a test certificate; and to stand as a permanent record of an individual instrument's performance at the beginning of its operating life.

Using the relatively simple power supply that has the circuit diagram shown in Figure 1.6, let us work through the

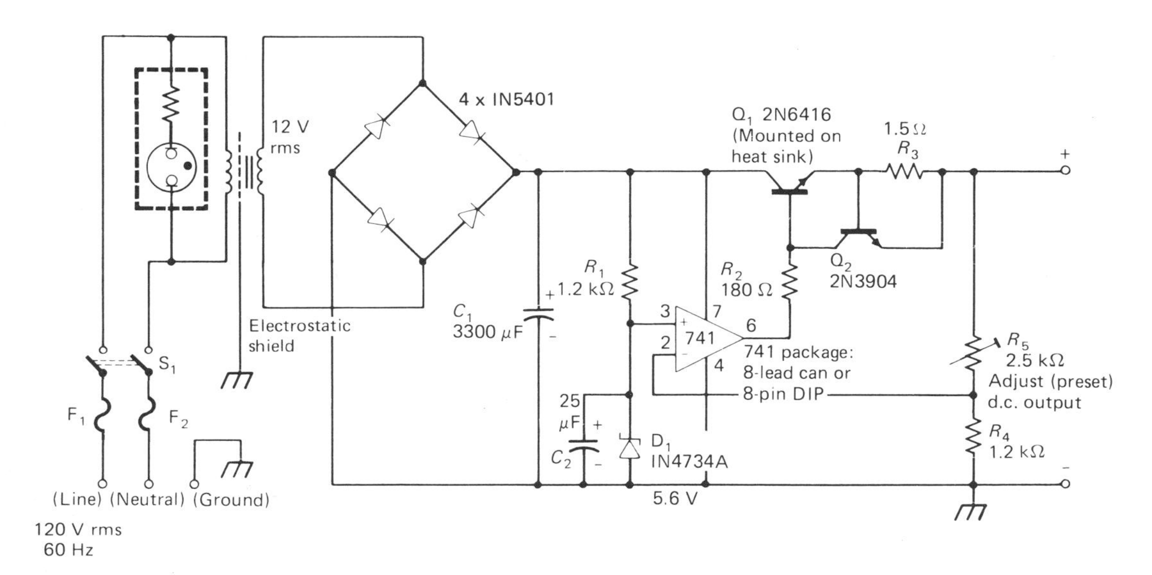

FIGURE 1.6 Regulated power supply; 10V at 0.3A.

various portions of its test specification. The power supply is designed to give the following performance:

Output voltage	10 V ± 50 mV
Maximum load current	300 mA
Load regulation	Better than 0.1% zero to full load
Line regulation	Not greater than 0.1% for 10% mains change
Ripple and noise	Less than 5 mV pk-pk
Overload protection	Between 360 mA and 440 mA (fixed)

The power supply is of standard design using a 5.6 V zener as the reference voltage, a 741 op-amp as the comparator (error amp), and a 2N6416 as the series control transistor (Q_1). Current limiting is provided by R_3 and Q_2.

In writing the Test Specification, only those measurements that are necessary should be requested.

EXAMPLE OF A TEST SPECIFICATION

1. Title

Additional comments

JP86 D.C. power supply
10 V at 300 mA
Issue 24 June 1981

(These would not normally be given.)

2. Test Equipment

2.1 **Variable a.c. mains source**
Output voltage 80 V to 140 V rms
Current rating 1 A
Frequency 60 Hz

An autotransformer is required so that the a.c. input can be varied for line regulation measurement.

2.2 **Multirange meter**
a.c./d.c./ohms
Moving coil type with a sensitivity of 20 kΩ/V on d.c. voltage ranges.

This will be used for measurement of resistance, a.c. input voltage, and for various d.c. voltages and currents.

2.3 **Digital voltmeter**
Display 5 digits
Accuracy ±0.1%
Resolution 1 mV in 10 V

Essential for the accurate setting of the d.c. output voltage and for the measurement of any small changes in d.c. output during the regulation tests.

2.4 **Cathode ray oscilloscope (single beam)**

Used for the measurement of any 100 Hz ripple present on the d.c. output.

Bandwidth: at least 1 MHz
Sensitivity: 1 mV/div.

2.5 **Load resistors**
20 Ω 10 W fixed ±10%
500 Ω 5 W variable ±10%

To enable the test engineer to adjust the load current taken from the supply.

2.6 **Insulation tester**
Output voltage 500 V d.c.

3. Continuity, Insulation and Resistance Tests

With no a.c. power applied, make the following tests and measurements.

A suitable continuity tester can be made up of two prods with a series-connected circuit of a battery and buzzer. The purpose of these tests is to ensure that a good ground connection exists.

3.1 **Continuity**
Using a suitable method,

(a) Check that the negative output terminal is connected to chassis.
(b) Check that the screen of the transformer is connected to the chassis.
(c) Check that the chassis is securely and electrically connected to the a.c. power line ground connector (G). Note that the a.c. power line ground is connected to earth.

3.2 **Insulation**
Using the insulation tester at 500 V, check that the insulation resistance from the high side of the a.c. line, connector (L), to chassis ground, connector (G), is greater than 50 MΩ. Repeat test with S_1 closed.

This should ensure that no breakdown or low leakage path exists between the a.c. line and chassis.

3.3 **Resistance tests**
Using the multirange meter on ohms, check resistances on unit as follows:

This simple test checks that no short circuits or low resistance paths exist before a.c. power is applied.

(a) Across output terminals (most positive prod of meter to positive output terminal). The resistance should be greater than 1 kΩ.

(b) Across C_1 (allow a short time for C_1 to charge). The resistance should be greater than 3 kΩ.

4. Voltage Measurements and Tests

4.1 Connect power supply to the a.c. line. Set a.c. voltage to 120 V rms and set load resistor to maximum value.

See the test setup in Figure 1.7.

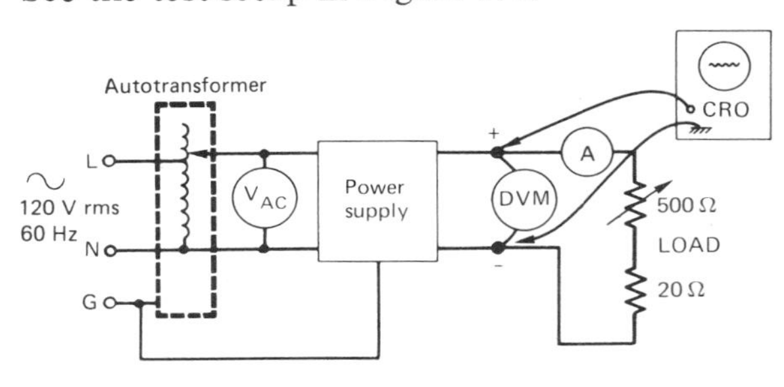

FIGURE 1.7 Test setup for power supply measurements.

4.2 Operate the power switch and check that the neon indicator comes on.

4.3 Using a multirange meter, measure and record the following d.c. voltages:

(a) Voltage across C_1: 16 V ± 1 V (record value)

(b) Voltage across D_1: 5.6 V ± 0.6 V (record value)

(c) Voltage across R_4: this should be identical to the reading obtained at **(b)**. (Record value.)

4.4 **Adjust output voltage**

Set load resistor to maximum value, i.e., load current approximately 20 mA. Connect a digital voltmeter (DVM) across d.c. output terminals and adjust R_5 until the output voltage is 10 V ± 10 mV.

4.5 **Load regulation**

Keeping the a.c. input constant at 120 V rms apply a load current of 300 mA by adjusting load resistor. Use the DVM to measure the change

This tests that the load regulation is within 0.1%.

in output voltage. Maximum change 10 mV. (Record value.)

4.6 Line regulation

Set load current to 300 mA.

Vary a.c. input from 108 V to 132 V rms. Measure the change in d.c. output voltage. Maximum change 10 mV. (Record value.)

This tests that the line regulation for a 10% mains change is within 0.1%.

4.7 Ripple

Keep load current at 300 mA.

Using the CRO, measure the peak to peak 100 Hz ripple at

(a) C_1: not greater than 1 V pk-pk.

(b) Output: not greater than 5 mV pk-pk. (Record value.)

4.8 Current limit

Gradually increase the load current from 300 mA while monitoring the d.c. output voltage.

The value cannot be precise, since R_3 has a tolerance of ±5%.

Record value of load current that causes the d.c. output to be 9 V. This should be between 360 mA and 440 mA.

Record the value of load current that causes the d.c. output to be 1 V. Not more than 20 mA greater than the previous reading.

This checks that the dynamic output resistance rises to at least 400 Ω when the current limit operates.

5. Burn-In Test

Operate the power supply under the following conditions for a period of not less than 50 hours.

A.C. power line input: 120 V
Load current: 300 mA.
Ambient temp.: +30°C

Check d.c. output voltage twice during test and at the end of the burn-in test.

D.C. output should be 10 V ± 30 mV.

The purpose of a burn-in test is to reduce risk of early component failures during normal use.

In some instances the temperature may be cycled during the burn-in period from a low to a high value.

This may appear lengthy in comparison with the power supply's makeup, but if you study the various points you will see that the tests follow a logical sequence and should ensure that all production units are fully tested, with a record made of all important parameters.

1.8 WORKING TO A TEST SPECIFICATION

In production test departments, the test specification or test instruction sheet is for the guidance of the test engineers. It is their job to ensure that production units meet all aspects of the agreed on performance; this requires skill in measurement and in rapid troubleshooting. When some part of an instrument is not functioning to specification, the test engineer must find the cause of the fault as quickly as possible and then return the instrument, or faulty subassembly, back to production for repair. In addition to measurements and troubleshooting, the engineer has to carefully record the required data from the instrument being tested.

Apart from knowledge of measuring techniques, testing procedures, instrument accuracies, and troubleshooting, the tester must also observe other basic points. They are

1. **Safety** The test bench should be supplied with a.c. power via an isolating transformer and the floor should be nonconducting. Test instruments should be carefully positioned so that they can be easily reached for adjustment and clearly observed for readings. Test equipment must be secure and never stacked dangerously. Cables and leads should not be trailing. The whole work area must be well illuminated.
2. **Test instruments** These must be well maintained and regularly calibrated. Most organizations recalibrate test instruments at least on a half-year or yearly basis. Any correction data must be kept with the instrument.
3. **Records** A clear space should be provided for writing reports and test data. If possible a small library of test instrument manuals and component data should be maintained for reference.

In general, a logical layout for any test setup should be used. Instruments used for measuring critical voltages and currents should be clearly marked. Any measuring leads should be kept short to avoid stray capacitance, inductance, and coupling between leads. Low capacitance probes should be used with oscilloscopes, especially when measurements of signals above audio are being made. More detail of test and measuring techniques is presented in later chapters.

1.9 SUMMARY

Performance Specification A detailed statement of the characteristics and parameters of a device, machine, structure, product, or process, when operating under stated environmental conditions.

Advantages:

(a) A well-defined product.

(b) Customers can select the most suitable device, instrument, or process for their needs at a price they are prepared to pay.

(c) Manufacturers know exactly the limits of each characteristic or parameter of their products. This enables them to forecast costs and selling price.

Test Specification A document used within a manufacturing plant which details the tests, with limits of measured values, that must be made on all production models.

Limits The maximum and minimum values of any parameter or characteristic.

Standard Specifications Those specifications issued by national, international and defense authorities for the guidance of manufacturers and users of components, equipment, and processes.

Standard specifications cover

(a) Glossaries of terms and symbols

(b) Dimensional standards

(c) Performance specifications

(d) Standard methods of test

(e) Codes of practice.

EXERCISES 1

1.1 Write performance specifications for the following instruments:

(a) A d.c. power supply, derived from the a.c. power line, suitable for supplying a TTL logic circuit. The power required is 5 watts.

(b) A signal generator capable of testing radio frequency systems.

1.2 List in short form the most important parameters and characteris-

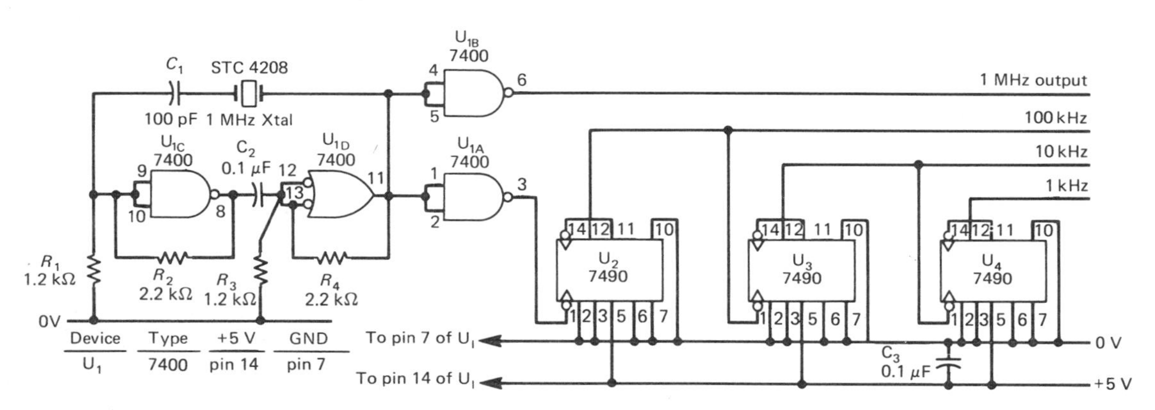

FIGURE 1.8 Frequency standard.

tics, with typical values, for
(a) Unijunction transistors.
(b) General purpose medium power thyristors.
(c) LEDs suitable for indicators. (See Figure 1.3.)

1.3 Convert the following limits to maximum and minimum values: (a) 10 V ± 3% (b) 910 ± 2% (c) 160 kHz ± 4.5% (d) 22 MHz ± 10% (e) 100 000 μF − 20% + 50%.

1.4 Write a suitable test specification to cover tests for the oscillator and frequency divider circuit shown in Figure 1.8.

1.5 List the test instruments and equipment necessary for the production testing of
(a) Pulse generators (frequency 0.1 Hz to 10 MHz).
(b) Audio frequency sine/square wave signal generators.
(c) General purpose single beam CRO (bandwidth 10 MHz).

CHAPTER 2

Reliability

2.1 INTRODUCTION AND DEFINITIONS

When something, such as a machine, is described as being "reliable," what is the intended meaning? The words that first come to mind are probably "dependable" or "trustworthy." A question most of us would then ask is: just how dependable is it? Or in other words, what is its actual *reliability*? The answer could be graded as fair, good, or excellent, and this, although subjective, does convey more information. However, when reliability is used in a technical sense, it must have a clearly defined meaning, and also must be capable of expression as a figure or percentage.

For example, the reliability of the machine might be quoted as 0.99 for a 1000-hour operating period under well-defined operating conditions (temperature, humidity, etc.). This means that the probability of satisfactory operation, without any failure, is 99% during a 1000-hour period. Note that reliability is concerned with *probability*, for it would be impossible to state that the machine was certain to operate correctly for that period. What we actually get is a prediction of the *likely* success of operation. Electronic instruments are made up of several types of component, and it is by measuring small samples of the various items that manufacturers gain information on failure rates. Using these figures we can then begin to predict the overall reliability of the complete instrument.

A good definition of reliability may be stated as follows:

Reliability *is the ability of an item to perform a required function (without failure) under stated conditions for a stated period of time.*

Here an item means a component, instrument, or system. Note that a reliability figure cannot be predicted without specifying the operating time and the operating conditions.

To complete the picture, since reliability is concerned with "failure," this also has to be defined:

Failure *is the termination of the ability of an item to perform its required function.*

Failures may be yet further defined depending on:

1. The degree of failure—is the item just out of specification or has it broken down completely?
2. The cause of failure—misuse or inherently weak.
3. The rate or time of failure—sudden or gradual.

1. Degrees of Failure

(a) **Partial failure**—failures resulting from deviations in characteristic(s) or parameter(s) beyond the specified limits, but not such as to cause complete lack of the required function.

(b) **Complete failure**—failures resulting from deviations in characteristic(s) beyond the specified limits such as to cause complete lack of the required function.

2. Causes of Failure

(a) **Misuse failure**—failures attributable to the application of stresses beyond the stated capabilities of the item.

(b) **Inherent weakness failure**—failures attributable to weakness inherent in the item itself when subjected to stresses within the stated capabilities of the items.

3. Time of Failure

(a) **Sudden failure**—failures that could not be anticipated by prior examination.

(b) **Gradual failure**—failures that could be anticipated by prior examination.

4. Combinations of Failure

(a) **Catastrophic failure**—failures that are both sudden and complete.

(b) **Degradation failure**—failures that are both gradual and partial.

It is always important both to the user and the manufacturer that the type of failure of any component be accurately reported, since it is often information on faults in operating equipment that eventually leads to improvements in the reliability of future products. For example, future designs could reduce the incidence of misuse failure and eliminate components that are inherently weak.

The reason for the interest in reliability and reliability engineering is due in part to the increasing complexity of electronic equipment and systems, with the resulting increase in the number of components, and also to an awareness of the fact that poor reliability carried with it a cost penalty for the owner. A cheap product may appear attractive because of its low capital cost, but during operation the cost of maintenance, repairs, and spare parts may be so high as to far outweigh the initial financial advantage compared with the purchase price of a product with high reliability.

The development of miniature solid state electronic components, in particular integrated circuits, has made possible extreme reductions in size and weight of electronic equipment and, in addition, has greatly improved reliability. At first sight this seems odd, for the electronic instruments themselves are becoming more and more complex, but integrated circuits aid reliability in two important ways: (i) they enable complete circuits to be isolated from environmental stresses such as humidity, shock, and temperature, and (ii) the soldered and plug and socket interconnections are replaced by connections inside the encapsulation. Intensive research and development work has also led to increasing reliability in discrete components, and in mechanical parts such as printed circuit boards.

Naturally the reliability requirement for a piece of equipment depends very much on its intended application. For example, the reliability of communication satellites must be very high for it is generally not possible to carry out repairs. In any situation where maintenance is difficult and costly and where the equipment must be kept working, high reliability is essential. Even in a situation where faults can be repaired, there are examples of where high reliability is required when demanded: for example, medical diagnostic and life support machines, aircraft electronics while in flight, and industrial control systems where the cost of a shutdown or lost production through system failure could be very high. We shall see later, however, that increased capital cost must be paid for high reliability.

2.2 FACTORS AFFECTING EQUIPMENT RELIABILITY

The various stages in the life cycle of an electronic instrument can be separated into four parts (see Figure 2.1):

(a) Design and development.

(b) Production.

(c) Storage and transport.

(d) Operation.

At each stage the overall reliability will be affected by the methods used. If, as should be the case, good reliability is considered an essential part of the manufacturer's aim, then a reliability program has to be established to cover these four stages.

At the design stage, a reliability specification must be prepared that establishes a target figure for the reliability. Each step the design engineers take will then be geared to this target figure. These steps are

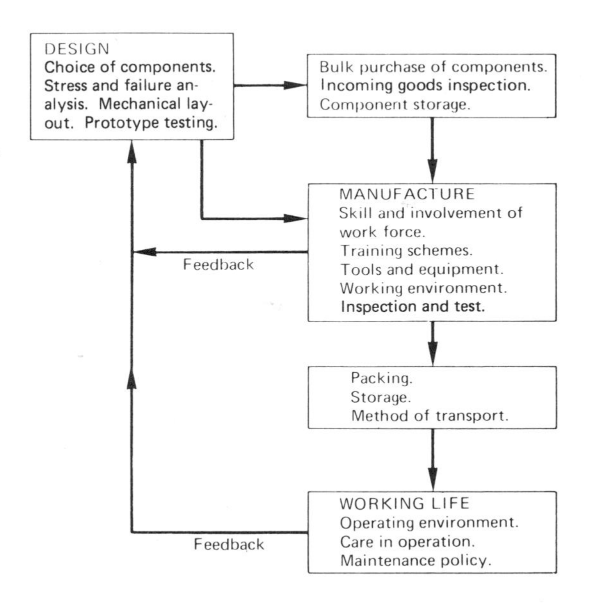

FIGURE 2.1 Factors affecting the reliability of electronic equipment and systems.

making the correct choice of components, and designing circuits so that no undue stresses act on components. This is called a *stress and failure analysis* and consists of investigating each component's chance of failure in the proposed circuit. Components will of course be derated for best reliability.

The mechanical layout of components, assemblies, and panels is also very important. Components must not be mechanically stressed by poor mounting, nor must excessive heat be allowed to develop inside the equipment housing. The effects of the environmental conditions in which the equipment will be operated must be taken into account, and protection must be built in to counter environmental hazards. These steps may include hermetic sealing; forced air cooling; antivibration mountings, or potting assemblies in some insulating compound.

Extensive prototype tests are essential to check that the design will be able to meet both the performance and reliability specifications.

What, then, are the factors affecting the instrument's future reliability when it is being assembled in production?

First, the Purchasing Department must ensure that the required components are ordered from a reputable source. Correct purchasing policy means obtaining the components at the right time and price, while at the same time ensuring that all the aspects of the item's specification can be met.

All component parts entering the factory should then pass through an Incoming Goods Inspection system. This may involve 100% test and inspection for small batches, or a sampling test and inspection method based on statistical analysis for larger quantities. The amount of checking required depends on the type of component. For example, items bought under a contract agreement will already have passed through a test and inspection system that is independently controlled, so further testing would be wasteful.

Another factor affecting the equipment's reliability is the method and length of storage of the individual components prior to assembly. Apart from the high costs of storing for long periods, all components change their characteristics while stored. This is often referred to as SHELF LIFE. It is, therefore, important to keep component storage time to a minimum. Stored components should have fairly constant values of ambient temperature (say 20°C ± 2°C) and medium values of humidity, to reduce aging effects.

Turning now to the actual process of production, it is fairly obvious that the skill, training, and sense of involvement of all the production personnel play a vital part in overall reliability. Each worker (toolmakers, production and methods engineers, assembly operators, test and inspection engineers, etc.) forms a link in the chain of production and can assist in producing a quality product that will meet the reliability specification.

Good training schemes will ensure that personnel are able to use the correct and most effective production techniques, and an information service should help in involving all in producing a reliable product. The importance of good communications between production and design in achieving rapid feedback of information cannot be overemphasized as this enables modifications to be made that will improve reliability and quality.

The tools and manufacturing equipment must be of the required standard that is adequate for the work, and be well maintained. It is also vital that the work environment is well ventilated, well illuminated, and generally kept at an even ambient temperature. For the highest reliability, assembly operations must be carried out in a dust-free assembly room.

Finally, as far as production is concerned, the inspection and test methods used will also affect the reliability of the final equipment. Printed circuit boards and assemblies should be inspected 100% if possible. Automatic test equipment, not subjected to human error, can be used to check for short or open circuits on copper traces and also for correct circuit function. A burn-in test when the complete instrument is operated for a specified period at elevated temperature or with temperature cycling will help in identifying components that are inherently weak.

Before the instrument is operating in the field, it has to be packed and stored, and the methods will affect operating reliability. Packaging must protect the instrument from corrosion and mechanical damage, and storage temperature and humidity levels must be controlled.

Following its sale, the equipment is transported and is subjected to vibrations, mechanical shocks, variations in temperature, humidity, and pressure. The packing and specified method of transport, therefore, have to be considered within the reliability specification.

Once installed, the equipment reliability is affected by the environmental conditions and the correctness of handling and operation. *Operability* is the name given to designing to reduce the possibility of operator error. A well-written operating instruction manual should ensure that failures through misuse do not occur. The environment hazards are dealth with more fully in a later section of this chapter.

2.3 THE COST OF RELIABILITY

The actual cost of ownership of a product is made up of the capital cost (the purchase price) and the cost of operation and maintenance. The latter which can, over a number of years, often exceed the capital cost is dependent to a large extent on reliability. Looking more closely at the breakdown of costs, it can be seen that, as far as the manufacturer is concerned (Figure 2.2*a*), as reliability is improved, design and production costs increase while the costs of repairs and free replacement under guarantee fall. The design costs increase for the obvious reason that more and more effort and time has to be taken in improving reliability. Adding the two graphs together gives the general shape of total manufacturing cost versus reliability. Note that this has an obvious minimum cost point. In other words, a firm that produces poor reliability products may be forced out of business simply because of the costs involved in providing an excessively large service organization.

Using the total cost graph of Figure 2.2*a* as a basis for customer purchase price, the graph of total ownership costs versus reliability can be obtained (Figure 2.2*b*). This is the sum of the purchase price and maintenance costs. The latter will naturally fall as reliability improves. The total ownership cost graph also has a minimum point which, however, is noticeably to the right

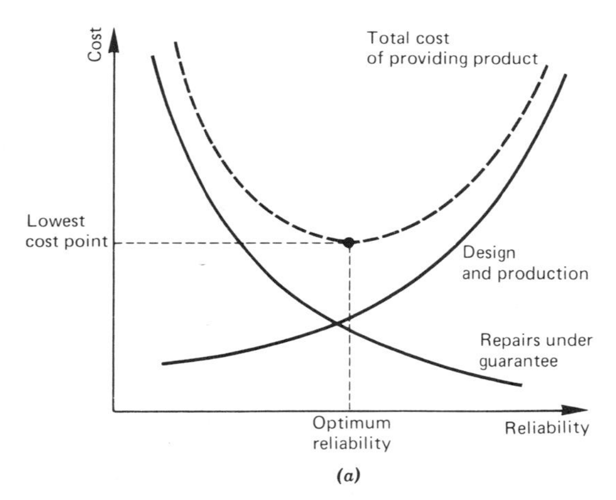

FIGURE 2.2*a* Manufacturing costs versus reliability.

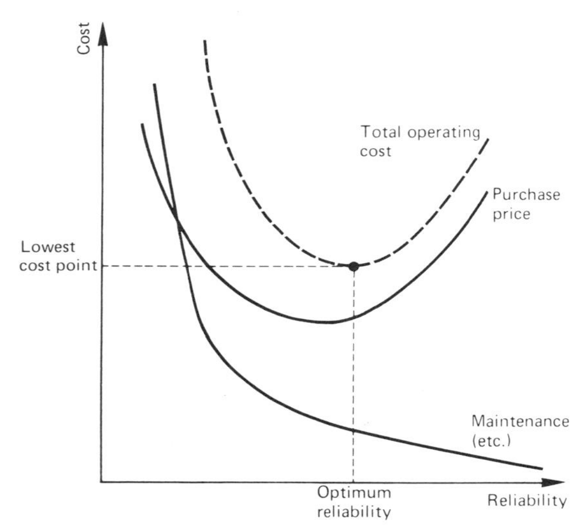

FIGURE 2.2*b* Ownership costs versus reliability.

of the manufacturer's minimum cost point. Hence, the customer may well be seeking a better reliability, for the price he or she is prepared to pay, than the manufacturer wishes to provide.

If the equipment is made to a specified order, some compromise must be made in the contract clauses relating to reliability. In most instances, however, equipment is offered for sale at a fixed price, and then it is up to customers to attempt some evaluation of their total costs.

The graphs are intended only as a guide to the variation of costs with realibility. The minimum-cost/optimum-reliability point is difficult to realize in practice because it involves a prediction of how the various cost factors with reliability are actually going to vary. This can be done with some accuracy if large numbers of similar instruments have been in service for a considerable time and all faults are accurately reported and analyzed. Unfortunately the pace of electronic development is so rapid that many completely new products using novel design features are being developed each year, and prediction of costs and reliability is then more difficult. It is the designer's job to evaluate very carefully new types of components before they are used in preference to those that are tried and tested.

Minimum total cost is not always the main factor in deciding the reliability required from a system. Many customers insist on, and are prepared to pay for, high values of operational reliability. These are, for example, the armed services, space authorities, and airlines.

2.4 FAILURE RATE, MTTF, MTBF

A study of reliability is essentially a study of the failures of components and systems. Field trials of equipment can be used to provide data on system failure rates, but for reliability prediction a designer needs to know with some confidence the failure rate of each type of component that goes to make up the system. We have already defined failure as the inability of a component or piece of equipment to carry out its specified function. In other words, it may be just out of specification limits (a partial failure) or totally nonfunctioning (a complete failure).

Why should components fail? The answer, of course, is that all man-made items have a finite life. Whatever the article, wear during use, and stresses acting on it, will at some time cause it to fail.

For electronic components the stresses are caused by:

(a) **Design operating conditions**
Applied voltage and current, power dissipated, and mechanical stress caused by the mounting method.

(b) **Environmental conditions**
High or low temperature. Temperature cycling. High humidity. Mechanical vibrations and shocks. High or low pressure. Corrosive atmospheres. Radiation. Dust. Attack by insects or fungi.

Some of these hazards are more damaging than others, for example, rapid and large changes in temperature coupled with high humidity could lead, within a short time, to the breakdown of some components.

As far as design is concerned, the life of a component can be greatly improved if it is operated well below the maximum rated values of current, voltage, and power. This design method, called DERATING, is used extensively to reduce failure rates. This means, for example, operating a 1 W resistor at 0.25 W maximum.

The FAILURE RATE of a component can be found by operating large numbers of the component for a long period and noting the number of failures that occur. Imagine a batch of small signal diodes taken straight from production and placed on test at their maximum rated power. Within a short time it is quite likely that several will fail, due to manufacturing techniques and material imperfections. Gradually, as the diodes that have inherent weaknesses fail and are withdrawn, the rate of failure will fall. This initial period of high failure rate, known as BURN-IN or EARLY FAILURE, is followed by a period where the rate of failure levels off to an almost constant value. This period is known as RANDOM FAILURE PERIOD or the USEFUL LIFE, and is the one of most interest since here the failures are entirely random, i.e., due to chance alone. Using the failure rate over this period enables predictions to be made of reliability by means of probability theory.

If the test were continued beyond the useful life period, a gradual increase in failure rate would be observed as the diodes failed one by one because of the aging process—this is called the WEAR-OUT PERIOD.

The variation of failure rate (FR) with time is shown graphically in Figure 2.3. Because of its shape it is often referred to as the *bathtub curve*.

If the diodes are to be used in a design where high reliability is required, the early failure effect could be eliminated by pre-aging all production for a short period (say, 100 hours), and using those that survive. Wearout could be delayed by im-

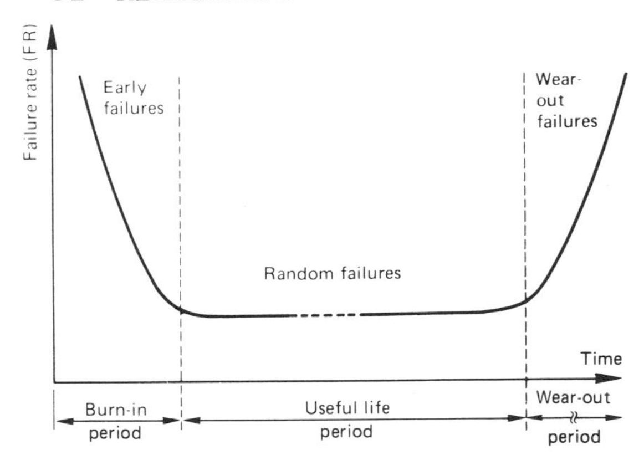

FIGURE 2.3 Failure rate variations with time.

proving the design, by changing materials, and by a closer control of production, but this may be very costly. The random failure rate during use could be reduced by operating the diodes at lower power (i.e., derate the power).

Returning to the test, suppose that after burn-in, out of 400 diodes left on test, five fail over a 1000-hour period, then the average failure rate is

$$FR = \frac{5}{400} \times 100\% \text{ per } 1000 \text{ h}$$

$$= 1.25\% \text{ per } 1000 \text{ h}$$

Often failure rate is written as a percentage as above, but it could just as well be expressed as failures per hour:

$$FR = \frac{5}{400} \times \frac{1}{1000} \text{ failures per hour}$$

$$= 1.25 \times 10^{-5} \text{ per hour}$$

or 0.0125 per 1000 hours.

This failure rate obtained from a sample of production is called the *best estimate*. Because sample sizes are relatively small, to reduce the risk of quoting too favourable a figure, a higher failure rate is usually given. The failure rate is then said to have a CONFIDENCE LEVEL of 60, 80, or 90%. For example, a confidence level of 90% means that, if an infinite number of samples were taken, 90% of the failure rates obtained would be below the figure for the 90% confidence level. To adjust the best estimate figure of failure rate a constant is added to give a confidence level figure. These constants C_{60}, C_{80}, C_{90} can be applied when the failure rate is itself constant. Tables are published giving these constants for various confidence levels and number of failures. Usually a 60% confidence level is used. The relationship between best estimate and 60% confidence levels for 2.5 million component hours is shown in Table 2.1. Here the failure rate is calculated using the formula:

$$FR_{60} = \frac{\text{Number of failures} + C_{60}}{\text{Number of component hours}} \quad (C_{60} = 1)$$

One reason component failure rates are often quoted with a 60% confidence level is that tests on some batches of components show no failures. The failure rate would then be zero, which we know cannot be true. Note that, as the number of recorded failures increases, the difference between best estimate FR and FR_{60} becomes smaller.

FAILURE RATE then is defined, over the useful life, as the number of failures per number of component hours. For the diodes in our example, the FR is 1.25×10^{-5}/h (best estimate). At first glance this appears a low figure and, in fact, if one diode is used, its MEAN TIME TO FAIL (MTTF) will be

TABLE 2.1

Number of Recorded Failures	Best Estimate FR ($\times 10^{-6}$/h)	60% Confidence Level FR_{60} ($\times 10^{-6}$/h)[a]
0	0	0.4
1	0.4	0.8
2	0.8	1.2
3	1.2	1.6
5	2.0	2.4
10	4.0	4.4
25	10	10.4

[a] 10^{-6}/h = per million hours.

$$\text{MTTF} = \frac{1}{\text{FR}}$$

$$\text{MTTF} = \frac{1}{1.25 \times 10^{-5}} = 80\,000 \text{ hours (3333 days)}$$

indicating a fairly long life.

But think of the case where 100 of the diodes are used to provide a decoder panel and where the failure of one diode constitutes a panel failure. Since each diode has a chance of failure of 1.25×10^{-5} in every hour, the chance of the panel failing is 100 times greater, i.e., 125×10^{-5} in every hour. Therefore, the MEAN TIME BETWEEN FAILURES (MTBF) of the panel is

$$\text{MTBF} = \frac{1}{125 \times 10^{-5}} = 800 \text{ hours } (33^1/_3 \text{ days})$$

The more components a system uses the greater is its chance of failure. Note that the figure for MTBF is an average, in other words, the time between failures might be found in practice to be 28 days, 36 days, 29 days, 40 days, 26 and 41 days over a 200-day period. What MTBF gives us is a prediction of the *average* time that a system will run before failing.

The term MTTF is normally applied to items that cannot be repaired whereas MTBF is used for repairable items, i.e., instruments and systems.

The MTBF of a complete system can be calculated by first finding the sum of the failure rates of all components. Consider a very simple circuit using three components, A, B, C. The system failure rate is

$$FR_{(\text{system})} = FR_{(A)} + FR_{(B)} + FR_{(C)}$$

Therefore,

$$\text{MTBF}_{(\text{system})} = \frac{1}{FR_{(\text{system})}}$$

The best way to see how system MTBF is calculated is to add the failure rates of components in a circuit, but before we can do that we must know the failure rates of individual components. Table 2.2 (page 36), *intended only as a guide*, shows failure rates for commonly used electronic components.

The failure rate for a component depends, of course, on the manufacturing method and its environment in use. Weighting factors must be applied if the environmental conditions are severe as, for example, in military mobile equipment.

As a further example consider the data presented in Figure 2.4, which shows the results of life tests on Mullard resistors and capacitors for 4800 hour endurance tests. This represents a total of about 436 000 000 component hours. For the resistors the test conditions are exclusively indicated by the hot spot temperature, which is the resistor body temperature caused by a combination of ambient temperature and load. For capacitors, the load is expressed as a percentage of the maximum permissible voltage. The normalized failure rate is then given for maximum load and 25°C.

The effect of derating on failure rate can be dramatically seen from Figure 2.5, which shows the results of long-term field tests in telephone transmission systems.

Component type	Hot-spot temperature (°C)	Number tested	Component hours ($\times 10^3$)	Catastrophic failures	Failure Rate 60% confidence level ($\times 10^{-6}$/h)
Carbon film resistors CR16, CR25, CR37					
≤1 MΩ	25 to 153	29 523	111 949	11	0·11
>1 MΩ	25 to 153	7 820	19 308	15	0·80
Metal film resistors MR25, MR30					
	50 to 170	26 160	159 514	1	0·012

FIGURE 2.4*a* Failure rates for resistors. *(Courtesy Mullard Ltd.)*

Component type			Test conditions Load (%)	Temperature (°C)	No. tested	Component hours ($\times 10^3$)	Failures Catastrophic	Degradation	FR 60% confidence level ($\times 10^{-6}$/h) Cat.+Degrad. Test	Normalized	Catastrophic Test	Normalized
Ceramic plate	629		150	55	920	4 051	4	7	3·1	0·19	1·3	0·08
	630		150	85	870	5 076	2	4	1·4	0·03	0·61	0·01
	632 (C333)		150	85	4 149	20 234	25	45	3·6	0·08	1·3	0·03
Polystyrene	424	63V	150	70	1 193	1 193	2	3	5·3	0·01	2·6	0·006
	425	125V	150	85	695	695	9	15	3·7	0·03	1·5	0·01
	435 (C295AH)	63V	150	70	740	7 082	8	—	—	—	1·3	0·003
	436 (C295AA)	125V	150	85	680	5 882	10	— }	—	—		
	438 (C295AC)	500V	150	85	600	5 687	6	— }			1·5	0·001
Polyester Film/foil	311 (C296AA)	160V	150	85	800	6 187	46	—	—	—	7·9	0·027
	311 (C296AC)	400V	150	85	320	3 020	34	—	—	—	1·2	0·041
	347		150	85	1 219	7 069	10	6	2·5	0·0086	1·6	0·0055
Metallized	341 (C281)		150	85	3 136	18 457	41	49	6·2	0·021	2·4	0·008
	342 (C280)		150	85	2 666	15 853	22	26	3·7	0·013	1·5	0·005
	344		150	85	2 559	15 059	10	24	2·9	0·010	0·8	0·003
Electrolytic Small G.P.	015 & 016		100	85	298	408	0	4	1·3	—	2·2	—
	017		100	85	529	987	1	2	4·2	—	2·0	—
Small long life	101 (C428)		100	70	1 566	5 076	1	5	1·4	—	0·40	—
	108		100	85	310	620	0	0	1·5	—	1·5	—
Large G.P.	071/2		100	85	380	760	0	20	28·8	—	1·2	—
Large long life	106/7		100	85	660	1 320	1	3	3·98	—	1·7	—
Solid aluminum	121		100	85	1 305	9 571	0	15	1·7	0·18	0·10	0·10
Solid tantalum	143 (C421)		100	25	985	9 649	0	1	0·21	0·21	0·09	0·09
			100	85	810	1 610	4	20	1·6	4·9	3·3	1·0

FIGURE 2.4*b* Failure rates for capacitors. *(Courtesy Mullard Ltd.)*

Component type	Number in operation (×10³)	Component hours (×10⁹)	Failures	FR 60% confidence level (×10⁻⁶/h)
Carbon film resistor	5725	130	3	0·00003
Metal film resistor	7·5	0·162	0	0·006
Polystyrene capacitor	2615	57·4	319	0·0057
Polyester capacitor Film/foil and metalized film types	516	11·8	4	0·00045
Electrolytic capacitor Long life, small type	641	15·8	28	0·0019
Electrolytic capacitor Small solid aluminum type	35	0·562	0	0·0016
Electrolytic capacitor Solid tantalum type	48	1·007	0	0·00091

FIGURE 2.5 Failure rates of components in an actual working system. *(Courtesy Mullard Ltd.)*

Loads were 10 to 20% typically but always less than 50%, and the ambient temperature was 40°C. Compare the failure rate in use with that indicated by the manufacturer's test; the failure rates in use are extremely low.

Now, using the figures in Table 2.2, let us estimate the MTBF for a typical system—say, a function generator with its own a.c.-derived power supply. It uses, say, 30 transistors; 75 metal film resistors; 45 capacitors; 5 potentiometers; 3 switches (30 contacts); a transformer (3 windings); 4 rectifiers; 20 small signal diodes; 2 linear ICs; 3 digital ICs; and has 750 soldered joints. The calculation for total system failure rate is made as follows:

Component	Average Failure rate FR (X 10^{-6}/h)	Number used (n)	n(FR) (X 10^{-6}/h)
Transistors	0.08	30	2.4
Resistors (metal film)	0.03	75	2.2
Capacitors (polyester)	0.1	45	4.5
Potentiometers	3	5	15.0
Switches	0.1	30 contacts	3.0
Transformer	0.4	3 windings	1.2
Diodes	0.05	20	1.0
Rectifiers	0.5	4	2.0
Linear ICs	0.3	2	0.6
Digital ICs	0.2	3	0.6
Soldered joints	0.01	750	7.5
			40.0

TABLE 2.2 Typical Failure Rates for Common Components

Components	Type	Failure Rate ($\times 10^{-6}$/h)
Capacitors	Paper	1
	Polyester	0.1
	Ceramic	0.1
	Electrolytic (Al. foil)	1.5
	Tantalum (solid)	0.5
Resistors	Carbon composition	0.05
	Carbon film	0.2
	Metal film	0.03
	Oxide film	0.02
	Wire-wound	0.1
	Variable	3
Connections	Soldered	0.01
	Crimped	0.02
	Wrapped	0.001
	Plug and sockets	0.05
Semiconductors (Si)	Diodes (signal)	0.05
	Diodes (regulator)	0.1
	Rectifers	0.5
	Transistor $<$ 1 watt	0.08
	$>$ 1 watt	0.8
	Digital IC (plastic DIP)	0.2
	Linear IC (plastic DIP)	0.3
Wound components	Audio inductors	0.5
	R.F. coils	0.8
	Power transformers (each winding)	0.4
Switches	(per contact)	0.1
Lamps and indicators	Filament	5
	LED	0.1
Vacuum tubes	(Thermionic)	5

Therefore FR(system) (symbol λ) is estimated to be 40×10^{-6}/hour.

So the MTBF(system) (symbol m) is

$$m = \frac{1}{\lambda} = \frac{10^6}{40} = 25\ 000 \text{ hours}$$

It must be pointed out that this is only an example of the way in which MTBF can be calculated. To obtain a reasonable degree of accuracy in the prediction, the actual failure rate of each type of component must be assessed correctly. It is the designer's job to obtain figures on FR from the component manufacturer and/or to carry out failure analysis on a batch of components. Since an analysis of this kind will involve millions of component hours, the cost can be high—one reason why high reliability is expensive.

2.5 THE EXPONENTIAL LAW OF RELIABILITY

If a constant failure rate applies, that is, failures are due to chance alone, the relationship between reliability (R) and system failure rate (λ) is given by the formula:

$$R = e^{-\lambda t}$$

where t is the operating time; λ, the system failure rate, is the sum of all component failure rates; and e is the base of natural logarithms. R is the probability of *zero* failures in the time t.

What the formula means is that the probability of no system failure occurring within a given time t is an exponential function of that time. In other words, the longer the system is operated the less reliable it becomes and the probability of failure (unreliability) rises.

$$\text{Unreliability } Q = 1 - R = 1 - e^{-\lambda t}$$

Now since MTBF or $m = 1/\lambda$, then $\lambda = 1/m$ and therefore

$$R = e^{-t/m}$$

It is useful to show the graph of R versus t, with time marked off in intervals of m (Figure 2.6). This shows that when $t = m$, i.e., the operating time is the same as MTBF, the probability of successful operation has fallen to approximately 0.37 or 37%. Only when the operating time is relatively short compared with MTBF does the reliability remain relatively high.

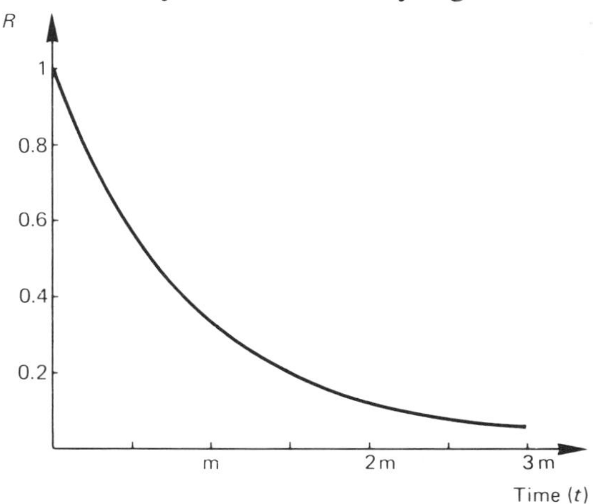

FIGURE 2.6 Plot of $R = e^{-t/m}$.

As an example, imagine a naval radar system with an estimated MTBF of 10 000 hours. What is the probability of it working successfully for mission times of 100, 2000, and 5000 hours?

Using the formula $R = e^{-t/m}$

$$\text{for } t = 100 \text{ hours}, R = e^{-0.01} = 0.99\ (99\%)$$

$$t = 2000 \text{ hours}, R = e^{-0.2} = 0.819\ (81.9\%)$$

$$t = 5000 \text{ hours}, R = e^{-0.5} = 0.6065\ (60.65\%)$$

Books of mathematical tables give values of e^{-x}, and they can be found with a calculator.

Note that e^{-x} is approximately equal to $(1 - x)$ for $x < 0.14$ within 1.1%.

The reliability for the 5000-hour period would certainly not be considered adequate. It is, of course, impossible to achieve perfect reliability, even for a very short operating period, since this implies that the chance of failure is zero. Reliability, the probability of success for a stated period of time under stated conditions, can approach unity (i.e., 0.999) but never equal it.

Good reliability depends on many factors such as choice of components, derating, protection from environmental stresses, operability, and maintainability.

Another method of improving reliability is to use what is called REDUNDANCY, where subunits or component parts are connected so that if one fails another takes over the function. A good example is the use of standby power sources that are switched in to supply the load if and when the a.c. power line fails. In redundancy design the units are usually paralleled. Consider the simple case of two units X and Y wired in parallel so that the whole system does not fail until *both* X and Y have failed. Let the individual units have reliabilities over an operating time of 1000 hours of

$$R_x = 0.85 \quad \text{and} \quad R_y = 0.75$$

Unreliability, or probability of failure, is given by

$$Q = 1 - R$$

Therefore,

$$Q_x = 1 - R_x = 1 - 0.85 = 0.15$$

and

$$Q_y = 1 - R_y = 1 - 0.75 = 0.25$$

The probability of failure of the units in parallel is the product of their unreliabilities:

$$Q_{xy} = Q_x \cdot Q_y = 0.15 \times 0.25 = 0.0375$$

So the system reliability over 1000 hours is

$$R_{xy} = 1 - Q_{xy} = 1 - 0.0375 = 0.9625$$

Note here the increase in reliability achieved by duplicating the circuits.

Alternatively since $Q_x = 1 - R_x$ and $Q_y = 1 - R_y$, it follows from

$$R_{xy} = 1 - Q_x \cdot Q_y$$

that

$$R_{xy} = 1 - (1 - R_x) \cdot (1 - R_y)$$
$$= 1 - (1 - R_y - R_x + R_x R_y)$$

i.e.,

$$R_{xy} = R_x + R_y - R_x \cdot R_y$$

This is covered by the probability theorem:

If a and b are two events that can occur together, the probability of either or both events taking place is $P_a + P_b - P_a \cdot P_b$.

Redundancy can be *active* where a standby unit is switched in following a failure, or *passive* where the elements share the load but are each capable of supplying the load or carrying out the function separately. The more units placed in parallel, the greater the overall system reliability, but the cost penalty is obviously high.

It should now be clear that, when units are placed in series such that the failure of one unit means failure for the whole system, then the reliability for the system will be lower than that of the individual units. The system reliability then depends on the product of the reliabilities:

$$R_{xy} = R_x \cdot R_y$$

If $R_x = 0.85$, $R_y = 0.75$, then

$$R_{xy} = 0.85 \times 0.75 = 0.6375$$

EXAMPLE

A power supply, oscillator, and amplifier are all used in a simple system. Calculate the individual and system reliabilities for a 1000-hour operating period if the MTBFs are 20 000 hours, 100 000 hours and 50 000 hours, respectively. Use the approximation for $e^{-x} = (1 - x)$ when $x < 0.14$.

Solution

Power supply reliability for $t = 1000$ hours is

$$R_p = e^{-t/m} = e^{-0.05} = 0.95$$

Oscillator reliability is

$$R_o = e^{-t/m} = e^{-0.01} = 0.99$$

Amplifier reliability is

$$R_a = e^{-t/m} = e^{-0.02} = 0.98$$

Since the units are in series the overall system reliability will be given by the product of the subunit reliabilities:

$$\begin{aligned} R_s &= R_p \cdot R_o \cdot R_a \\ &= 0.95 \times 0.99 \times 0.98 \\ &= 0.922 \end{aligned}$$

that is, a system reliability of 92.2% for a 1000-hour operating period.

2.6 ENVIRONMENTAL EFFECTS ON RELIABILITY

As already noted, the prevailing conditions of the environment in which an instrument or system is operating have a profound effect on its reliability (Figure 2.7). This is, of course, true whether the instrument is operating (active), switched off (static), or stored. In fact, when switched off, the effects of the environment may be even more damaging, since there is no self-generated heat to counteract the effects of moisture. Hence, many equipments are rarely switched off. The study of the various effects can be quite involved and what follows in this section is an introduction. We shall look at the major environmental hazards in turn, consider their most obvious effects, and look at ways in which these effects can be minimized by design and manufacture.

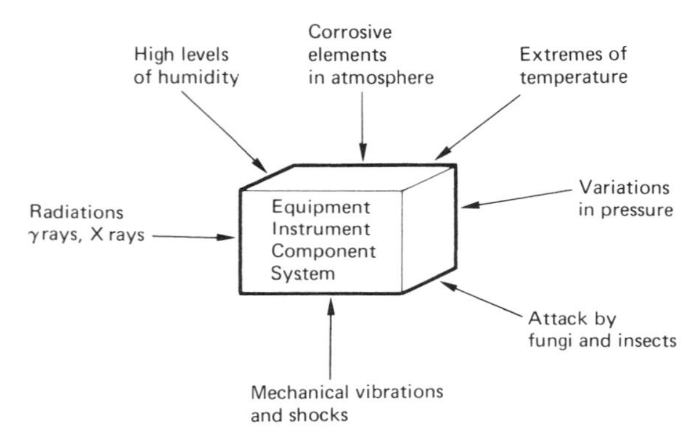

FIGURE 2.7 Environmental hazards.

Temperature

To achieve highest reliability it would be best to operate any electronic system at a fairly constant and medium value of temperature, say 20°C ± 5%, the sort of temperature conditions in an air-conditioned office.

This is not always possible and equipment such as portable communications receivers for armed services usually has to be designed to withstand large changes in ambient temperature. The temperature variations on the earth's surface are typically:

Deserts	-15°C to +60°C
Tropics	+20°C to +45°C
Polar regions	-40°C to +30°C
Seas	-10°C to +30°C

All electronic equipments have some self-generated heat, so, if the outside temperature rises to say +60°C, the internal temperature may be 70°C or 80°C and the temperature inside components that are generating this heat may rise to well above 100°C. Thus a component, operating well within its maximum rated power dissipation at 25° ambient, may well be almost out of specification at 60°C ambient.

The other major effects of high temperatures are the stresses caused by the expansion of materials and the acceleration of the chemical actions that age components. As a rough rule of thumb, the failure rate of many components can be considered to double for every 10° C rise in temperature. For example:

Estimated life times of electrolytic capacitors with ambient temperature:

50°C	40 000 hours
60°C	23 000 hours
70°C	14 000 hours
90°C	5 000 hours

At very low temperatures, materials harden and become brittle and are therefore more liable to break. Transistor current gains fall. Some components, in particular electrolytic capacitors, can become almost ineffective, because of freezing of the electrolyte and the resulting increase in series resistance.

The temperature effects that may prove most damaging are those of rapid temperature cycling. Where the ambient temperature varies between a low temperature and a relatively high value during a day or part of a day, the continual expansions and contractions of materials and components will accelerate failure.

The various effects of temperature and methods to minimize these effects are shown in Table 2.3.

Humidity

Humidity, the level of moisture content in air, is another environmental condition that

TABLE 2.3 Environmental Effects

Environment	Main Effects	Design Action
High temperature	Exceeding power rating of components. Expansion and softening. Increased chemical action resulting in rapid aging.	Adequate heat sinks and/or ventilation or forced air cooling. Choice of components with low expansion and temperature characteristics.
Low temperature	Contraction. Hardening and freezing. Brittleness. Loss of gain and efficiency.	Possibly heating to controlled temperature. Correct choice of materials and components.
Temperature cycling	Severe stressing. Fatigue failure.	Introduction of large thermal delays to prevent rapid changes affecting internal components.

has to be taken into account. It may vary from as little as 3% RH for deserts and arid regions up to nearly 100% RH in the tropics. Relative humidity (RH) is the ratio of the amount of water vapor present in the air at the same temperature.

The main effects of high levels of humidity are

1. To reduce values of insulation resistance leading to possible electrical breakdown.
2. Corrosion resulting from moisture forming an electrolyte between dissimilar metals.
3. Promotion of fungi growths that reduce insulation.

The worst sort of conditions are those of high temperatures together with high levels of humidity as is common in the tropics.

To minimize the effects, insulating materials must be used that do not absorb moisture or support a water film and if possible resist growth of fungi. Typical materials that do not absorb moisture are silicones and polystyrene.

Obviously the best and most effective method is to hermetically seal (totally close) any sensitive components or in some cases whole assemblies or instruments. For components this may mean encapsulation in a plastic resin and, for assemblies, using sealing rings and gaskets. Desiccants, or drying agents, can be included inside the sealed area to absorb any moisture leaking through the seal. Silica gel is a commonly used desiccator which can also be used as an indicator of moisture, since it changes color (becomes pink) when it has absorbed water.

Mechanical Vibrations and Shocks

All instruments when transported and handled (even when moved from one bench to another) will experience some degree of mechanical shock and vibration. Portable instruments and those used in mobile equipment naturally suffer the worst. Vibrations may be in the frequency range of d.c. to several kHz depending on whether the method of transport is road, rail, sea, or air (including space vehicles). The actual vibration amplitude can vary from centimeters at very low frequencies to fractions of millimeters at the high frequencies. In addition, shocks of up to 20*g* or greater may be experienced if the instrument is dropped. The effects are to weaken supports, loosen wires and connections, bend and possibly fracture components, and to set up stresses that lead to fatigue failure. All of these can be minimized by careful design techniques such as using antivibration mountings, shake-proof washers, locking nuts and varnishes, and by the encapsulation of sensitive components in some protective material. Silicone rubber compounds are often used to "pot" assemblies to provide an absorber for the mechanical energy caused by vibrations and shocks.

Variations in Pressure

Atmospheric pressure variations are important at high altitudes where the pressure falls to a low value, and can cause leaking of seals in components and subassemblies. This can occur if equipment is transported by air in an unpressurized area. In working instruments, low pressure causes a decrease of electrical breakdown voltage between contacts using air insulation. Therefore, for equipment subjected to low pressure the distances between electrical conductors have to be increased, and care must be taken in maintaining a dust and dirt-free path.

Combinations

Usually the conditions around a working piece of equipment are a combination of the several environmental factors, and the effects of all must be carefully assessed during the design and development stage.

Finally other hazards that can occur are those of radiation (X-rays, γ-rays, etc.), corrosive elements contained in the surrounding air (salt spray for example), and the possibility of dust and other unwanted elements entering the equipment. Insects may even decide that the equipment housing makes a suitable living space!

2.7 AVAILABILITY

The one thing most customers require from any equipment or system is maximum use, or maximum "up-time." A system may possess excellent reliability, i.e., have a low chance of failure during operation, but if and when a failure does occur the repair time (or "down-time") must be short. No customer wishes to wait weeks for the repair to be carried out, and in some instances even a few hours can be costly.

$$\text{Availability} = \frac{\text{MTBF}}{\text{MTBF} + \text{MTTR}}$$

where MTTR is the mean time taken to repair any fault and includes the time taken to diagnose, locate, and then repair the fault.

Availability can be expressed as a fraction or as a percentage. Take, for example, the case where an instrument has an MTBF of 1000 hours and an MTTR of 100 hours:

$$\text{Availability} = \frac{1000}{1000 + 100} \approx 0.91 \text{ or } 91\%$$

If the average repair time increased to say 500 hours, the availability of the instrument would fall.

$$\text{New value of Availability} = \frac{1000}{1000 + 500} \approx 0.67$$

$$\text{or } 67\%$$

To keep Availability high, the MTTR must be as short as possible. Given that the necessary test gear and spares are available, much depends on the skill of the service technician in effecting a rapid return to operation. This is why training in troubleshooting and testing is so important. However, other factors that can reduce MTTR are concerned with the overall design, the location of the instrument or system, and the maintenance policy.

During design, the repair time for any possible fault should be considered, and the constructional and layout methods used should make repair as easy as possible. The designer should ensure that the equipment is easily dismantled so that all components are accessible and easily changed. Such methods include using plug-in circuit cards, plug-in components, hinged panels, and clearly labeling component locations. Test points, internal test signal sources, self-checking facilities, and power failure indicators can also be designed in, provided, of course, that these additions (including plug and socket connectors) do not seriously reduce the overall reliability. The provision of a comprehensive service manual is vital. This must contain easy-to-read circuit and layout diagrams; spare part lists with possible equivalents; specifications and test instructions; troubleshooting guides; and dismantling instructions.

The other factors affecting MTTR can be listed as follows:

1. Location of instrument or system relative to service facilities.
2. Availability of spare parts.
3. Availability of diagnostic instruments and test gear.
4. Skill and level of training of the service technicians.

Maintenance policy is discussed later in this book.

2.8 RELATIONSHIP BETWEEN QUALITY AND RELIABILITY

Most of this chapter has been concerned with an introduction to reliability. Another term used in manufacture is quality. QUALITY can be defined as the ability of an item (component, instrument or system) to meet its specification. Note that this definition does not include a statement of time, so reliability is really an extension of quality in time. Reliability could be restated as the probability of quality being maintained for a stated period of time. It's important to realize also that quality when used in an engineering sense does not necessarily imply perfection or excellence—just the ability to meet the specification. Quality and quality control are growing and important areas in manufacture.

2.9 SUMMARY

Failure The termination of the ability of an item to perform its required function.

$$\textit{Failure Rate (FR)} = \frac{\text{Number of failures}}{\text{Number of component hours}}$$

Typical values of FR for electronic components are 10^{-6}/h and 10^{-9}/h. The latter figure represents components with very low FR.

Mean Time to Failure (applies to non-repairable items)

$$\text{MTTF} = \frac{1}{\text{FR}} \text{ hours}$$

Mean Time Between Failures (applies to items that can be replaced)

$$\text{MTBF} = m = \frac{1}{\lambda} \text{ hours}$$

where λ is the system failure rate and is the sum of failure rates of all component parts.

Reliability The ability of an item to perform a required function under stated conditions for a stated period of time.

Where a constant failure rate applies, i.e., failures are random,

$$R = e^{-t/m} \quad \text{or} \quad R = e^{-\lambda t}$$

Unreliability $Q = 1 - R = 1 - e^{-t/m}$.

Product Law of Reliability For units in series, the failure of one means failure of whole system,

$$R_s = R_x \cdot R_y \ldots R_n$$

Redundancy Used to improve system reliability by placing units in parallel.

$$R_p = 1 - Q_p$$

where Q_p is parallel system unreliability

$$Q_p = Q_A \cdot Q_B \ldots Q_n$$

For the special case of two units.

$$R_p = R_x + R_y - R_x \cdot R_y$$

Factors Affecting Reliability

Design
Choice of components
Derating
Layout

Manufacture
Workforce
Tools and equipment
Test and inspection method

Operating environment
Temperature
Humidity
Mechanical vibrations and shock
Pressure
Dust and corrosive elements in atmosphere

$$\text{Availability} = \frac{\text{MTBF}}{\text{MTBF} + \text{MTTR}}$$

EXERCISES 2

2.1 During a long-term test on a batch of 5000 components a total of 3 failed in a 10 000-hour period. Calculate the failure rate and quote it both in failures per hour and percentage failures per hour.

2.2 Calculate the MTTF for one component in Exercise 2.1.

2.3 Calculate the MTBF of an electronic system that contains the following components:

	Quantity	FR ($\times 10^{-6}$ per hour)
Resistors	200	0.004
Transistors	300	0.05
Digital ICs	290	0.02
Thyristors	4	0.5
Capacitors	20	0.02
Connections	1000	0.001

2.4 For an electronic instrument the total failure rate is estimated to be 2.5×10^{-6} per hour. Calculate the MTBF and the reliability for a 10 000-hour operating period.

2.5 In an electronic system comprising a power supply, two amplifiers, an oscillator, and a modulator, the individual MTBFs are as shown below. If the failure of one unit means a total system failure, calculate the reliability for a 1000-hour operating period.

Unit	MTBF
Power supply	30 000 h
Pre-amplifier	100 000 h
Oscillator	60 000 h
Modulator	75 000 h
Power amplifier	40 000 h

2.6 Comment on the environmental conditions affecting the reliability of electronic equipment used in:

(a) Space vehicles

(b) Mines

(c) Geophysical research

(d) Car assembly plant.

CHAPTER 3

Electronic Components

3.1 INTRODUCTION

The various types of electronic components currently available are now so numerous that it is impossible for anyone to have a thorough knowledge of them all. However, it is important that the basic and most common types of component are known, and this chapter covers the constructional details, types and causes of failure, and functional testing, of these basic components. The approach is from the user's point of view, and in particular the test and service technicians.

A good understanding of components and their limitations is an essential part of troubleshooting. For example, knowing that it is highly improbable that a resistor of any type will have a short circuit failure eliminates many components in a circuit when a short is suspected. Another point worth noting is that many failures of components are due to misuse, estimated to be perhaps as high as 40%. Misuse failure results either from operating the component outside its ratings or from poor handling. The designer must ensure that every component is used correctly, that is to say operated within its voltage, current, power, and frequency ratings under all specified conditions and preferably underrated for greater reliability. Test and service technicians need to understand component limitations in terms of handling, soldering, and measurement to reduce the likelihood of premature failure. For example, unplugging while the power is still applied often causes damage to some components by the electrical stresses set up. Test and service technicians should act as a part of a feedback information loop to the designer, enabling weaknesses in design and misuse of components to be identified and then reduced. Accurate reporting of failures is very important, since it assists in sorting out failures caused by misuse from those that can be attributed to the component manufacturer.

3.2 FIXED RESISTORS

The various types of fixed resistor were discussed in terms of their specification in Chapter 1. They are:

Carbon composition.

Carbon film.

Metal oxide.

Metal film.

Metal glaze.

Wire-wound.

It was shown that, for many applications, metal film, metal oxide, or cermet (metal glaze) are now the types usually chosen; mainly for the reasons that these types offer a reasonable resistance range (typically 10 Ω to 1 MΩ), have low temperature coefficients (better than ±250 ppm/°C), and good stability both when stored and under operating conditions. The carbon composition type, of which hundreds of millions have been used, is now less common because of its inferior stability and relatively poor temperature coefficient (−1200 ppm/°C).

Before considering the construction of the various types of resistors, let us look at the methods used for coding resistors and their preferred values. The standard color code is well publicized and will not be repeated here. Note that resistors with a 5, 10, or 20% tolerance are color coded with two significant-digit bands followed by a number of zeros (or decimal multiplier) band and a tolerance band; however, resistors with a 1 or 2% tolerance are color coded with three significant-digit bands, a number of zeros (or decimal multiplier) band, and a tolerance band. The value and the tolerance of the resistor printed on the resistor body are sometimes in a straightforward manner such as 1.82k 1% (1820 Ω ±1%) or in a coded form such as 1821F. For values of 100 ohms and above three significant digits are shown followed by a fourth digit that specifies the number of zeros to follow. For values below 100 ohms the letter R represents the decimal point with all digits significant.

After the value code, a letter will be added to indicate the tolerance:

F = ±1%   G = ±2%   J = ±5%   K = ±10%

M = ±20%

Thus,

R33M is a 0.33 Ω ± 20% resistor
4701F is a 4700 Ω ± 1% resistor
6804M is a 6.8 MΩ ± 20% resistor
2202K is a 22 000 Ω ± 10% resistor.

Resistance Value Code

Examples

Resistance	Marking
0.47 Ω	R47
1 Ω	1R0
6.8 Ω	6R8
68 Ω	68R
100 Ω	1000
1 kΩ	1001
4.7 kΩ	4701
100 kΩ	1003
10 MΩ	1005

It is possible to purchase a resistor of any particular value provided, of course, one is prepared to pay the price. However, in the majority of cases a suitable value can be found within the preferred ranges. Take, for example, a resistor, in a noncritical application, whose value is calculated to be 9.4 Ω, with a tolerance of say ±20%. The nearest preferred value in the E12 series is 10 Ω, which with a ±10% tolerance can lie anywhere between 9 Ω and 11 Ω. A 9.1 Ω value does not appear in the E12 series, since it is covered by the limits of the 8.2 Ω and the 10 Ω as shown below.

Resistance values E12:

8.2 Ω ± 10%	min. value	7.38 Ω
	max. value	9.02 Ω
10 Ω ± 10%	min. value	9 Ω
	max. value	11 Ω
12 Ω ± 10%	min. value	10.8 Ω
	max. value	13.2 Ω

The E24, E12, E6, and E96 series of preferred values of resistance are shown in Table 3.1. In the E96 series much closer values can be obtained (for ±1% tolerance).

Construction of Fixed Resistors

Typical constructions of fixed resistors are shown in Figure 3.1. The **carbon composition** type is made by mixing finely ground carbon with a resin binder and an insulating filler. The resulting mixture is compressed, formed into rods, and then fired in a kiln. The ratio of carbon to the insulating filler determines the final value of the resistance. Silver-plated end caps, which have tinned copper leads attached, are then pressed onto the rod. Alternatively some manufacturers mold the carbon around the leads so that the connection is embedded. This method is claimed to be more mechanically sturdy and to reduce the risk of electrical noise resulting from a poor connection. Finally, the whole resistor is either molded in plastic or given several coats of insulating lacquer to provide electrical insulation and protection from moisture.

FILM RESISTORS are manufactured by depositing an even film of resistive material onto a high-grade ceramic rod. The resistive material may be pure carbon (carbon film); nickel chromium (metal film); a mixture of metals and glass (metal glaze); or a metal and an insulating oxide (metal oxide). The choice of the ceramic rod is important, since this will enhance or degrade the properties of the final resistor. For example, its thermal expansion has to be similar to that of the film material to prevent the film from cracking. Alumina is commonly used. The required resistance value is then obtained by cutting a helical track through the film material, in other words, spiraling off part of the resistive film. By using a close pitch the resistance value can be increased by as much as 100 times or more. This spiraling technique has the added advantage that the final value of the resistor can be trimmed to a very close

TABLE 3.1 Preferred Values (resistors)

Series																								
E24 Series	10	11	12	13	15	16	18	20	22	24	27	30	33	36	39	43	47	51	56	62	68	75	82	91
E12 Series	10		12		15		18		22		27		33		39		47		56		68		82	
E6 Series	10				15				22				33				47				68			
E96 Series	100	102	105	107	110	113	115	118 etc.																

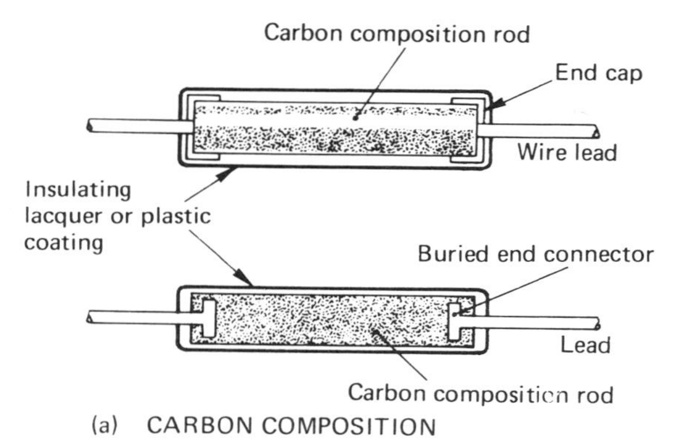

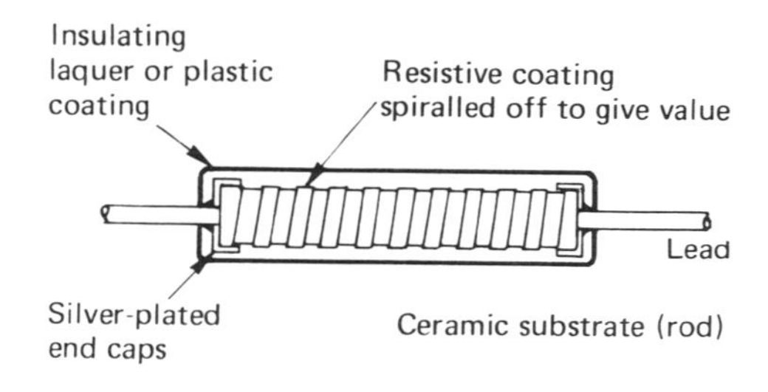

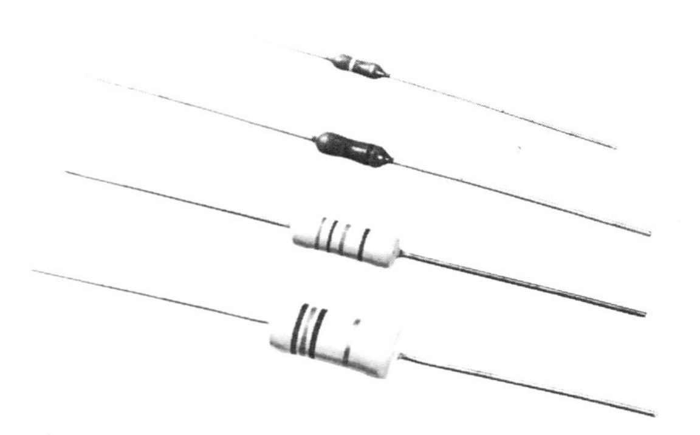

FIGURE 3.1 Construction of fixed resistors.

tolerance of say ±1% or better. The process can be made automatic with the pitch of the diamond cutting wheel being controlled by the output of a bridge circuit which is monitoring the value of the resistor being trimmed.

With film resistors the material used and the thickness of the film will determine the initial resistance value. For example, a metal film of nichrome, approximately 150 Å thick (0.015 μm), will give a resistance of about 125 ohms per square.* The term ohms per square is commonly used for the following reason:

$$R = \rho \frac{l}{a}$$

where ρ is material resistivity, l = length of material, a = cross-sectional area.

Therefore,

$$R = \frac{pl}{dw}$$

where d = depth of film, w = width of film.

Therefore,

$$R = \rho' \frac{l}{w}$$

where ρ' is termed resistivity per unit depth or surface resistivity.

For a square sheet when $l = w$,

$$R = \rho'$$

regardless of the size of the square. Sheet resistance values can therefore be written in ohms per square.

Looking more closely at one of the film resistors, the metal glaze type, the process

---

*One angstrom (Å) is equivalent to one ten-thousands of a micron (μm). Hence,

$$\text{Å} = \frac{1}{10000}\ \mu\text{m}$$

is as follows. First a metal, either chromium, tungsten, thallium, tantalum, or some other type, is ground down into small particles of micron size. This metal powder is then mixed with glass powder of similar size and an organic solvent. The percentage of glass to metal will determine the sheet resistance of the glaze, or resistive ink as it is sometimes called. The glaze is then coated onto a ceramic rod and fired at a temperature of about 1150°C for, say, 30 minutes. This causes the glass to melt and reflow, thus fixing the glaze onto the ceramic rod and producing a very stable component. End caps are fitted with tinned copper leads attached, and then the resistance value is adjusted by cutting the helical track through the glaze with a diamond wheel. The finished resistor is molded in a plastic material to provide electrical and environmental insulation.

The majority of resistors discussed thus far have power ratings of typically 250 mW, although by increasing their physical size power ratings up to 2 W can be manufactured. For dissipating more power, WIRE-WOUND RESISTORS (Figure 3.2) are available with power ratings ranging from 1 W up to, and above, 25 W. A wire-wound resistor is manufactured by winding resistance wire onto an insulating former. Common materials used are nickel chromium (Nichrome), copper nickel alloys (Eureka), and alloys of nickel and silver. The wire is produced by drawing the material through a suitable die and then annealing it. The wire must have good uniformity, be ductile, resist corrosion, and have fairly high resistivity. Ductility is important otherwise the wire might crack or break after winding. On the larger resistors, terminations are provided by a band of conducting metal (ferrule) with a lug. Smaller wire-wound resistors may have end caps or a lead brazed onto the ends of the resistance wire. The connecting lead is

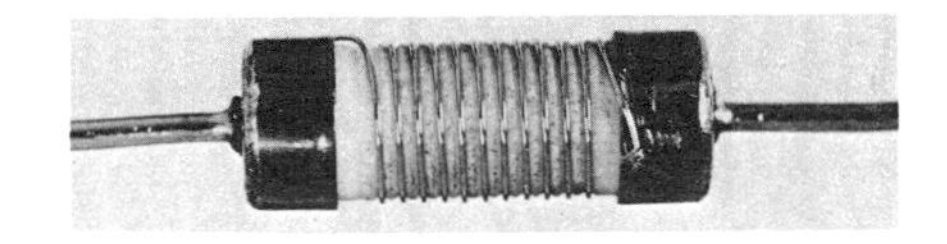

(a)

FIGURE 3.2*a* A miniature wire-wound resistor showing the method of assembly. The final processes of vitreous enamelled coating has still to be applied. The rod is high-quality ceramic with nickel-plated mild steel preformed end caps pressed over the ceramic rod, and the terminating wires are welded to the end caps. These wires are nickel or nickel alloy, capable of withstanding the high firing temperature during the coating process. The helical wire element is a nickel alloy chosen for its comparatively high resistivity and small temperature coefficience. (*Courtesy Welwyn Electronics*)

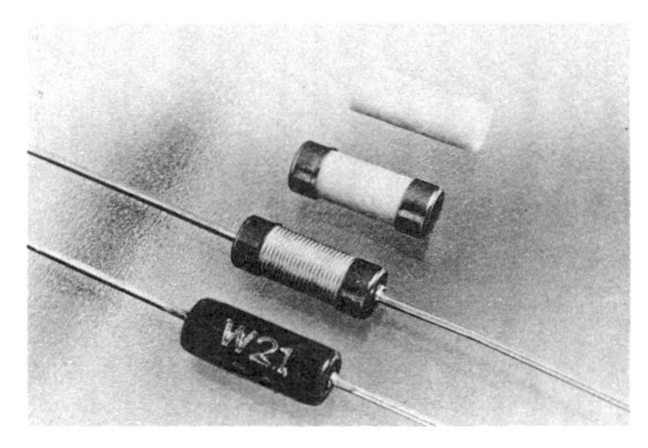

(b)

FIGURE 3.2*b* Showing the process sequence in the manufacture of a vitreous enamelled wire-wound resistor.
(*a*) Ceramic rod.
(*b*) Pressed steel endcaps force fitted.
(*c*) Wound and termination leads fitted.
(*d*) Enamel coating fired and component marked.
(*Courtesy Welwyn Electronics*)

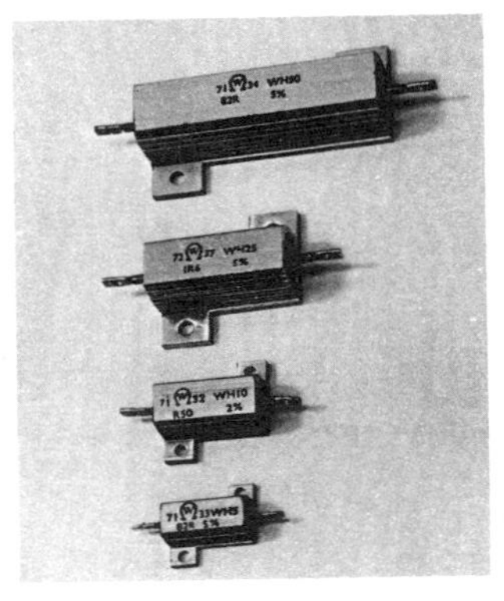

(c)

FIGURE 3.2*c* A range of wire-wound resistors enclosed in aluminum housings. This type of resistor is particularly useful for dissipating comparatively large wattage in a confined space, when it can be mounted on a heat sink such as a chassis. (*Courtesy Welwyn Electronics*)

(d)

FIGURE 3.2*d* A range of high dissipation vitreous enamelled wire-wound tubular resistors. (*Courtesy Welwyn Electronics*)

then folded and inserted inside the ceramic former. This firmly retains the wire leads and prevents mechanical strain being applied to the actual resistance wire winding. The whole resistor is then covered in some insulating material. Usually, wire-wound resistors are vitreous-enamelled, which gives excellent protection against moisture, allows good heat dissipation, and is noninflammable; other types may be cement-coated, which naturally is not impervious to moisture.

To achieve relatively high values of resistance, the wire must be thin and many turns must be used; thus, the maximum value for wire-wounds is limited to about 100 kΩ. General purpose wire-wounds are used where a relatively large power dissipation is required, but another important application is in providing a precision resistor with a very low temperature coefficient and excellent stability. Tolerances can be ±0.1%, temperature coefficients ±5 ppm/°C, and long-term stability better than 40 ppm/year. Other types of resistors cannot yet approach this performance.

3.3 FAILURES IN FIXED RESISTORS

The failure rate and failure mode of a resistor will depend on its type, the method of manufacture, the operating and environmental conditions, and the resistance value.

Any resistor, while operating, has to dissipate power. High powers can be dissipated at low ambient temperatures, but lower dissipation will give improved stability and lower values of failure rate. Since most resistors are of uniform construction, the temperature rise caused by the power being dissipated will be a maximum in the middle of the resistor body. This is called the *hot spot temperature*. Stability, the percentage change of resistance value with time, can be determined experimentally, and some manufacturers publish nomographs showing predicted stability with power dissipation and ambient temperature (Figure 3.3). Note that the hot spot temperature is the sum of the ambient temperature and the temperature rise caused by the power being dissipated. Naturally as the resistor value drifts out of specification, a partial and gradual failure results. It has to be stressed that resistors, in general, exhibit very low failure rates or, in other words, they are very reliable. However, failures do occur because, during the passage of time, mechanical stress and vibrations, heat, applied voltage, and humidity all act to cause chemical or other changes and lead to gradual deterioration. Failures and their possible causes are listed in Table 3.2.

3.4 VARIABLE RESISTORS (POTENTIOMETERS)

These useful components consist basically of a track of some kind of resistive material to which a moveable wiper makes contact. The most simple construction is shown in Figure 3.4. Naturally the methods of manufacture vary considerably, but potentiometers can be grouped under three main headings depending on the resistive material used:

(a) **Carbon** Either molded carbon composition giving a solid track, or a coating of carbon plus insulating filler onto a substrate.

(b) **Wire-wound** Nichrome or other resistance-wire wound onto a suitable insulating former.

(c) **Cermet** A thick film resistance coating on a ceramic substrate.

Several different types of resistors are made, from single turn either open or enclosed slide type, up to multiturn. The component may be intended only as a resistance that requires to be preset and, therefore, adjusted only a few times during its operational life, or as a control that is required to be continually varied over its specified range. The latter must be sturdy, stable, and capable of many thousands of rotations before failing. Generally the requirement for a potentiometer falls into one of the following three categories:

Preset or trimmer.

General purpose control.

Precision control.

Typical examples together with required performances are given in Table 3.3.

A further constraint, not included above, is that of the expected environment. Should the component be enclosed or fully sealed? What performance can be expected under severe stress? For example, some cermet trimmers give the following worst-case changes in resistance when tested to the following specifications.

		Percentage Change
Load	0.5 W at 70°C, 1000 h	±2%
Moisture	95% R.H., 40°C, 21 days	±2.5%
Temperature	250 h at −25°C	±1%
	250 h at +125°C	±1.5%
Thermal shock	5 cycles −25°C to +125°C	±2%
Mechanical vibration	20 *g*	±1%
Soldering	3 sec at 350°C	±0.5%
Rotations	200 cycles at full load	±2%

This specification shows that the multiturn cermet trimmer exhibits good stability and explains why it is now so popular. A typical construction for this type is shown in Figure 3.4. The metal/ceramic resistance track is housed in a dust-proof plastic case and the end leads and wiper connections brought out for mounting on a plastic-

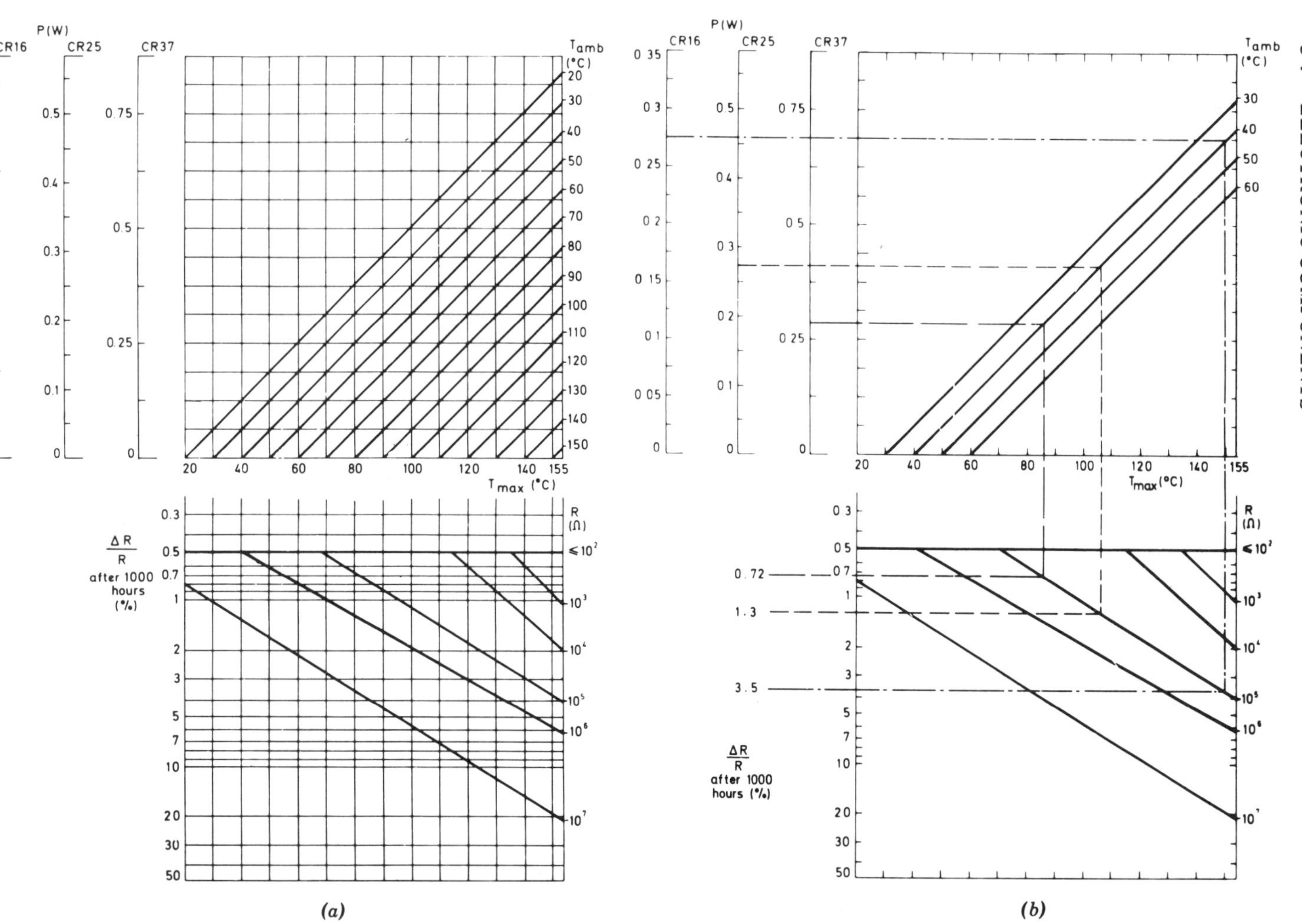

FIGURE 3.3 Nomograph of resistor stability with power dissipation and ambient temperature. (*Courtesy Mullard Ltd.*)

(*a*) Performance nomogram for CR16, CR25 and CR37 showing the relationship between power dissipation (P), ambient temperature (T_{amb}), hot-spot temperature (T_{max}) and stability ($\Delta R/R$) for 1000 hours.

(*b*) A 100 kΩ resistor, dissipation 0.275 W, operating in an ambient temperature of 40°C is required. Under these conditions CR16 will have a hot spot temperature of 150°C with a stability of 3.5%; the CR25 will have a hot spot temperature of 107°C with a stability of 1.3%; and the CR37 will have a hot spot temperature of 86°C with a stability of 0.72%.

TABLE 3.2 Failures in Fixed Resistors

Resistor Type	Failure	Possible Cause
Carbon composition	High in value	Movement of carbon or binder under influence of heat, voltage or moisture. Absorption of moisture causes swelling, forcing the carbon particles to separate.
	Open circuit	Excessive heat burning out resistor center. Mechanical stress fracturing resistor. End caps pulled off by bad mounting on a circuit board. Wire breaking due to repeated flexing.
Film resistors (Carbon, metal oxide, metal film, metal glaze)	Open circuit	Disintegration of film with high temperature or voltage. Film may be scratched or chipped during manufacture. With the higher values (greater than 1 MΩ) the resistance spiral has to be thin and therefore open circuit failure is more likely.
	High noise	Bad contact of end connectors. Usually the result of mechanical stress caused by poor assembly on a circuit.
Wire-wound	Open circuit	Fracture of wire, especially if fine wire is used. Progressive crystallization of wire because of impurities–leads to fracture. Corrosion of wire due to the electrolytic action set up by absorbed moisture. Failure of welded end connection.

(a) WIRE-WOUND

Wiper

Spindle (insulated from wiper)

Card former

Resistance wire wound on former

(b) SKELETON TRIMMER (CARBON)

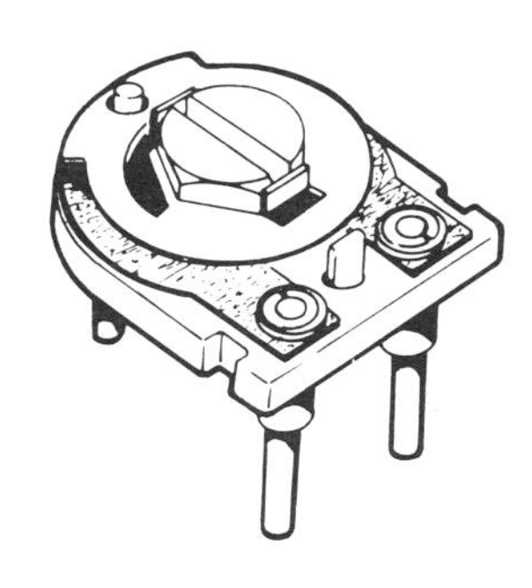

(c) CERMET MULTITURN POT

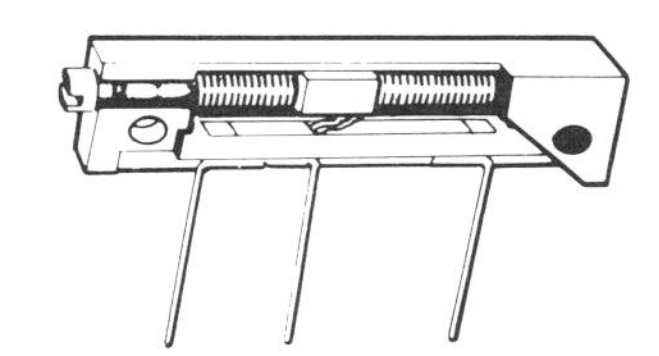

FIGURE 3.4 Basic potentiometer constructions.

TABLE 3.3 Applications of Variable Resistors

Type	Example of application	Selection tolerance	Linearity	Stability	Expected number of rotations	Turns
Preset or trimmer	Adjustment of fixed pulse width from a monostable	±20%	Not important	High ±2%	Less than 50	Single or multiturn
General purpose control (panel mounting)	Brilliance control on CRO	±20%	±10%	Medium ±10%	10 000	Single
Precision control (panel mounting)	Calibrated output voltage control from a laboratory power supply	±3% or better	±0.5%	High ±0.5%	50 000	Single or multiturn

coated base. The wiper, which may be multifingered, is usually a precious metal of say gold-plated bronze, and travels along the resistance track via a stainless steel lead-screw. A slipping clutch action is included so that the wiper assembly idles at both ends of the track to prevent damage from overadjustment.

Apart from specifying such characteristics as tolerance, temperature coefficient, stability, and power rating, designers may require information on:

(a) Starting torque—how much effort is required to cause the wiper to move.

(b) Rotational noise—the change of resistance between wiper and track as the wiper moves (100 equivalent noise resistance).

(c) End or (terminal) resistance—the resistance left between wiper and track at maximum and minimum travel (2% or 1% of track).

(d) Insulation resistance—resistance of the case ($10^9 \Omega$)

(e) Breakdown voltage—maximum voltage between track and case and track and insulated control spindle (400 V).

Figures for a typical cermet trimmer are shown in parenthesis.

Table 3.4 gives information on the comparative performance of the three types of resistive tracks.

3.5 FAILURES IN VARIABLE RESISTORS

Any component that has moving parts and depends for its operation on a good electrical contact between a wiper and a resistance track is bound to exhibit a higher

TABLE 3.4 Variable Resistors—Comparison Table

Type	Construction	Law	Value	Tolerance	Power Rating	Temperature Coefficient	Stability	Life Expectancy Typical
Carbon composition (molded or coated track)	Single turn or preset. Slide type can be ganged.	Linear or log	100 to 10 M	±20%	0.5 W to 2 W	±700 ppm below 100 k. ±1000 ppm above 100 k.	±20%	20 000 rotations
Wire-wound General purpose	Single or multi-turn control or preset ganged.	Linear sine cosine	10 to 100 k	±5%	3 W	100 ppm/°C	±5%	20 000 to 100 000 rotations
Precision	Multiturn (Helipot)			±3%		50 ppm/°C	±2%	
Cermet	Single or multiturn preset	Linear	10 to 500 k	±10%	1 W	±200 ppm/°C	±5%	500 cycles

failure rate than a fixed type. For general purpose variable resistors a failure rate of about 3×10^{-6} per hour is common, but this figure varies depending on the method of manufacture. Partial failures, such as a rise in wiper contact resistance giving increased electrical noise, are common, and so also is intermittent contact. This may be caused by particles of dust, grease, or abrasive matter trapped between the wiper and track. With an open-type construction it is possible to attempt to remove the substance with cleaning fluid. Complete failures will be open circuit, either between track and the end connections or the wiper to track. This can be caused by the corrosion of metal parts by moisture, or swelling and distortion of plastic parts such as track moldings by moisture or high temperature.

3.6 CAPACITORS

A capacitor is created by two parallel conducting plates separated by an insulating DIELECTRIC. The familiar formula for capacitance C is

$$C = \epsilon_0 \epsilon_r \frac{A}{d} \quad \text{(neglecting any edge effects)}$$

where ϵ_0 is the absolute permittivity (vacuum)
ϵ_r is the relative permittivity (or dielectric constant)
A is the area of the plates
d is the distance between the plates, i.e., the thickness of the dielectric.

To achieve any reasonable value of capacitance the area of the plates must be large, the relative permittivity high, and the dielectric thickness small. For the manufacturer this means that the conductors and the dielectric must be thin in order to create a component that has a reasonably small volume. The capacitance-to-volume ratio is important, since the space available for a capacitor on a printed circuit board (p.c.b.), for example, is usually limited. With many types (plastic, paper) long strips of thin conducting foils separated by a thin dielectric are rolled together to make the capacitor (Figure 3.5). Values up to a few microfarads are then possible. Another limitation is that the thin insulating dielectric must be able to withstand reasonable d.c. voltages without breakdown, so the dielectric strength is also important. Often the CV product (capacitance times voltage) is used as a measure of the efficiency of a capacitor type as it gives the total charge Q that can be stored. Electrolytic capacitors have the highest CV ratio available.

From the preceding paragraph we can see that the characteristics tend to be highly dependent on the type of dielectric material. Modern capacitors can be broadly classified by dielectric as in Table 3.5.

Each of these types has particular advantages and areas of use. Applications for capacitors are widespread and include:

Welding, Photoflash } Energy stored in capacitor and then discharged rapidly.

Spark suppression contacts on thermostats, relays, etc.

Smoothing filters in power supplies.

Decoupling and coupling in amplifiers.

Tuning resonant circuits.

Timing elements for monostables, delay circuits, and multivibrators.

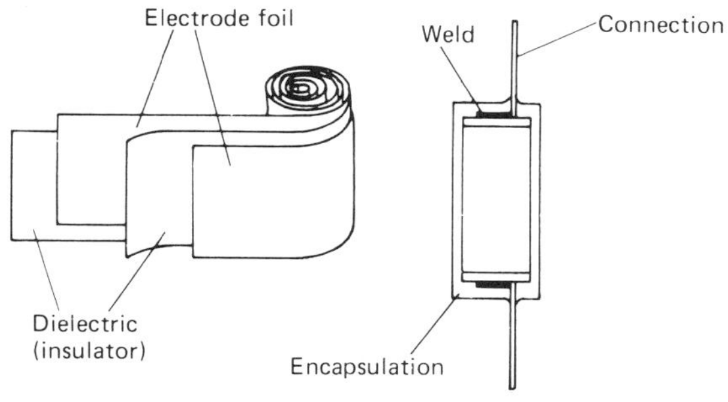

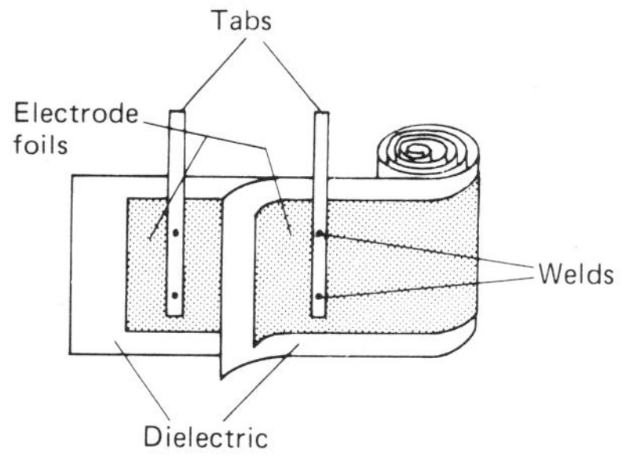

FIGURE 3.5 Basic capacitor.

TABLE 3.5 Capacitors Classified by Dielectric

Dielectric	Dielectric Constant $(K = \frac{\epsilon_r}{\epsilon_0})$
Air	1
Ceramic	
(a) Low-loss types	7
(b) Temperature compensating	90
(c) High permittivity (high K)	1000 → 50 000
Mica (silver mica)	4 to 6
Paper (waxed), gradually being replaced by polypropylene	4
Plastic film	
Polycarbonate	2.8
Polystyrene	2.4
Polyester	3.3
Polypropylene	2.25
Aluminum Oxide electrolytics	7 to 9
Tantalum Oxide electrolytics	27

Filters and waveform shaping and oscillators.

Power factor correction.

Motor start and run, etc.

Later the types which are particularly suited for various applications will be discussed, but before that let us consider the general features and characteristics of a capacitor. The capacitance is not constant, but will vary with temperature, applied voltage, and frequency. With an increase of temperature, the expansion of materials and changes in dielectric permittivity cause the capacitance to change. Polystyrene, mica, and some ceramic types exhibit the lowest changes with temperature (see Table 3.6).

The equivalent circuit for a capacitor (Figure 3.6) shows that it consists of a series inductance, a series resistance (equivalent to losses in the dielectric), and a capacitance with a parallel leakage resistance. Neglecting the leakage for the moment, the rest form a series resonant circuit where

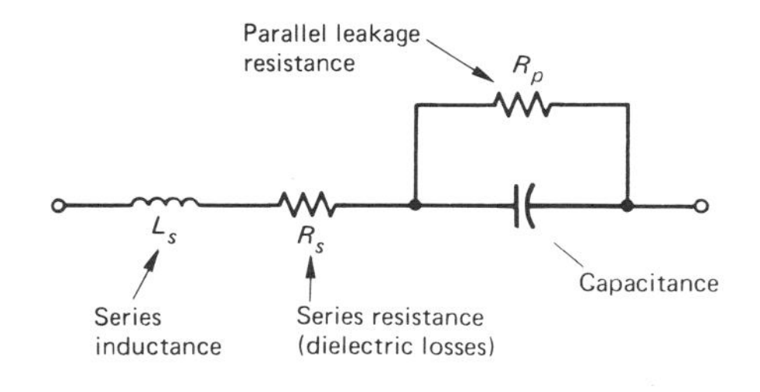

FIGURE 3.6 Equivalent circuit for a capacitor.

TABLE 3.6

Type		Range	Tolerance	Typical a.c. Voltage	Typical d.c. Voltage
PAPER	foil metallized	10 nF to 10 μF	±10%	250 V/500 V rms	600 V
SILVER MICA		5 pF to 10 nF	±0.5%	–	60 V to 600 V
CERAMIC	low-loss high K monolithic	5 pF to 10 nF 5 pF to 1 μF 1 nF to 47 μF	±10% ±20% ±10%	250 V	 60 V to 10 kV 60 V to 400 V
POLYSTYRENE		50 pF to 0.5 μF	±1%	150 V	50 V to 500 V
POLYESTER	foil metallized	100 pF to 10 nF 1 nF to 2 μF	±5%	400 V rms	400 V
POLYPROPYLENE		1 nF to 100 μF	±5%	600 V	1250 V
ALUMINUM ELECTROLYTIC	plain foil etched foil	1 μF to 22 000 μF 1 μF to 100 000 μF	 -20% +50%	Polarized Polarized	 6 V to 100 V
TANTALUM ELECTROLYTIC	foil solid/wet	1 μF to 1000 μF 1 μF to 2000 μF	±10% ±5%	Polarized Polarized	1 V to 50 V

$$Z = \sqrt{R_s^2 + (X_L^2 - X_c^2)} \;;$$

$$X_L = 2\pi f L, X_c = \frac{1}{2\pi f C}$$

Below RESONANCE the impedance is capacitive, at resonance it is resistive, and above resonance it becomes inductive. It is obviously important to keep the inductance low to increase the working frequency range of the capacitor. Low inductance connections are used during manufacture for this purpose. Typical values of resonant frequency are also included in Table 3.6. Above resonance the impedance rises as the inductive reactance predominates, and the capacitor loses its effectiveness.

Electrolytic capacitors may have resonant frequencies well below 100 kHz, and this is the reason why they cannot be used to remove short-duration switching spikes

Temperature Coefficient	f_R	tan δ	Leakage Resistance	Stability	Typical Application
300 p.p.m./°C	0.1 MHz	0.005 0.01	$10^{10}\ \Omega$ $10^{9}\ \Omega$	Fair	Motor start and run Line interference suppression
100 p.p.m./°C	10 MHz	0.0005	$10^{11}\ \Omega$	Excellent	Tuned circuits Filters
±30 p.p.m./°C Varies Varies	10 MHz 10 MHz 100 MHz	0.002 0.02 0.02	$10^{8}\ \Omega$ $10^{8}\ \Omega$ $10^{10}\ \Omega$	Good Fair Good	Temp. compensating Coupling and decoupling
− 150 p.p.m./°C	10 MHz	0.0002	$10^{12}\ \Omega$	Excellent	Tuned circuits Filters Timing
400 p.p.m./°C	1 MHz 0.5 MHz	0.005 0.01	$10^{10}\ \Omega$ $10^{11}\ \Omega$	Fair	General purpose Coupling and decoupling
− 170 p.p.m./°C	1 MHz	0.0003	$10^{10}\ \Omega$	Fair	Line suppression Motor start and run
1500 p.p.m./°C	0.05 MHz	0.08	Specified by leakage current	Fair	Decoupling L.F. Smoothing in power supplies
500 p.p.m./°C 200 p.p.m./°C	0.1 MHz	0.01 0.001	Specified by leakage current	Good Excellent	Coupling and decoupling at L.F. Timing, etc.

from power supply lines. When, for example, digital logic integrated circuits (ICs) switch from one state to another, a relatively large current pulse (14 mA) of short duration (10 nsec for standard TTL) is demanded from the power supply line. Usually a 0.01 μF to 0.1 μF ceramic or other low inductance type of capacitor is wired very near the IC between the positive supply line and ground. The capacitor stores charge during the static state of the logic IC and delivers the extra current required when the gate switches. This, therefore, prevents a voltage noise pulse from being developed on the power line. On the p.c.b. a small tantalum electrolytic, usually 2 μF to 20 μF, may well be used to remove power supply ripple.

Other important characteristics for a capacitor can be derived from the series resistance losses and the leakage resistance. Looking again at the equivalent circuit, for frequencies below resonance we can neglect the effect of the series inductance. A phasor diagram can be drawn showing the effect of the series resistance losses (Figure 3.7). From this

Loss angle $= \delta$ where $\tan \delta = R_s/X_c$.

Phase angle $= \theta$ where $\cos \theta = R_s/Z$.

Impedance $Z = \sqrt{X_c^2 + R_s^2}$.

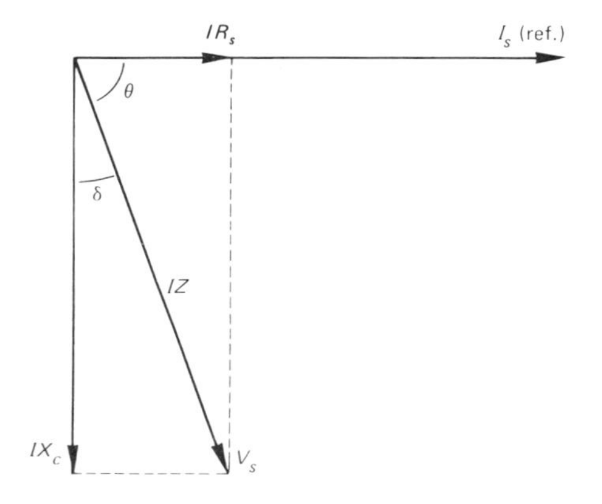

FIGURE 3.7 Phasor diagram of capacitor δ = loss angle θ = phase angle $\tan \delta = R_s/X_c$ (dissipation factor) $\cos \theta = R_s/Z$.

The loss angle δ is a measure of the size of the series resistance compared to the capacitive reactance. For a capacitor δ must obviously be small. The DISSIPATION FACTOR (d.f.) or $\tan \delta$, quoted at a particular frequency (60 Hz for electrolytic capacitors, 1 kHz for other types) is a measure of how "lossy" a capacitor is:

Dissipation factor $= R_s/X_c = \tan \delta$

Values range from as low as 2×10^{-4} for polystyrene capacitors (at 1 kHz) to above 0.3 for large-value aluminum electrolytics (at 60 Hz).

Alternatively manufacturers may quote the POWER FACTOR, which is given by

Power factor $= R_s/Z = \cos \theta = \sin \delta$

Note that for small angles (less than 5°), $\sin \delta = \tan \delta$.

The leakage resistance for most non-electrolytic types of capacitors is very high, typically greater than 10^{10} Ω. For electrolytic capacitors the actual leakage current is usually quoted, which for a very large size type may be several milliamps. Leakage resistance and leakage current are, of course, dependent on both temperature and applied voltage. The resistance decreases as voltage and temperature are increased.

Another method of comparing capacitors is the *RC* product, the TIME CONSTANT formed by the actual capacitor and its parallel leakage resistance. The capacitor is charged and then disconnected from the supply. The charge will leak away through the parallel leakage resistance, and the time for the voltage to fall by 63% will be equal to one time constant. Polystyrene capacitors have one of the highest *RC* products, perhaps of several days, while at the other extreme the *RC* value for aluminum electrolytics may be as low as a few seconds. The *RC* product is used extensively as an indication of capacitor deterioration during life tests.

3.7 CAPACITOR CONSTRUCTION

Basically the constructions of many capacitor types is very similar. The major types are as follows.

Paper Capacitors

Thin sheets of paper, impregnated with oils

or waxes to prevent moisture being absorbed and also to increase the dielectric strength, are wound with thin aluminum foils as shown in Figure 3.5. Contact to the metal foils is made either by welding on tabs giving what is termed "buried foil" design, or by extending the foil at either end. With the extended foil type, the end connection is made by soldering or welding an end cap to the exposed foil. Finally the capacitor is sealed inside a metal can, or encapsulated in resin.

Foil capacitors tend to be rather large for their capacitance values, since the foils themselves may be 5 μm thick and the paper, say, 10 μm. Physically smaller capacitors are made by metallizing the dielectric where the aluminum is vacuum-deposited onto the paper. Metallized capacitors have a free margin (Figure 3.8), so that end connections can be made. Another advantage of the metallized type is that they possess what is called "self-healing" properties. If the dielectric breaks down at one point for some reason, the heat generated by the arc will rapidly vaporize the thin metallization around the breakdown and this clears the short circuit (see Figure 3.8). With the foil type capacitors "self-healing" does not occur, and a defect in the dielectric will cause an arc that destroys more of the dielectric and rapidly leads to complete breakdown. The advantage of the foil over the metallized type is that the foil can dissipate more heat, and has better pulse and overload characteristics.

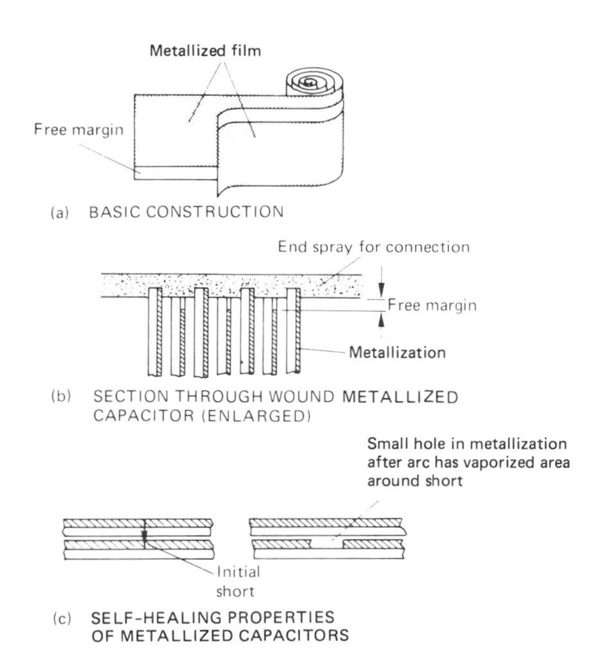

FIGURE 3.8 Metallized capacitor.

Plastic Film Capacitors

These are very similar in construction to paper capacitors and both foil and metallized types are produced. In the foil type a number of thin films of plastic material are interleaved with aluminum foils and rolled into a coil by a winding machine. The coil is then fitted with end caps and encapsulated either in resin or in insulating lacquer. Polystyrene film/foil capacitors were the first plastic capacitors to be manufactured and exhibit excellent stability, high insulation resistance, and low temperature coefficient. They are manufactured to fairly close tolerances, but tend to be rather bulky because metallization of the polystyrene film is not possible, since it has a rather low melting point. The plastic dielectric capacitor that is used extensively in electronic circuits for noncritical applications is the metallized polyethylene, i.e., terphthalate film. They are more commonly referred to as polyester types.

Mica Capacitors

Mica is a naturally occurring substance which, because of its platelike crystal structure, can be laminated into very thin sheets. It is a very stable material with a high permittivity; thus capacitors made from it give good performance. Silver electrodes are metallized directly onto the sheets of mica and several of these are stacked together to make the complete capacitor. The assembly is then either molded in plastic or dipped in resin. The latter have higher reliabilities, since less stress is applied to the capacitor during the sealing process.

Ceramic Capacitors

These can roughly be divided into two classes: the low-loss low-permittivity types and the high-permittivity types (high K). The low-loss types are usually made from steatite, which is a natural mineral. It is finely ground, compressed, and then heated to about 900°C to remove impurities. After being reground it is then re-formed at about 1300° C. Ceramic capacitors are made in disc, tubular, and rectangular plate form. For example, a thin plate is metallized on both sides and connecting leads are soldered to the metallization. The body is then given several coats of insulating lacquer. Such capacitors have low temperature coefficients (±30 ppm/°C) and are usually designated NPO types.

High permittivity ceramics (high K) have the advantage of achieving a relatively large capacitance in a small volume. The common material used is barium titanate. The permittivity can be as high as 10,000. Such capacitors are useful for general purpose coupling and decoupling where fairly wide variations in capacitance value due to temperature, frequency, voltage, and time can be tolerated.

A relative newcomer is the so-called monolithic ceramic or "block" type. Originally it was intended as a chip component that could be attached direct to film circuits, headers, or p.c.b.'s. In this application it had the advantage that the inductance is low because no leads are required. Because of its success this type is now sold as an encapsulated unit with connecting leads attached. The component is made of alternate layers of thin ceramic dielectric (medium to high K) and electrodes. These are compressed and fired to form a monolithic block having the appearance and properties of a single piece of ceramic. A value of, say, 4.7 μF with a working voltage of 50 V d.c. has a size of only 12 mm × 12 mm × 5 mm. These types are now competing with small electrolytics for general purpose coupling and decoupling applications.

Electrolytics

These types have one of the highest CV products of all capacitors and are commonly used as smoothing elements (reservoirs) in power supplies and for coupling and decoupling in audio frequency (a.f.) amplifiers. The large value of capacitance is achieved by the fact that the dielectric formed by the electrolytic action is extremely thin, only a few nanometers.

Electrolytics can be divided into the following subclasses:

Aluminum	plain foil
	etched foil

	solid
Tantalum	solid
	wet sintered.

In the ALUMINUM ELECTROLYTICS, an aluminum foil of very high purity (99.9%) is immersed in a bath of electrolyte and moved through this bath at a constant velocity while a fixed voltage is applied. This voltage causes a forming current to flow which gradually falls in value as aluminum oxide grows on the surface of the foil. This thin covering of aluminum oxide, an insulator, is to be the dielectric of the final capacitor. The anode foil with its thin coating of oxide is wound together with another foil and tissue paper spacers on a winding machine. The paper spacers are saturated with an electrolyte such as ammonium borate or ethylene glycol and this electrolyte, being a conductor, is the true cathode of the final capacitor (Figure 3.9). By chemically etching the aluminum foil before the forming process, the effective surface area of the plates is increased. The "etched foil" type can achieve up to four times the capacitance value of a plain foil because of the effective increase in the surface area of the plates.

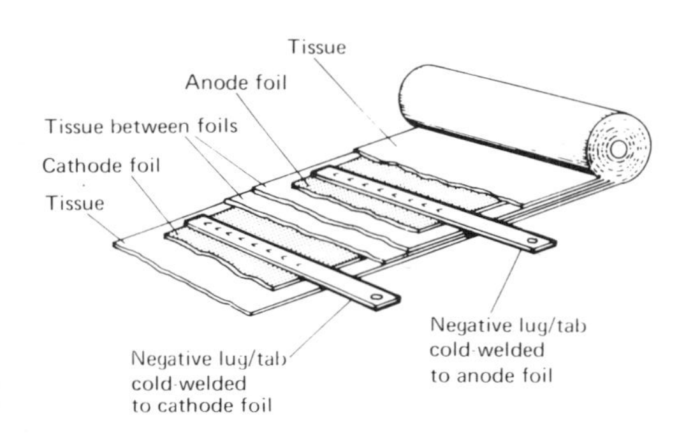

FIGURE 3.9 Aluminum electrolytics.

A few further points on aluminum electrolytics are worth noting:

1. The capacitor is usually polarized and the voltage across it should not be reversed. If a reverse voltage is applied, the dielectric will be removed from the anode and a large current will flow as oxide is formed on the cathode. The gases released from the electrolyte may build up and cause damage to the capacitor, or cause the capacitor to explode and damage other components.

2. The tolerance of the capacitance value is wide, typically −20% to +50%.

3. Leakage current can be high and is usually specified at maximum rated voltage and 20°C five minutes after the voltage is applied. The value of leakage current is due in part to the quantity of imperfections on the surface of the oxide film. A small current has to flow because of the electrolytic action as oxygen is made available from the electrolyte to renew the oxide dielectric on the anode film. Thus leakage current at a fixed temperature depends on the values of the capacitance and the working voltage.

A typical value can be calculated from the following:

$$I_L \approx 0.006CV \text{ amps}$$

(where C is in farads and V in volts). Thus

I_L for a 10 000 μF at 100 V is 6 mA

I_L for a 50 μF at 25 V is 7.5 μA

These are only approximate figures, and manufacturer's data should be consulted for the leakage current of a particular type.

When the capacitor is operating in a circuit the value of the leakage current falls and will usually be much smaller than the specified maximum value.

4. The equivalent series resistance (ESR) is made up of the resistance of the leads, aluminum electrodes, electrolyte, and the series losses of the dielectric. The value of ESR is often quoted instead of dissipation factor (tan δ):

$$\text{ESR} = X_c \tan \delta \text{ ohms}$$

Typical values of ESR are in hundreds of milliohms. Any alternating voltage across the capacitors will cause a "ripple current" to flow through this series resistance. Power loss (I^2R) will result in a temperature rise inside the capacitor, and this heat must be effectively removed by radiation from the surface area of the capacitor case. A maximum value of ripple current I_R at 70°C is quoted for electrolytics used in smoothing applications, and this maximum value must not be exceeded, otherwise the internal temperature rise will cause the capacitor to fail and possibly explode.

Values of max. I_R depend to a great extent on the size of the capacitor: for example, Computer power supply electrolytic 47 000 μF (size 64 mm dia. × 115 mm long)

$$I_R \approx 18 \text{ A rms}$$

Miniature electrolytic for coupling and decoupling (size 12.9 mm dia. × 25 mm long)

$$200\ \mu\text{F at } 10\text{ V} \qquad I_R \approx 125 \text{ mA rms}$$

5. Prolonged storage of an electrolytic may result in the gradual loss of the dielectric film from the anode. When taken from store the film can be re-formed by connecting the capacitor to a d.c. voltage via a high value resistor (10 MΩ).

TANTALUM ELECTROLYTICS use tantalum oxide as the dielectric. This has a higher permittivity than aluminum oxide, and therefore a very high capacitance is available in a small size. Apart from this, tantalums have lower leakage, higher reliability, and better selection tolerance than the aluminum types. The cost is therefore higher.

The first type is similar in construction to the aluminum foil whereas solid types are manufactured by sintering powdered tantalum around a tantalum anode. The resulting "slug" contains tantalum particles separated by an insulator (the dielectric) of tantalum pantoxide.

The cathode is formed by dipping the slug in a solution of manganese nitrate which, when heated to 300°C, gives a semiconductor layer of manganese dioxide. This is then coated with a mixture of graphite and silver. After the anode and cathode leads are welded on, the capacitor is encapsulated either by a resin dip process or in a metal case.

Wet sintered tantalum capacitors use a similar anode slug as in the solid types, but the electrolyte is in liquid form. Typical electrolytes are sulfuric acid or aqueous phosphoric acid. The construction of a tantalum capacitor is shown in Figure 3.10.

3.8 FAILURES IN CAPACITORS

Capacitors are in general reliable components exhibiting low failure rates especially

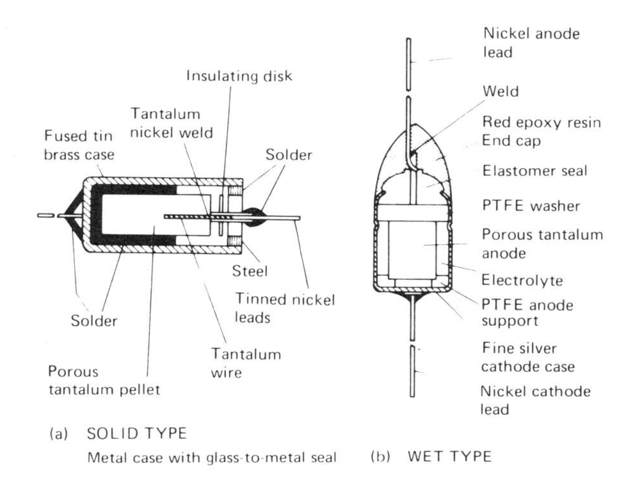

FIGURE 3.10 Tantalum electrolytic capacitor.

when derated. Life can be prolonged by operating well within the rated voltage and at low ambient temperatures, a 10°C reduction in temperature almost giving double the life.

Since all capacitors are essentially metal foils or electrodes separated by an insulating dielectric, the possible general failures are:

Catastrophic

1. Short circuit: dielectric breakdown (except for metallized types).
2. Open circuit: end connection failure.

Degradation

1. Gradual fall in insulation resistance or a rise in leakage current for electrolytic types.
2. Rise in series resistance, i.e., an increase in dissipation factor (tan δ).

Some of the causes of failure are

(a) **Manufacturing defects.** Included impurities, for example, trace chloride contamination in electrolytics causing corrosion of the internal connections. Mechanical damage to the end spray of metallized capacitors leading to overheating and open circuit.

(b) **Misuse.** Capacitor subjected to stresses beyond its stated capability, i.e., exceeding ripple current rating for electrolytics, or poor assembly techniques causing excessive mechanical stress to end connections and seals.

(c) **Environmental.** Mechanical shocks and vibrations. High and low temperatures. Thermal shock. Humidity.

Table 3.7 lists some typical faults and their possible causes for various capacitor types.

3.9 INDUCTORS

The trend in electronics is to use inductors only in those applications where no other good solution exists. This is partly because inductors are generally expensive, bulky, less ideal than capacitors, and not capable of being integrated like capacitors are integrated in one piece of silicon.

3.10 SEMICONDUCTOR DEVICES

The almost explosive growth of the elec-

TABLE 3.7 Failures in Capacitors

Capacitor Type	Failure	Possible Cause
Paper foil	(*a*) Loss of impregnant leading to eventual short circuit. (*b*) Intermittent or open circuit.	(*a*) Seal leak. Mechanical, thermal shock or variations in pressure. (*b*) Damaged during assembly or mechanical/ thermal shock.
Ceramic	(*a*) Short circuit. (*b*) Open circuit. (*c*) Fluctuations in capacitance value.	(*a*) Fracture of dielectric from shock or vibration. (*b*) Fracture of connection. (*c*) Silver electrode not completely adhering to ceramic.
Plastic film	Open circuit.	Either damage to end spray during manufacture or poor assembly.
Aluminum electrolytic	(*a*) Leaky or short (*b*) Fall in capacitance value (*c*) Open circuit.	(*a*) Loss of dielectric. High temperature. (*b*) Loss of electrolyte via leaking seal caused by pressure, thermal or mechanical shock. (*c*) Fracture of internal connections.
Mica	(*a*) Short circuit. (*b*) Intermittent or open circuit.	(*a*) Silver migration caused by high humidity. (*b*) Silver not adhering to mica.

tronic industry in the last few decades can be attributed to the discovery, development, and exploitation of semiconductors. The history of development from the discovery of the point contact transistor in 1948, the bipolar junction transistor in 1950, the planar process in 1959, the first integrated circuit in the early 1960s, the MOS transistor in 1962, through to the large-scale integrated devices such as microprocessors, large RAM memories, etc., of the 1970s is really fascinating.

It would be impossible to cover all the important steps and breakthroughs in manufacturing techniques in this brief section. Instead we shall concentrate on the planar process, since this forms the basis of the manufacture of the vast majority of modern semiconductors from diodes, bipolar and unipolar transistors, to ICs and large scale integrated (LSI) devices. For this reason an understanding of the process is important. In this section ICs have been included with discrete semiconductor components since, as far as test or service is concerned, an IC must be regarded as an individual component that has to be replaced if proved faulty.

The PLANAR PROCESS resulted from the early research work carried out on the diffusion of impurities into silicon, and it solved many of the problems concerned with the manufacture of transistors and diodes in volume, at lower cost, and with reliable characteristics. Since the process is precise and accurate, more refined transistor designs became possible and transistors with cutoff frequencies (f_T) of 1 GHz could be manufactured. The basis of the process depends on the fact that a thin coating of silicon dioxide (SiO_2), grown on the surface of a silicon wafer, forms a very effective seal against moisture and impurities. If selected areas of this oxide are removed, in other words, "windows" are created, then impurities could be forced into exposed silicon by diffusion to create individual p or n regions. Planar processing is a continuous repetition of three basis sequences:

(a) Oxidation—the formation of silicon dioxide over the whole surface of the silicon wafer.

(b) "Window" opening by photo-resist techniques via an accurate mask.

(c) Solid-state diffusion to make selected areas p or n type.

Consider the manufacturing steps required to make discrete transistors (npn) (Figure 3.11).

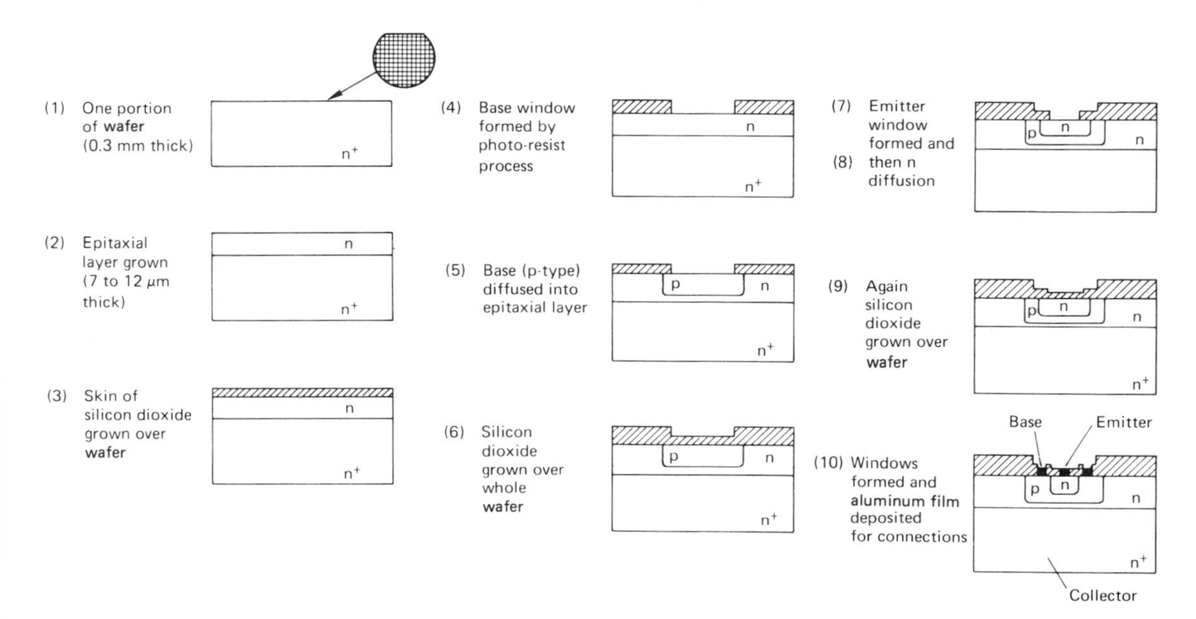

FIGURE 3.11 Discrete planar transistor.

1. Wafers of heavily doped n-type silicon are cut, lapped, and chemically polished. The wafer diameter can be from 25 mm up to 75 mm and is typically 0.3 mm thick. On a large wafer several thousand transistors can be made at the same time.

2. An epitaxial layer of lightly doped n-type silicon is deposited onto the surface of the wafer. It is in this epitaxial layer that the active parts of the transistor will be created. This layer, typically 7 μm to 12 μm thick, is deposited by exposing the wafer to an atmosphere of silicon-tetrachloride at a temperature of about 1200°C.

3. The wafer is heated to about 1100°C in steam for 1 to 2 hours and a skin of silicon dioxide forms over the surface of the wafer. The thickness of the oxide is typically 0.5 μm to 1 μm.

4. Photo-resist stage. Wafers are spun-coated with a film of resist material, and this is then dried by baking. The first mask is brought into contact with the wafer and ultraviolet light is exposed to those areas of the photo-resist not covered by the mask. The unexposed photo-resist is removed with a solvent and the silicon dioxide thus revealed is chemically dissolved away, leaving a "window" to the expitaxial layer. Finally the rest of the photo-resist is stripped off. All in all five process steps.

5. Base diffusion—an impurity, in this case boron since the base is to be p-type—is first deposited on the surface of the wafer and is then diffused into the silicon via the exposed window by heating the wafer in a furnace at 1100°C for between 1 and 2 hours. Since the process of solid state diffusion is slow, it is readily controllable and therefore precise. When the required base is achieved the wafers are removed from the furnace and the excess boron removed chemically from the surface.

6. Silicon dioxide is again grown over the whole wafer, thus covering the newly created base region.

7. The emitter window is defined by a photo-resist process and a second mask. Note that this mask must be accurately aligned with respect to the first mask.

8. Emitter diffusion using phosphorus as the impurity.

9. Silicon dioxide is again grown over the whole wafer, thus covering the newly created emitter region.

10. Access to the transistor is achieved by a third photo-resist and masking stage to define the contact areas. The contacts are made by evaporating aluminum onto these exposed areas. The back surface of the wafer will be the collector contact area.

11. Each transistor on the wafer is now tested, usually automatically, and devices that fail are marked

and discarded. Transistors that pass, i.e., are within specification, are separated from the wafer by scribing with a diamond stylus and breaking into individual units called dies.

12. Each die is then mounted on a header, usually gold plated, which then forms the collector contact. Connections are made to the base and emitter metallizations, and the transistor is then sealed either inside a metal can or encapsulated in plastic (Figure 3.12).

13. Final test of complete transistor.

During the manufacturing process great care has to be exercised at each step. Cleanliness and careful handling are essential factors in avoiding unwanted effects due to small particles of dust which can cause pinholes during the photo-resist stage. Photo-resist and wafer preparation processes have to be carried out in "superclean" rooms to avoid any possibility of contamination. This means achieving dust counts for 0.5 μm size particles of less than 35 000 particles per cubic meter in rooms and 3500 particles per cubic meter in processing cabinets. For comparison the dust count in an uncontrolled room for 0.5 μm particles could be approximately 2×10^7 particles per cubic meter. All liquids used in the various stages such as water, solvents, photo-resist solutions, and acids must also be filtered to remove any contaminating particles.

The quality of the masks is also of the utmost importance. These must be very accurate in detail and dimensions and have as few defects as possible.

As described this process is used for manufacturing diodes, bipolar transistors, unipolar transistors (FETs), and integrated circuits (both bipolar and MOS). It can be seen that many separate processing steps are necessary for even a single transistor and so for bipolar integrated circuits the process is quite lengthy. At each step there is the possibility of a failure or "yield loss," for example, imperfections in the silicon wafer, incomplete wafer cleaning, contamination of diffused areas with unwanted impurities, mechanical damage to the wafer or die, and so on. Thus for bipolar ICs the yield may be quite low, say 20% or less, and for this reason, especially for LSI devices, simplified processes are being used. These include MOS and CDI (Collector Diffusion Isolation). IC manufacture is discussed later.

Returning to the planar transistor, apart from the improvement in manufacture, the device itself possesses several advantages over the alloy-junction type, notably:

(a) Low leakage current.

(b) Accurate control of base width giving high f_T.

(c) Low collector resistance and low saturation voltage $V_{CE(sat)}$.

These transistors can, then, be used for fast switching applications and for digital logic circuits.

As stated previously the FET family of transistors is also manufactured using the planar process, although the number of process steps is fewer. Unipolar transistors, so called because the current flow is carried by only one type of charge carrier (either

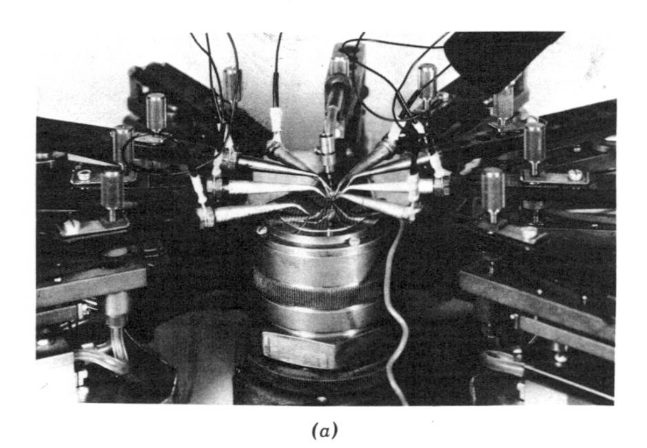

(a)

FIGURE 3.12 Transistor testing and mounting (*a*) Probing machine for checking breakdown voltage and direct current gain. The machine checks four transistors at a time and marks any faulty one with a tiny spot of ink. The picture shows the many devices made from a wafer of silicon a little over an inch in diameter.

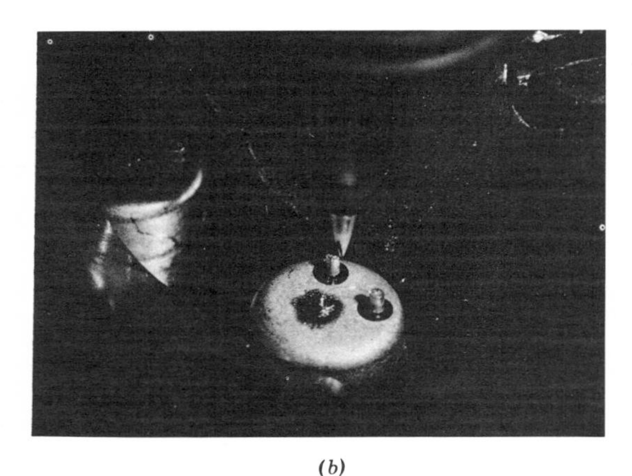

(b)

(*b*) Transistor just after the base and emitter have been connected to the leadout wires. The aluminum wire, only 25 μm thick, is connected to the transistor by thermocompression bonding. The transistor is 0.5 mm square and mounted on a TO-18 type header. (*Courtesy Mullard Ltd.*)

holes or electrons), depend for their operation on an electrostatic field to control the flow of current.

It is easy to be confused by all the various types, especially as different names have been given to the same device. The types are:

Junction Field Effect Transistors JFET
for which both p and n channels are produced

and

Metal Oxide Semiconductor Field Effect Transistors MOSFET
sometimes referred to as MOST—Metal Oxide
or MOS
or IGFET—Insulated Gate FET

With MOSFETs, in addition to n and p channels, the transistor can be made to be either a depletion or an enhancement mode type.

The structures of these FETs are shown in Figure 3.13. With the enhancement MOSFET, which is important to understand since this is the type that is used in CMOS logic and other MOS ICs, the gate electrode is insulated from the silicon substrate by silicon dioxide, and there is no conducting path between source and drain. This is because for a p-channel type the substrate is actually doped n-type. With the drain held negative with respect to the source, no current flows. However, if the gate electrode is made negative with respect to the substrate (normally connected to the source), some holes are attracted from the n-substrate to the region just beneath the gate, and electrons are repelled from this region. If the gate to substrate voltage is above what is called the threshold voltage

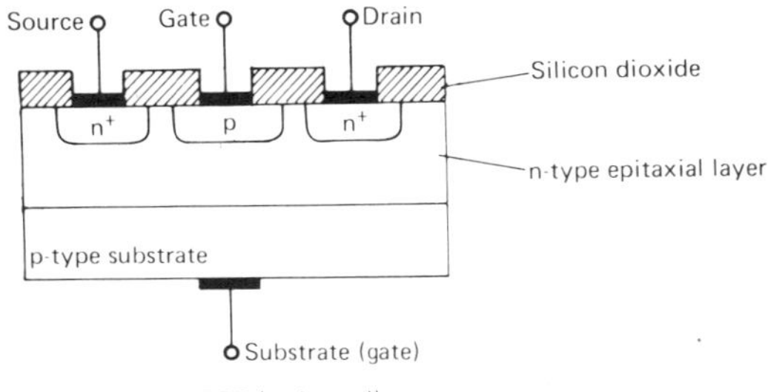

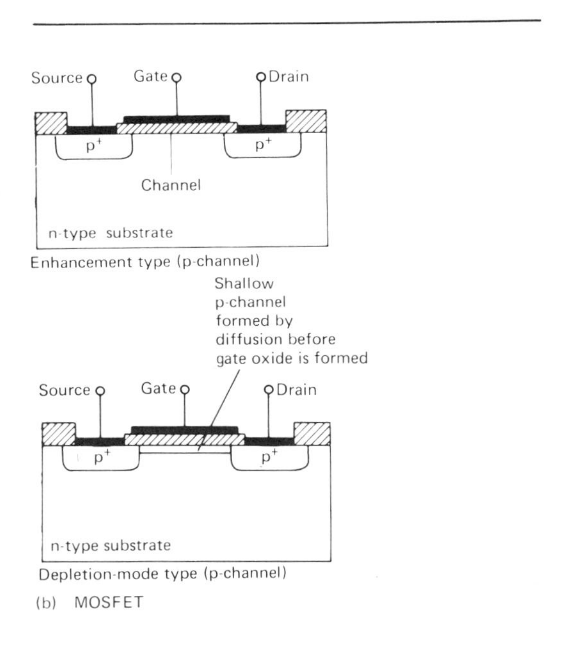

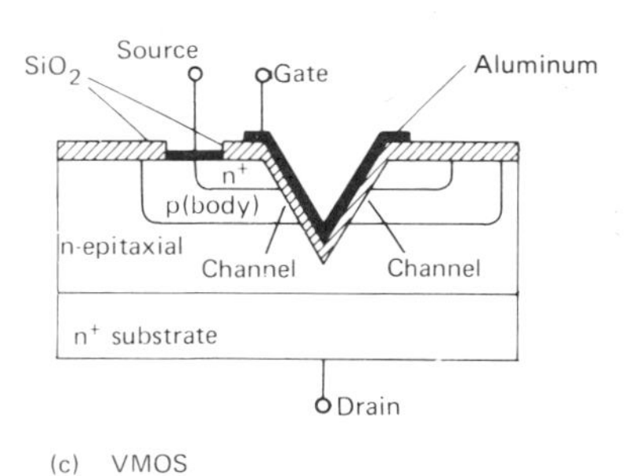

FIGURE 3.13 Field effect transistors.

($V_{GS(th)}$), a conducting channel is induced below the gate between source and drain and current flows. This then is the enhancement operation of a MOSFET.

In the depletion type of MOSFET, manufactured as a discrete component, a thin conducting channel is diffused in just beneath the gate so that there is current between the source and drain with zero gate to substrate bias. If the bias gate to substrate is made positive, the conducting channel (for p-type) is reduced, i.e., depleted, causing the drain current to decrease. Conversely if the gate to substrate bias is made negative, the conducting channel is increased, i.e., enhanced, and drain current increases. This causes part of the confusion because this type can be used either in depletion or enhancement mode. The big advantage of MOS devices whether as discretes or integrated is that the input impedance is very high (typically $10^{14}\Omega$). However, the thin insulation between gate and substrate (1000Å) can be easily damaged by electrostatic fields, and for this reason special precautions in handling have to be observed. In most cases diode protection circuits are built in to protect the gate insulation from punch-through; this can be seen in Figure 4.9*d* (p. 105).

VMOS Power FETs

These devices are a relatively recent innovation and represent a significant advancement of MOS transistors in power applications. For example, with a VMOS transistor it is possible to switch 1 A in less than 4 nanoseconds. Figure 3.13*c* shows that the structure of the VMOS is made of a heavily doped n^+ substrate and a lightly doped epitaxial layer. A p-region for the source and then a further n^+ region are diffused into the epitaxial layer. Following this, a V-groove is etched through the source region (p) and through into the epitaxial layer (n^-). Oxide is grown over the groove and aluminum deposited to form the gate. Connections are made to the p-region for the source and to the substrate for the drain. The operation is similar to a conventional MOSFET in which a conducting channel is induced from source to drain by a controlling voltage on the gate, but unlike the conventional MOSFET the VMOS uses vertical, rather than lateral, current flow. In other words electrons flow down from source to drain in a VMOS. With both gate and drain positive with respect to the source, the gate induces an n-type channel in both surfaces of the p-type source in the regions facing the groove. Current then flows as electrons are attracted from the source to the drain.

There are several advantages resulting from this type of construction:

(a) The lengths of the induced channels are determined by the diffusion process not by masking, and are therefore more controllable. Channel lengths are relatively short (1.5 μm) giving greater current density (ratio of channel width to length determines current density).

(b) Since each V-groove has two channels, the current density is doubled.

(c) The substrate forms the drain contact, and this enables heat to be removed efficiently especially as the drain can be mounted directly on the header.

(d) Since drain metallization is not required on the top of the die, the die area can be reduced and this gives lower capacitance.

(e) As with a bipolar transistor, the substrate is heavily doped, therefore the saturation resistance of the VMOS is low.

(f) The VMOS epitaxial layer absorbs the depletion region from the reverse biased body to drain pn diode and this therefore increases the breakdown voltage rating.

These are the advantages of the VMOS compared with the conventional MOSFET, but the VMOS being a MOS device possesses several advantages over the power BJT. It has very high input impedance, requiring input currents of less than 100 nA. This means the device can be directly interfaced with CMOS logic or other high-output impedance circuits. The VMOS being a unipolar device exhibits negligible charge storage effects and this gives it its fast switching speed.

Finally the temperature coefficient of the VMOS drain-to-source on voltage is positive, the VMOS current decreasing as temperature rises. There are then less problems of thermal runaway than in BJTs. The latter point can cause problems with high voltage supplies when the drain current is held relatively constant, since a rise of $V_{DS(on)}$ caused by the temperature will cause a rise of V_{DS} and increase dissipation.

Integrated Circuits

Some of the main reasons for the growth of ICs are that they are small and light in weight; more reliable than discrete circuits; have fewer connections; can be mass produced and are therefore cheap; and circuits can be created that would otherwise be uneconomical. The trend is towards packaging more and more of the circuit functions inside the IC so that future instruments will consist of maybe only one LSI circuit plus a few discrete components such as variable resistors.

The term IC generally means a monolithic integrated circuit, or in other words an element in which all the necessary diodes, transistors, capacitors, and resistors for a particular circuit function are diffused and interconnected in one piece of silicon. The word "monolithic" is derived from the Greek, "mono" meaning single and "lithos" meaning stone (the stone in this case being silicon). Other types of ICs are thin and thick film which will be discussed later. In general, film circuits are used where the ratio of passive to active devices is high.

In monolithic IC production the devices are produced several hundred at a time on one 50 mm or 75 mm diameter silicon wafer, using the planar process. There are two important differences between bipolar ICs and discrete semiconductors. For the IC,

(a) Each transistor, diode, resistor, etc., must be isolated from adjacent elements.

(b) All connections must be made to the top surface; this includes collector leads.

Several methods of isolation exist:

Junction isolation.

Dielectric or oxide isolation.

Collector diffusion isolation.

With junction isolation, p-regions are diffused through the thin n-type epitaxial layer to the p-substrate. Each element is then surrounded by a pn junction. During operation the substrate is connected to zero volts, thus reverse-biasing all the isolation junctions. In dielectric isolation, a layer of silicon dioxide is formed around each element. This is a more costly method than junction isolation, and it is only used on special circuits.

Collector diffusion isolation (CDI) is a relatively new process that simplifies the total number of steps required to make bipolar ICs. This method is already being used in the manufacture of LSI devices. The process is really an extension of the BURIED COLLECTOR REGION, and this will be explained first to show how a transistor is created in IC form.

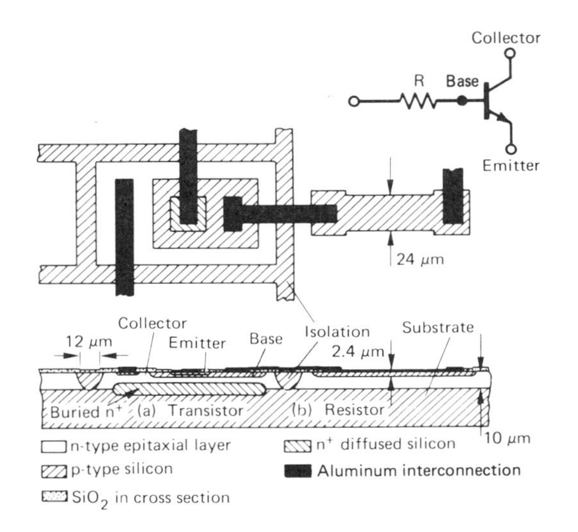

FIGURE 3.14 IC transistor with resistor load.

An npn IC transistor construction is shown in Figure 3.14. Since the collector as well as base and emitter connections must be brought out to the top surface, it was soon discovered that the collector resistance and $V_{CE(sat)}$ were relatively high if the collector region consisted only of the relatively lightly doped n-epitaxial region, whereas in the discrete planar transistor a low resistance path resulted from the heavily doped n-substrate. To reduce the collector resistance, a highly doped n-region is diffused into the p-substrate before the epitaxial growth, and this is called the buried n-region. To create the npn IC transistor the process steps required are:

1. Preparation of silicon wafer p-type.
2. Oxidization of wafer.
3. First photo-resist to define buried collector n-regions.
4. Buried collector diffusion n-type. Arsenic 5 hours at 1300° C.
5. Removal of oxide.
6. Epitaxial layer grown, lightly doped.
7. Oxidization of wafer.
8. Photo-resist to define isolation.
9. Isolation diffusion p-type. Boron 16 hours at 1180°C.
10. Oxidization.
11. Photo-resist to define base regions.
12. Base diffusion.
13. Oxidization.
14. Photo-resist to define emitter regions.
15. Emitter diffusion.
16. Oxidization.
17. Photo-resist to define contact areas.
18. Metallization of aluminum onto contacts.
19. Wafer probe test.
20. Scribe and break into dies.
21. Mounting and wire bonding.
22. Encapsulated and final test.

As noted before, this is a continual repeat of the planar process, and naturally during the process other elements such as diodes and resistors have to be created to make a complete microcircuit. A resistor is usually a p-type area as shown diffused into the epitaxial region.

COLLECTOR DIFFUSION ISOLATION is a simpler process and requires a much smaller area for the transistor and is, therefore, being exploited for LSI devices. In this process the buried n-layer is used as the actual collector. On this is grown a thin p-type epitaxial region which forms the base. A single selective n-diffusion then gives both the collector and isolation as shown in Figure 3.15, since this n-diffusion reaches through the thin p-region to the buried n-layer. A shallow nonselective p-type layer is next diffused over the whole wafer to give resistors their desired value and then the n-region for the emitter.

MOS ICs are manufactured using fewer process steps than bipolar, and the resulting MOSFET structure can be created in an

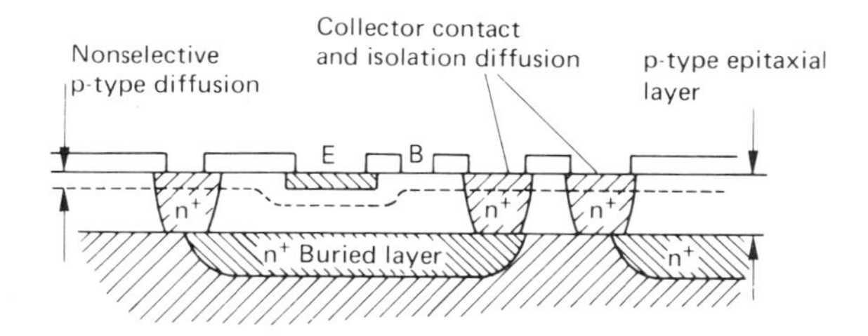

FIGURE 3.15 Collector diffusion isolation.

area about 60 μm by 40 μm. In addition, MOS transistors can be used as load resistors and are self-isolating. This is why the packing densities of MOS ICs can be so large. Logic circuits using MOS are described in the next chapter.

Film or hybrid ICs fall into two main classes:

THIN FILMS A film of appropriate conducting material is deposited onto an inert substrate such as borosilicate glass or glazed alumina. Film materials, such as nickel-chromium, gold, tantalum, etc., are either evaporated or sputtered from a cathode, and the required pattern is obtained either by masking the substrate or by etching after the film has been applied.

THICK FILMS A silk screen process is used, where the film material is deposited in paste form onto a substrate via the screen. The unit is then fired in a furnace at about 1000°C. Cermets and resistive inks are used as the film materials, and the substrates are commonly alumina or beryllium.

For both types of film circuits, adjustments of resistor values can be made to very close tolerance, which is not possible in monolithic circuits. Trimming to better than ±0.5% can be achieved with air abrasion, and with laser beams to within ±0.1%.

Low-value capacitors and inductors (spiraling film) can be made, but more usually these components and active devices such as "flip-chip" transistors and diodes are added to the film network. Connections are made by ultrasonic bonding, and finally the unit is encapsulated.

3.11 FAILURES IN SEMICONDUCTORS

Semiconductor devices can be broadly classified into two areas:

1. *Bipolar*	Transistors Diodes, unijunctions Thyristors Logic ICs such as TTL, ECL Linear ICs
2. *Unipolar*	Junction FETs MOSFETS VMOS power FETs CMOS logic Some linear ICs

With the bipolar group the action of the devices depends on the flow of the type of charge carrier across forward or reverse biased pn junctions. In an npn bipolar transistor, for example, the flow of electrons across the reverse biased collector/base junction is controlled by the forward biased base/emitter junction. As electrons cross into the base, some recombine with holes, but the majority diffuse across the base and are swept up by the collector.

Unipolar devices, on the other hand, use only majority carriers for current flow, and this current is controlled by an electrostatic field between gate and source or gate and substrate.

What the two groups have in common is that they can easily be destroyed if subjected to an overload, and also the fact that the manufacturing processes are very similar.

Failure mechanisms inherent in the manufacturing process include:

Diffusion and Epitaxial Growth Processes

(a) Flaws and imperfections in basic crystal wafer.
(b) Incorrect resistivity.
(c) Incorrect diffusion.
(d) Contamination.
(e) Epitaxial "washout" of pattern or "spike" formation.

Photo-Resist Processes

(a) Mask imperfections, scratches or pinholes.
(b) Poor mask alignment.
(c) Poor mask definition.
(d) Contamination of photo-resist.
(e) Insufficient cleaning or etching.

Metallization processes

(a) Poor ohmic contact.
(b) Metal adhering badly or too thin.
(c) Microcracks or voids over oxide steps.

Mechanical Processes

(a) Chipping, cracking or fracture during scribe and break.
(b) Weak bonds, sagging leads, microcracks in leads during wire bonding.
(c) Poor seal.

Many of these defects will be picked up during manufacture or by "burn-in." However, some units, whether discrete or IC, may possess inherent weaknesses from any of these sources when delivered and may fail prematurely in equipment.

Of much greater importance are failures caused by misuse, by bad handling during assembly or test, or by exceeding the maximum rated values of voltage, current, and power. Most ICs, for example, will be permanently damaged if the maximum supply voltage is exceeded or if the device is removed from or inserted into a test socket while the power is applied.

Apart from the main environmental hazards, electrical interference is another cause of premature failure in semiconductors. Voltage surges carried by the a.c. power lines, caused by heavy machines or relays being switched, can easily cause breakdown of semiconductor junctions.

Failures are mostly those of an open or short circuit at a junction. In other words, a bipolar may fail short or open between base and emitter or collector and base. In addition it is possible for a short circuit to occur between collector and emitter. With digital ICs it is usually only possible to identify which gate in the package has failed, but accuracy is important, since reporting circuit failures to the design engineer can help to reduce future failures.

3.12 PRECAUTIONS WHEN HANDLING AND TESTING COMPONENTS

Any component can be damaged by careless handling during assembly and test or while equipment is being serviced. By following a few sound principles, most damage can be avoided. First of all it is important to be aware of the type of component in question and any special precautions necessary when handling or testing it. This means being familiar with component specifications. Most manufacturers supply excellent data in books or sheets for this purpose.

The major causes of component damage are the following.

Lead Bending

A few sharp back and forth bends in a component lead can easily cause it to break or crack. Bending a lead too close to the component encapsulation can generate excessive stress where the lead enters and cause cracks in the encapsulation. Such cracks offer openings for moisture and therefore cause gradual degradation of the component and premature failure. As a general rule, allow at least 3 to 5 mm clearance between the component and the start of a lead bend.

Mechanical Shock and Damage

Dropping components, particular semiconductors, can cause damage by the high impact shock. For example, dropping a transistor from table height onto a concrete floor can subject it to an impact shock of several hundred *g*. Even cutting a lead can cause shock waves which may damage delicate or brittle components. Apart from this, always avoid scratching or cutting surfaces of components by the careless use of tools or sharp test probes. Printed circuit boards should not be stacked one on top of another without protection by, for example, thin sheets of plastic foam between them. Both the print and the components can be scratched and damaged very easily.

Overheating and Thermal Shock

If, during soldering, the maximum temperature for the component is exceeded, it may be permanently damaged, weakened, or be caused to drastically change its value or characteristics. These effects are not usually noticed during assembly or test but, unfortunately, may show up later when the equipment is in use.

Heat transmitted along connecting leads can cause unequal expansion between leads and packages and crack hermetic seals. In general, for hand soldering, a low wattage iron (20 to 50 W) should be used, and for iron temperatures of between 300° to 400°C the tip of the iron should be in contact for not more than 5 seconds. Other safety precautions are to remove ICs and transistors, if they are in sockets, to use heat sinks, and most important of all to solder quickly and cleanly.

When removing a faulty component, solder wick should be used to absorb the molten solder from each joint in turn until the component is entirely free. Alternatively a desoldering tool that operates by suction can be used. The tool is primed and the nozzle placed over the molten solder, the plunger released, and the surplus solder sucked up into the nozzle of the gun where it solidifies. When the plunger is operated again, the solder pellet is ejected. Obviously during any repair work it is well worth taking time and care so as not to damage or lift copper trace from the printed circuit board. The p.c.b. is usually a very expensive item.

HANDLING MOS DEVICES

Though all our MOS integrated circuits incorporate protection against electrostatic discharges, they can nevertheless be damaged by accidental over-voltages.
In storing and handling them, the following precautions should be observed.

1. Store and transport the circuits in their original packing. Alternatively, use may be made of a conductive material or special i.c. carrier that either short-circuits all leads or insulates them from external contact.

Caution Points 2 and 3 call for special attention to personal safety.
Personnel handling MOS devices should be connected to earth via a resistor.

2. Work on a conductive surface (e.g. metal table top) when testing the circuits or transferring them from one carrier to another. Electrically connect the person doing the testing or handling to the conductive surface by, for example, a metal bracelet and a conductive cord or chain. Connect all testing and handling equipment to the same surface.
3. Mount MOS integrated circuits on printed circuit boards after all other components have been mounted. Take care that the i.c's, the metal parts of the board, the mounting tools, and the person doing the mounting are kept at the same electric potential. If it is not possible to earth the printed circuit board, the person mounting the circuits should touch the board before bringing MOS circuits into contact with it.
4. Soldering iron tips or soldering baths should also be kept at the same potential as the MOS circuits and the board.
5. After the MOS circuits have been mounted on the board, proper handling precautions should still be observed. Until the sub-assemblies are inserted into a system in which the proper voltages are applied, the board is no more than an extension of the leads of the devices mounted on the board. To prevent static charges from being transmitted through the board wiring to the device, it is recommended that conductive clips or conductive tape be put on the circuit board terminals.
6. To prevent permanent damage due to transient voltages, do not insert or remove MOS devices from test sockets with the power on.
7. Beware of voltage surges due to:
   - switching electrical equipment on or off.
   - relays.
   - d.c. lines.
8. Signals should not be applied to the inputs while the device power supply is off.
9. All unused input leads should be connected to either the supply voltage or earth.
10. If possible, personnel should wear anti-static clothing (no wool, silk or synthetic fibers).

FIGURE 3.16 Example of special-care precautions. (*Courtesy Mullard Ltd.*)

Apart from these general precautions MOS devices need special care. A list of precautions that should be observed when handling MOS devices is provided in Figure 3.16. Note that for test purposes points 6, 7, and 8 apply to almost all types of active devices. Also when making any measurements, especially on ICs, large test probes should not be used because they may short out adjacent pins. Always observe the maximum ratings of the device and in *no circumstances* apply test signals or voltages that exceed the absolute maximum voltages. It is also wise not to remove or replace any component while the power is on. This may well produce voltage or current surges that could damage the component itself and other sensitive components in the circuit.

3.13 TEST CIRCUITS FOR COMPONENTS

Component testing can be divided approximately into three main areas:

1. Simple verification of device operation: for example, checking that a resistor value is approximately as

stated and not high in value or open circuit, or checking that a saturated switching transistor will turn off when the base/emitter junction is shorted.

2. Go/No-go tests to determine that some parameter or characteristic of a device is within specification limits. For example, testing small signal diodes with a constant value of forward current and monitoring the value of forward voltage.
3. Relatively accurate measurement of a component parameter.

Usually in the test and service of equipment, the object is to find any fault quickly, and therefore the first two methods are used much more frequently than the third. However, there will be occasions when fairly accurate measurement of a device is required, and it is then useful to be aware of the general principles involved. For good accuracy (±0.1%), bridge methods are used to compare the unknown against a standard.

The Wheatstone bridge arrangement (Figure 3.17) can be used for resistance measurements and is balanced when $R_a/R_b = R_x/R_s$. The detector (D) indication is then a minimum. This is because the voltage drop across R_b is then the same as that across R_s. The balance point is independent of the value of the supply voltage, and any sensitive null indicator can be used. The accuracy depends on the tolerance and stability of the ratio resistors R_a, R_b, and the standard R_s.

At balance, when R_a and R_b have been adjusted for a null indication

$$R_a/R_b = R_x/R_s$$

It follows that

$$R_x = R_s \cdot \frac{R_a}{R_b}$$

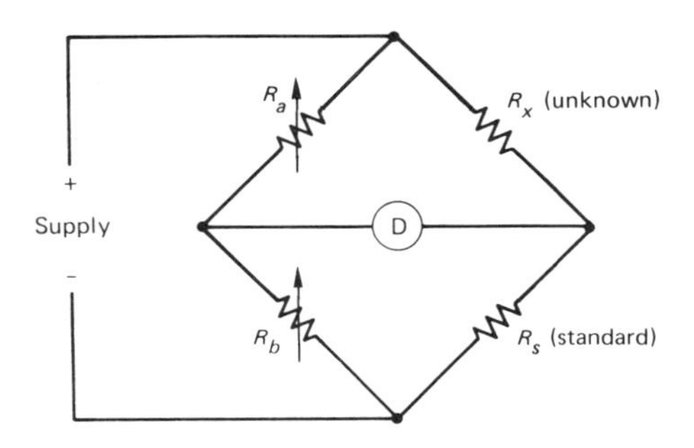

FIGURE 3.17 Wheatstone bridge.

In commercial LCR universal bridges, three a.c. bridge circuits are used (Figure 3.18). The frequency of the bridge supply is typically 1 kHz, and the highly sensitive detector is often an a.c. amplifier tuned to 1 kHz with its output feeding a moving coil meter via a recitifier. At balance the value of the component is usually presented in digital form for easy readout. A typical performance specification for a general purpose universal bridge is:

Inductance	1 μH to 100 H
Capacitance	1 pF to 1000 mF
Resistance	10 mΩ to 10 MΩ
Q-factor (coils)	0 to 10 at 1 kHz
Dissipation factor (capacitors)	0 to 0.1 at 1 kHz
Accuracy on all measurements	0.5%

Apart from bridges, which are not often required in the service situation,

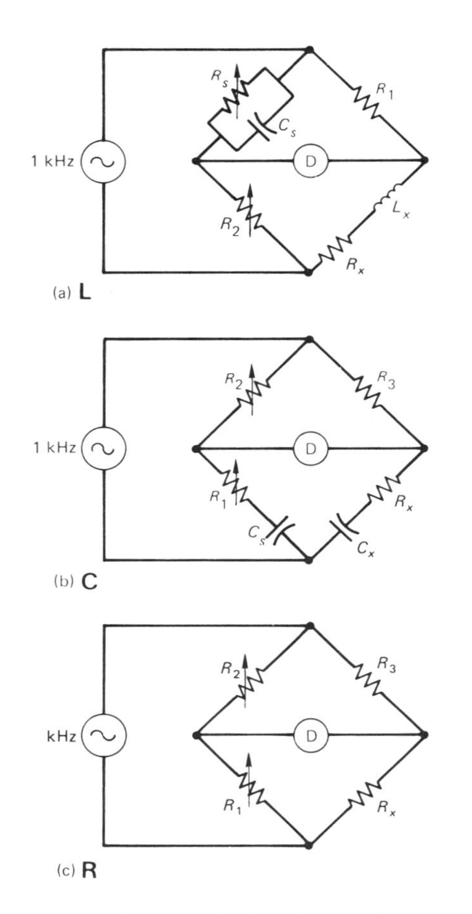

FIGURE 3.18 A.C. bridge circuits for L, C, R.

there are several other good and quick methods for component measurement. Two things have to be kept in mind:

(a) The effect of any measuring currents or voltages on the component. For example, too high a current may cause excessive power dissipation in the device being measured, or a high test voltage may cause breakdown.

(b) The sources of error that are present in the measurement. That is, such things as meter inaccuracies and loading effects, lead inductance, capacitance, and resistance. In general, all test leads should be kept as short as possible, especially when low values are being measured, and particularly if the test is at a high frequency.

Measurement of Resistance

To start with, consider the measurement of resistance in the range 1 Ω to 10 MΩ. Some basic circuits are shown in Figure 3.19. One method commonly used is the ohmmeter ranges or moving coil multirange meters. Normally three switched ranges are provided:

$R \times 1$	0 to 2 kΩ
$R \times 10$	0 to 200 kΩ
$R \times 100$	0 to 20 MΩ

Before the resistor is measured, the two meter leads are shorted together and the appropriate set zero control is adjusted to give zero ohms indication (this being maximum meter current). When the leads are connected across the unknown resistor, the current through the meter falls, and the resistance value can be read from the scale. Unfortunately, the scale is nonlinear and accuracy is not high, being about ±3% zero to midscale, ±5% midscale to two-thirds full-scale deflection, and ±10% two-thirds to full scale deflection (FSD).

A linear ohmmeter using an op-amp can be constructed as shown. A constant voltage circuit applies a stable reference level (1 V) to a series of range resistors, and the unknown resistor is connected as the feedback element of the inverting amplifier. A

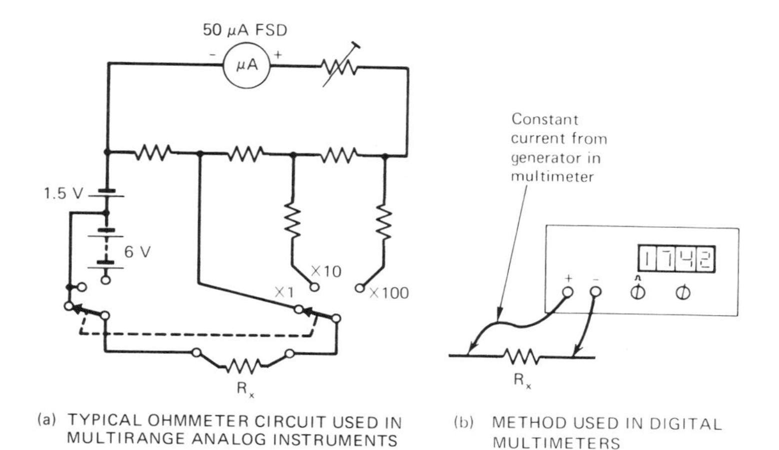

(a) TYPICAL OHMMETER CIRCUIT USED IN MULTIRANGE ANALOG INSTRUMENTS

(b) METHOD USED IN DIGITAL MULTIMETERS

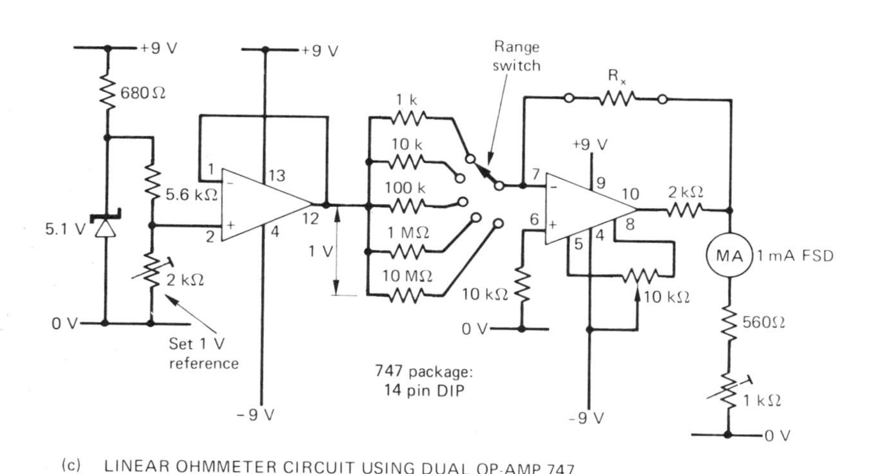

(c) LINEAR OHMMETER CIRCUIT USING DUAL OP-AMP 747

FIGURE 3.19 Circuits for measurement of resistance in the range 1 Ω to 10 MΩ.

moving coil meter together with a series resistor indicates the output voltage. The gain of the amplifier circuit to the meter is given by R_X/R_S. Therefore if $R_X = R_S$, the output voltage will be equal to the input reference. If $R_X = \frac{1}{2}R_S$ then the output will fall by a half, and so on, so that a linear scale results. With range resistors of ±0.5% tolerance and a reasonable quality meter, an accuracy of a few percent can be achieved.

In digital multimeters a constant current source is used for resistance measurement. When a known current is passed through an unknown resistor, the measured voltage across the resistor will be proportional to its value. The constant and accurately known current from the digital multimeter causes a voltage to be developed across the resistor. This voltage is measured by the digital voltmeter section of the instrument. An accuracy of ±1% is common. Currents vary from 1 mA on the 100 Ω range to, say, 1 μA on the 1 MΩ range.

This method can also be used to measure very low resistance values such as the contact resistance of plug and socket connectors, or switch and relay contacts. Essentially the four-terminal (kelvin) method is used. A constant current somewhere in the range of 1 mA to 100 mA is passed through the contacts and the voltage developed across the contacts is measured using a digital voltmeter. It is, of course, essential that the voltage probe measures

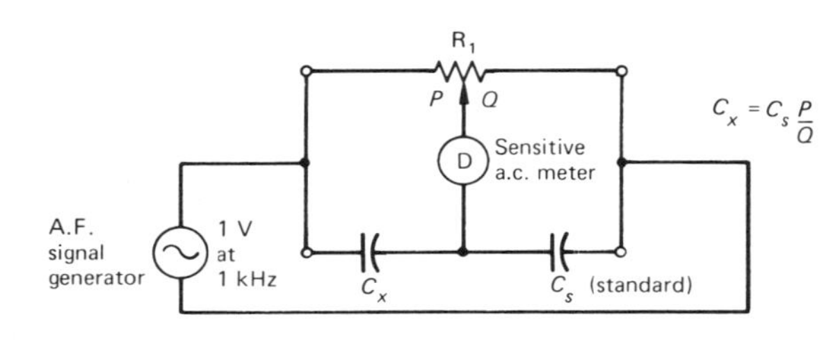

(a) SIMPLE COMPARISON BRIDGE (DE-SAUTY)

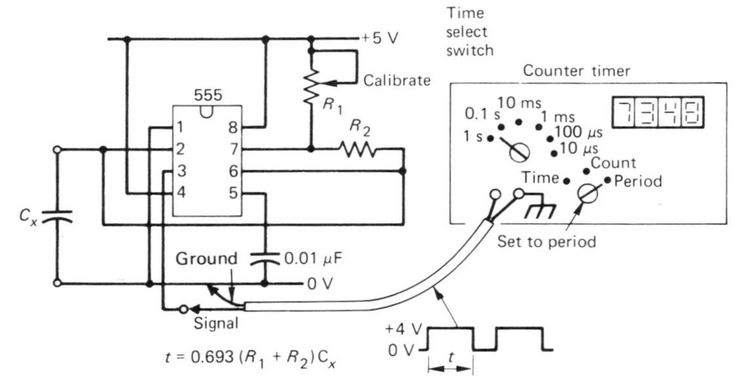

(b) DIGITAL METHOD USING 555 TIMER IC AND COUNTER TIMER

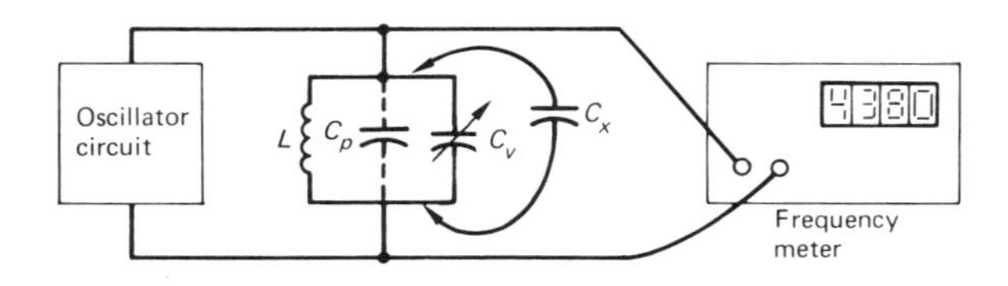

(c) SUBSTITUTION IN RESONANT CIRCUIT (for small value of C)

FIGURE 3.20 Measurement of capacitance.

directly across the contacts and does not, therefore, include any of the volt drop in the measuring leads carrying the current.

Measurement of Capacitance

The measurement of capacitance can be made with a simple comparison bridge (Figure 3.20). This is suitable for the rapid checking of capacitors against a standard unit. The Schering bridge is used for the measurement of capacitance and loss angle δ, since it enables a balance to be made for the series losses of the capacitor. (See Figure 3.18.)

Fairly accurate measurement of capacitance value from 500 pF up to 10 000 μF can be made using a simple circuit based on the 555 timer. In this circuit the 555 (IC) is used as an astable multivibrator. Its output is connected directly to the input of a digital counter via a short piece of shielded coaxial cable. The counter is set to read Period, so that the time for the unknown capacitor to charge from $^1/_3 V_{CC}$ to $^2/_3 V_{CC}$ is displayed. The system can be calibrated by standard close tolerance capacitors and different period ranges can be used to display μF, nF, and pF. (For a description of the 555 IC see Chapter 5.)

The measurement of small capacitance values (1 pF to 500 pF) is more difficult, since at low frequencies the impedance of the capacitor will be very large, and if a higher frequency is used the effects of stray capacitance in the circuit cannot be ignored. One reliable method is to substitute the unknown low-value capacitor in an active resonant circuit. A standard variable capacitor Cv with a calibrated dial, together with an inductor L, and any stray capacitance Cp, form the tuned circuit for an oscillator. With Cv set to a known value Cv_1, the frequency of the oscillator is measured with a frequency meter. When the unknown capacitor is connected in parallel with Cv the frequency will fall. Cv is then adjusted until the previous frequency is indicated. The new value Cv_2 is noted and the difference between Cv_1 and Cv_2 will be the value of Cx. Note that the value of the stray capacitance Cp does not affect the result, since it is always included in the measurement. Nor is the accuracy of the frequency meter important; the measurement depends on the calibration of Cv.

Measurement of Inductance

Measurement of inductance is usually made by one of three bridge circuits. For coils with low series resistance and, therefore, high Q, the Hay bridge is commonly used. Coils with relatively high resistance and, therefore, low Q are measured on the Maxwell Bridge, which has the advantage of being nonsensitive to the bridge supply frequency. The accuracy of both these arrangements can be as high as ±0.1%. The Owen bridge is a circuit that is useful for testing iron-core inductors with a superimposed d.c. current. All three circuits are shown in Figure 3.21.

Measurement of Active Components

Turning now to active components such as diodes, transistors, thyristors, and ICs, it would obviously not be practical to discuss in detail all the possible measurements. In the test and service situation we are not concerned so much with all device parameters but rather with one or two that are of

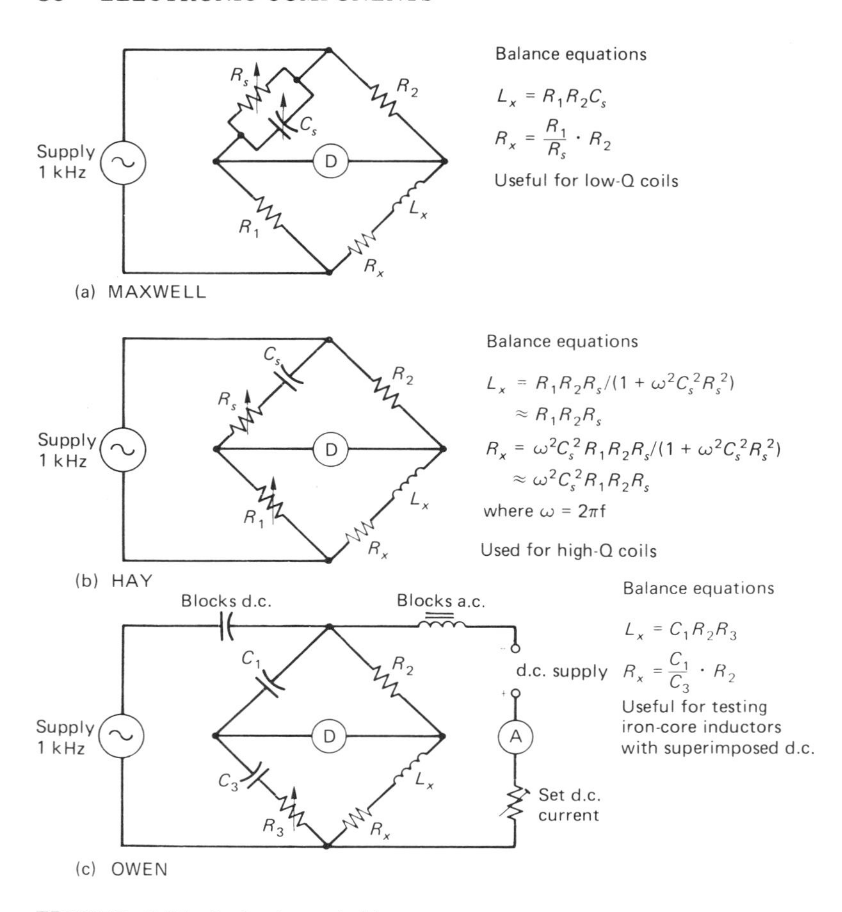

FIGURE 3.21 Inductance bridges.

vital importance for correct circuit operation. For various discrete semiconductors the most important parameters are given in Table 3.8.

For **diodes**, simple tests to check that values of V_F and $V_{(BR)}$ are within limits can be made using constant current sources. In nearly all measurements of this kind the current must be kept constant to avoid overheating and possible destruction of the component. A typical diode characteristic is shown in Figure 3.22. Measurement of V_F is simple. A fixed value of forward current, say, 5 mA, is passed

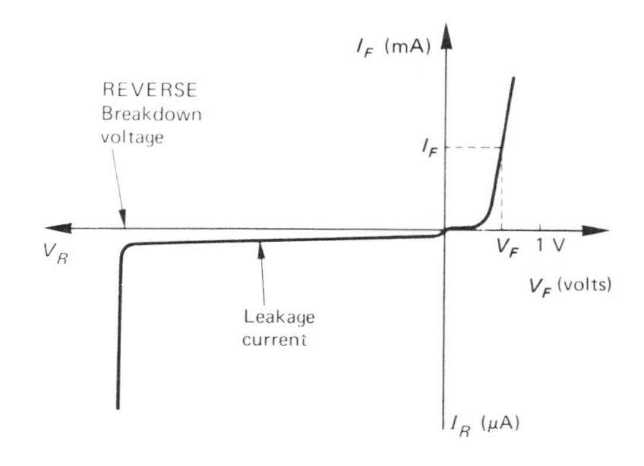

FIGURE 3.22 Typical semiconductor diode characteristics.

TABLE 3.8 Important Parameters of Discrete Semiconductors

Diodes	Zeners or Reference Diodes	Bipolar Transistors	FETs	Thyristors
V_F forward voltage drop	V_Z breakdown voltage	h_{FE} d.c. current gain	Y_{fs} transconductance	V_T foward voltage drop
I_R reverse leakage current	z_z dynamic impedance	$V_{CE(sat)}$ collector/emitter saturation	$V_{GS(off)}$ gate source voltage that turns device off; practical measurement of pinch-off (V_P)	I_{GT} gate trigger current V_{GT} gate trigger voltage I_H holding current
$V_{(BR)}$ reverse breakdown voltage		$V_{(BR)CEO}$ collector/emitter breakdown (base open circuit)		
For switching diode: t_{rr} reverse recovery time		I_{CBO} leakage current (emitter open circuit)	I_{DSS} drain current with $V_{GS} = 0$	V_{DRM} repetitive peak off-state voltage
		I_{CEO} leakage current (base open circuit)	$r_{DS(on)}$ drain to source resistance with $V_{GS} = 0$	I_R reverse current

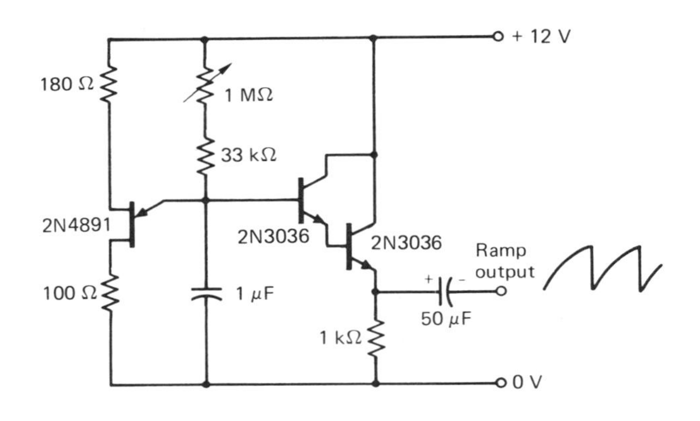

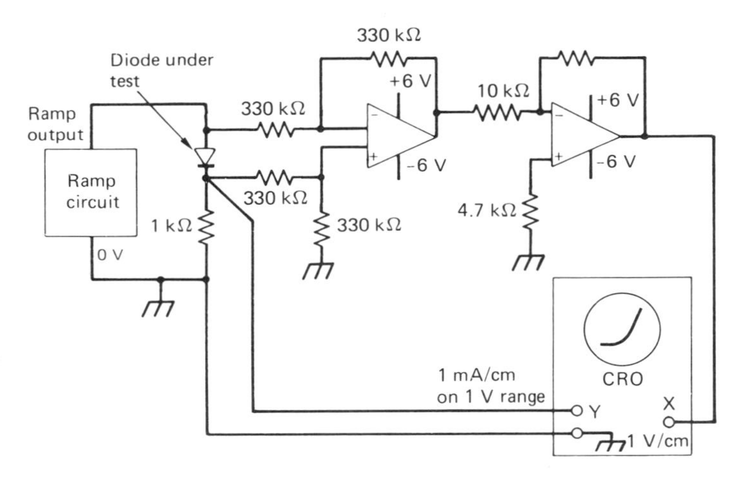

FIGURE 3.23 Ramp circuit for test circuits, and use of CRO to display diode forward characteristics.

through the diode and V_F is read off a voltmeter. If the I_F/V_F characteristic is required, a circuit can be used to display this either on an oscilloscope or an XY plotter. In this case some form of ramp generator must be used, such as that shown in Figure 3.23.

Breakdown voltages of semiconductors must always be measured with a constant current source. At breakdown, which is mostly an avalanche effect, a rapid rise in current occurs for only a small increase in voltage. A breakdown test circuit (Figure 3.24) can be used to nondestructively test for $V_{(BR)}$, V_Z, $V_{(BR)CEO}$, etc. The circuit is essentially a constant current generator formed by Q_1 circuit. Q_1 base is held at 5.6 V by the zener diode so that V_E is approximately 5 V. The emitter current, and therefore the collector current, can be adjusted by varying the emitter resistance R_5. It will then remain fairly constant over

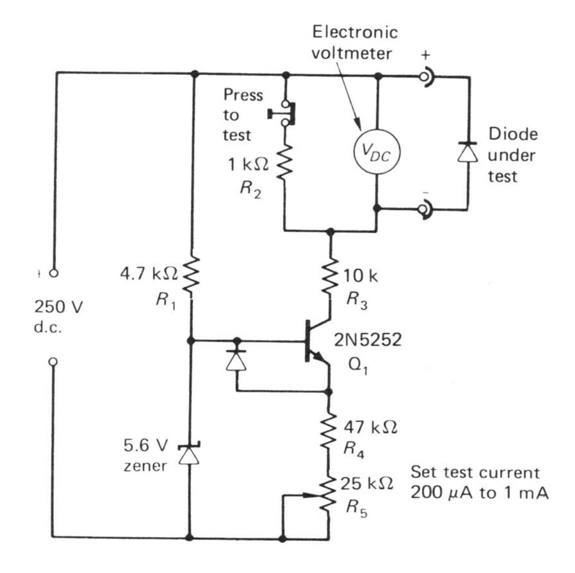

FIGURE 3.24 Reverse breakdown test circuit.

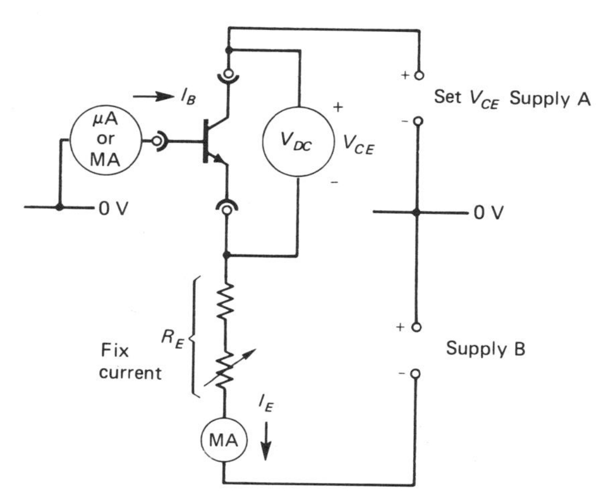

FIGURE 3.25 Circuit for measuring $h_{FE} = \frac{I_C}{I_B} \approx \frac{I_E}{I_B}$.

a variation of collector voltage from 200 V to 10 V. Note that the maximum current is approximately 1 mA, low enough not to cause failure. When a component is being checked for breakdown, the test switch is pushed and the voltage across the component will rise to the breakdown value where the current will then be limited. The voltage across the device under test can be read using a multirange meter.

Tests for h_{FE} are commonly used as an indication of transistor operation, and a simple circuit for measuring this is given in Figure 3.25. Note that h_{FE} is the large signal d.c. common emitter current gain, i.e.,

$$h_{FE} = \frac{I_C}{I_B}$$

at some specified value of V_{CE} and I_C.

Elaborate circuits can be set up for exact measurement of h_{FE} and h_{fe} as well as all the other h parameters but, in practice, since these vary so widely from one transistor to the next, the usefulness of the exercise is questionable. It is perhaps better to produce a set of output characteristic curves using, say, an XY plotter to automatically produce the results (Figure 3.26).

$V_{CE(sat)}$ is usually specified with an I_C/I_B ratio of 10:1. Thus for switching transistors a go/no-go test circuit such as that in Figure 3.27 can be easily constructed and values of $V_{CE(sat)}$ at particular values of I_C measured using a digital voltmeter.

For **field effect transistors** the most accepted parameters for device verification are

$$Y_{fs} \approx \frac{\Delta I_D}{\Delta V_{GS}} \quad \text{with } V_{DS} \text{ constant}$$

and

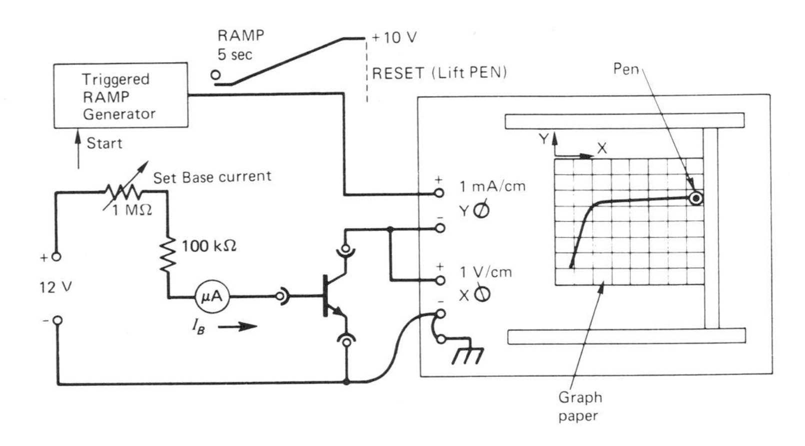

FIGURE 3.26 Use of an XY plotter to obtain transistor characteristics.

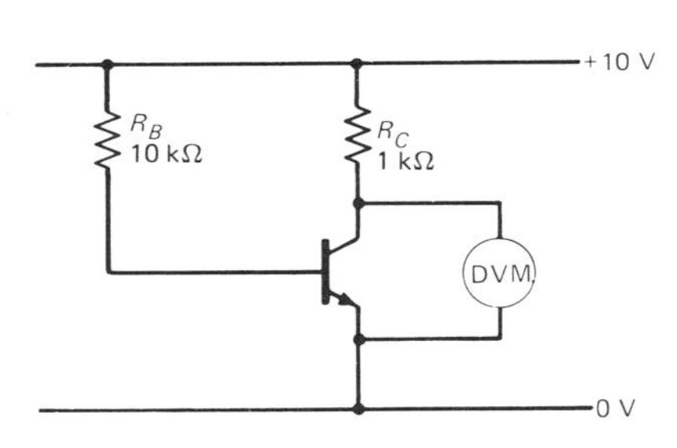

FIGURE 3.27 Measurement of $V_{CE(sat)}$.

In this case I_C = 10 mA.
Note that ratio R_B:R_C = 10:1.
Measurement of $V_{CE(sat)}$ at other values of I_C are obtained by changing values of R_B and R_C.

I_{DSS}, the value of drain current with
$V_{GS} = 0$ and $V_{DS} = V_P$

Circuits for checking these are shown in Figure 3.28. For Y_{fs}, the transconductance, the circuit has a fixed bias level set up so that an operating point is defined. Then V_{GS} is varied by a signal from the a.c. source and the resulting change in drain current is noted. A typical value for Y_{fs} will be 2 milliSiemens.

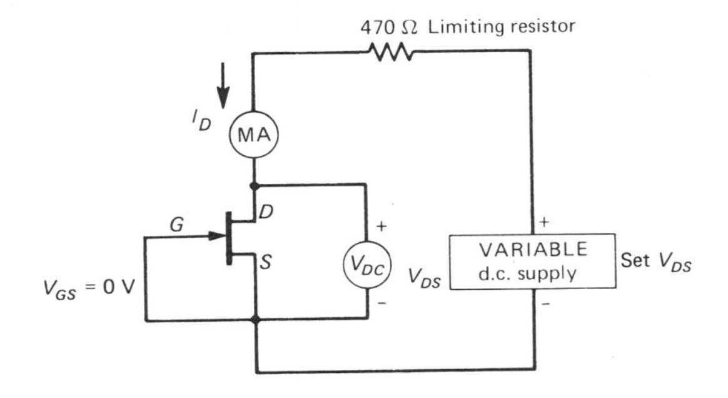

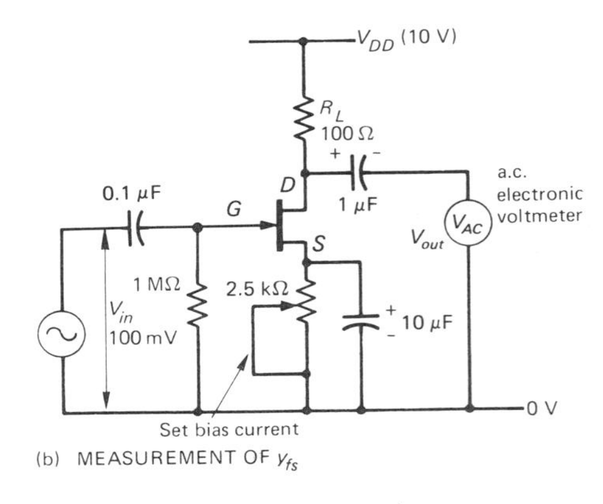

FIGURE 3.28 FET measurements

$$Y_{fs} = \frac{V_{out}}{V_{in} \cdot R_L} \text{ in } (b)$$

Finally, for discrete components a test circuit for **thyristors** is given in Figure 3.29. This can check correct operation by applying the specified values of I_{GT} and V_{GT} to the gate of the thyristor. Initially, with R_2 set to minimum, S_1 is closed. The current through meter 1 should be very low (50 μA) and the voltmeter should indicate nearly 24 V. This is because the thyristor is forward blocking and, therefore, nonconducting. By momentarily operating S_2 the thyristor should be triggered into conduction. M_1 should indicate approximately 100 mA, and M_2 about 1 V. This is

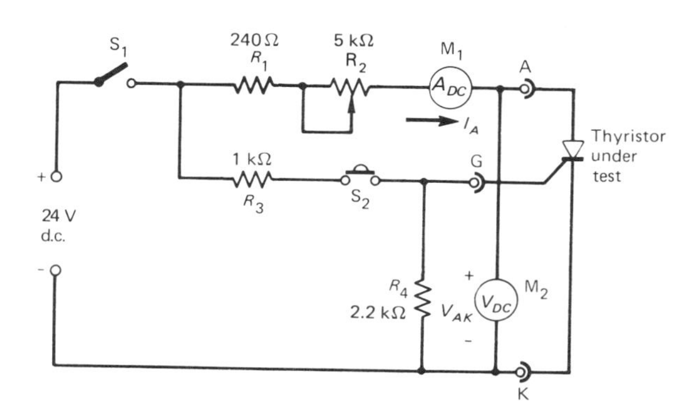

FIGURE 3.29 Thyristor test circuit.

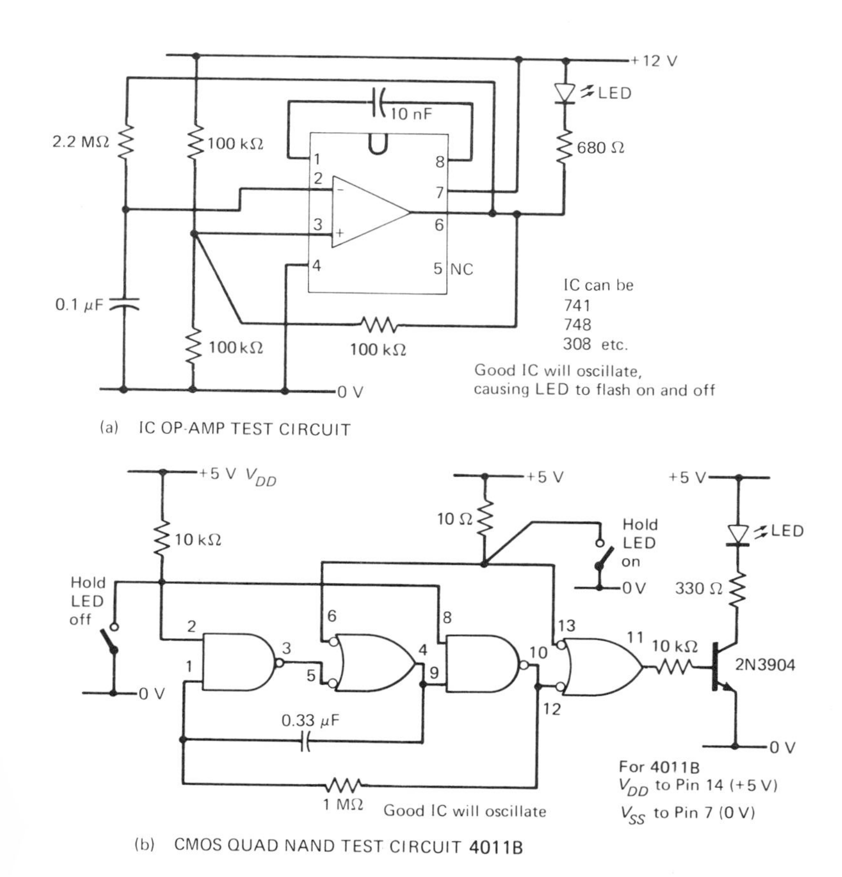

FIGURE 3.30

the forward conducting voltage. Then as R_2 is increased in value the current will gradually fall until a point is reached where the thyristor turns off. The value of current indicated just before turnoff is the holding current I_H.

Tests on **linear and digital ICs** can, of course, also be made to accurately measure all parameters, but more usually indication of a circuit function is required. In other words, does an op-amp have gain or will a counter IC divide correctly? By placing the IC into a test jig that requires the device to oscillate or perform some logic function, the good devices can be sorted from those that are faulty. This procedure can also be used to test any active device such as transistors, unijunctions, and thyristors.

Two examples of this method are illustrated in Figure 3.30. The first shows how linear op-amps of the 8-pin DIP variety can be given a functional check. The components connected around the IC socket will form a low frequency (2 Hz) oscillator. When the IC is inserted correctly in the test jig, the LED should switch on and off. A CMOS Quad 2 input positive (I/P) NAND gates (4011B) can also be checked by wiring components around a socket (14 pin) so that low-frequency oscillation occurs. Additional checks on internal gates can be made by operating the two inhibit switches S_1 and S_2.

EXERCISES 3

3.1 Devise simple test circuits for checking the functional operation of **(a)** unijunction transistors, **(b)** light-emitting diodes (LEDs), and **(c)** reed relays.

3.2 Derive the balance equations for the de Sauty capacitance bridge and the Maxwell inductance bridge.

3.3 Sketch a circuit layout to show how the common source characteristics of an n-channel JFET could be obtained.

3.4 Devise a circuit that could be used to measure the leakage current of large value (1000 μF–10 000 μF) aluminum electrolytic capacitors. What is the expected accuracy of your measurement?

3.5 Suggest five reasons for the premature failing of silicon transistors in electronic instruments and systems.

CHAPTER 4

Digital Logic Circuits

4.1 INTRODUCTION

Digital ICs are used extensively in all branches of electronics from computing (for which they were originally designed) to industrial control, electronic instruments, and communications systems. In fact, there does not seem to be any area where digital circuits in some form are not or will not be used. The basic reason for this is that digital circuits operate from defined logic levels; in other words, a signal is either high or low. Logic levels, for a positive convention, are usually a few volts positive (high or logic 1) or nearly zero (low or logic 0), and this reduces any uncertainty about the resulting output of a circuit. Whatever form the allowed states take, a logic gate will give a certain output as long as a particular set or combination of states exist on its input leads. In industrial control, for example, this can mean that a safety guard on a machine is either shut or open, *never* nearly shut.

The basic elements of all digital circuits are logic gates that perform logical operations on their inputs. To describe these operations, Boolean algebra is used. This type of algebra, based on logical statements that are either true or false, is a very useful tool in the design and troubleshooting of digital logic circuits.

The identities and properties of Boolean algebra are shown in Table 4.1, together with some logic gate symbols. It is useful to think of logic being represented by switches. The OR function will be two switches in parallel; closing A or B or A and B gives the output. The AND function will be two switches in series, so that an output is obtained only when both switches A and B are closed.

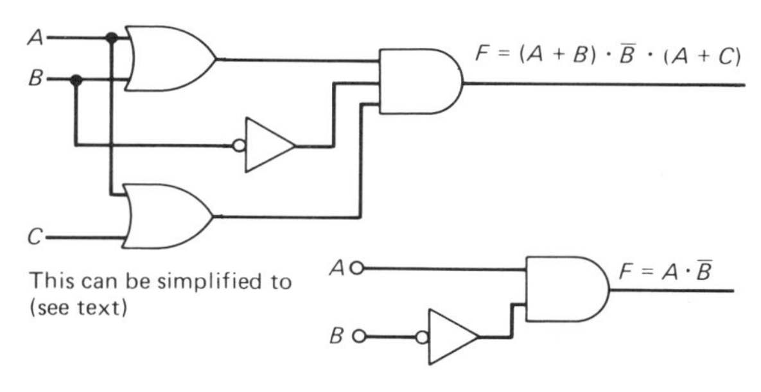

FIGURE 4.1 Logic diagrams for $F = (A + B) \cdot (A + C) \cdot \overline{B}$.

The use of Boolean algebra can be seen in the simple example of Figure 4.1 where

TABLE 4.1 Commonly Used Combinational Logic Gates for Two Inputs and Identities and Properties of Boolean Algebra

FUNCTION	ANSI SHAPE DISTINCTIVE SYMBOL AND EXPRESSION	ANSI SHAPE DISTINCTIVE EQUIVALENT SYMBOL	IEC NON-SHAPE DISTINCTIVE SYMBOL	IEC NON-SHAPE DISTINCTIVE EQUIVALENT SYMBOL
OR (INCLUSIVE-OR)	A, B — $F = A + B$		≥ 1	&
AND	A, B — $F = A \cdot B$		&	≥ 1
INVERTER (NOT)	A — $F = \overline{A}$		1	1
NOR	A, B — $F = \overline{A + B}$		≥ 1	&
NAND	A, B — $F = \overline{A \cdot B}$		&	≥ 1
EXCLUSIVE-OR	A, B — $F = A \oplus B$; $A \oplus B = A \cdot \overline{B} + \overline{A} \cdot B$		=1	

Bubble and directional triangle indicators are both used in industry but not in the same circuit.

Examples: o— bubble indicator; ▷— directional triangle indicator.

Identities	Proof by Switch Circuit
$A + 0 = A$	
$A + A = A$	
$A + 1 = 1$	
$A + \bar{A} = 1$	
$A \cdot 0 = 0$	
$A \cdot 1 = A$	
$A \cdot A = A$	
$A \cdot \bar{A} = 0$	

Rules

$A + B = B + A$, $A \cdot B = B \cdot A$ — Commutative rules

$(A + B) + C = A + (B + C)$, $(A \cdot B) \cdot C = A \cdot (B \cdot C)$ — Associative rules

$A \cdot (B + C) = (A \cdot B) + A \cdot C$, $A + (B \cdot C) = (A + B) \cdot (A + C)$ — Distributive rules

$A + A \cdot B = A$, $A \cdot (A + B) = A$ — Absorption rules $\left[A + A \cdot B = A \cdot (1 + B);\ A \cdot (A + B) = A \cdot A + A \cdot B = A + A \cdot B \right]$

$A + \bar{A} \cdot B = A + B$

$\overline{A + B} = \bar{A} \cdot \bar{B}$, $\overline{A \cdot B} = \bar{A} + \bar{B}$ — De Morgan's rules

the logic arrangement gives the output as

$$F = (A + B) \cdot (A + C) \cdot \overline{B}$$

Using Boolean algebra this can be simplified as follows:

combining brackets

$F = (A \cdot A + A \cdot C + B \cdot A + B \cdot C) \cdot \overline{B}$

$F = (A + A \cdot C + B \cdot A + B \cdot C) \cdot \overline{B}$ since $A \cdot A = A$

$F = (A + B \cdot C) \cdot \overline{B}$ since $A + A \cdot C + B \cdot A = A$ by absorption rule

$F = A \cdot \overline{B} + B \cdot C \cdot \overline{B}$ removing parenthesis by ANDing $\overline{B}$ with A and then with $B \cdot C$

$F = A \cdot \overline{B} + 0 \cdot C$ since $\overline{B} \cdot B = 0$

Finally $F = A \cdot \overline{B}$ since $0 \cdot C = 0$

Apart from simplification methods like those illustrated above, the Karnaugh mapping technique can be used; it will not be covered here but is well worth looking up.*

Logic diagrams can be drawn in two forms. The classical form is drawn with bubbles placed on outputs only. The classical form requires engineers and technicians to recall truth tables or manipulate Boolean expressions to trace signal flow through a circuit. The functional form is drawn with matching indicators (i.e., choosing necessary equivalent gates so that output bubbles, or directional triangles, on logic symbols are connected to input bubbles, or directional triangles; and outputs without bubbles, or directional triangles, are connected to inputs without bubbles, or directional triangles). Functional forms of logic diagrams are most often preferred by engineers because signal flow is easier to trace. Technicians also appreciate the functional form because logic diagrams are drawn with matching indicators which facilitate troubleshooting. Since the emphasis of this book is on troubleshooting, the functional form for drawing logic diagrams has been used. The equivalent gates shown in Table 4.1 can be used to convert a diagram drawn in classical form into the easier to use functional form. For positive logic, an active low output is an output with a bubble, or directional triangle, while an active high output has no bubble, or directional triangle, on its output. Also a bar over the name of an output implies an active low output (i.e., a bubble, or directional triangle, is present on the output) while no bar over the name of an output implies an active high output (i.e., no bubble, or directional triangle, is present on the output). To make the output expression indicate the active low level performed by the INVERTER, the NOR, or the NAND in the first column of Table 4.1, invert both sides of each expression (e.g., $F = \overline{A + B}$, becomes $\overline{F} = A + B$ and $\overline{F}$ now indicates an active low level output as expressed by the presence of the bubble on the output of the NOR gate). The expression $\overline{F} = A + B$ may be written more descriptively as F, $L = A + B, L$ to indicate the active low level at the output caused by an active high level at input A or B.

The validity of a Boolean statement can also be checked out by drawing a truth table. This table is simply a list of all the possible states of the inputs together with the resulting output. For example, the

*R. S. Sandige, *Digital Concepts Using Standard Integrated Circuits*, McGraw-Hill, New York, 1978.

truth table for the example of $F = A \cdot \overline{B}$ is

Input States		Output
A	B	$F = A \cdot \overline{B}$
0	0	0
0	1	0
1	0	1
1	1	0

This is just another way of stating that the output from the circuit in Figure 4.1 will only be 1 when $A = 1$ and $B = 0$. Table 4.2 shows the truth tables for the commonly used combinational logic gates for two inputs. The tables can easily be extended for three or more inputs but tend to become rather large. The exclusive-OR circuit has the function $F = A{\cdot}\overline{B} + \overline{A}{\cdot}B$. It is a circuit that is often used, and the ways in which it is achieved with the various logic types are shown later.

Quite often the designer prefers to use mostly NAND-type or NOR-type gates because: (1) all possible types of logic circuits can be generated by using either all NAND gates or all NOR gates; (2) these are more available or cheaper in the logic family chosen. De Morgan's rules can be used to show how the equivalent logic symbols are obtained for the OR, AND, NOR, and NAND functions in Table 4.1.

De Morgan's rules, easily proved by truth table, are

$$\overline{A + B} = \overline{A}{\cdot}\overline{B} \quad \text{and} \quad \overline{A{\cdot}B} = \overline{A} + \overline{B}$$

It follows that

$$\overline{\overline{A}{\cdot}\overline{B}} = \overline{\overline{A}} + \overline{\overline{B}} = A + B \quad \text{the OR function}$$

This shows that the two ANSI (or IEC) symbols in the first row of Table 4.1 are, in fact, equivalent and

$$\overline{\overline{A} + \overline{B}} = \overline{\overline{A}}{\cdot}\overline{\overline{B}} = A{\cdot}B \quad \text{the AND function}$$

This shows that the two ANSI (or IEC) symbols in the second row of Table 4.1 are also equivalent. Figure 4.2 shows how NAND gates can be used to give the OR, NOR, and exclusive-OR functions. In Figure 4.2*a*, when input A is high or B is high, output F is high as seen by tracing the signals through the functionally drawn logic circuit; therefore, $F = A + B$, which may be written more descriptively as $F, H = A + B, H$. In Figure 4.2*b*, when input A is low (i.e., $\overline{A}$) and input B is low (i.e., $\overline{B}$), output F is high as seen by tracing the signals, thus $F = \overline{A}{\cdot}\overline{B}$; but, $\overline{A + B} = \overline{A} \cdot \overline{B}$ by De Morgan's rule, so $F = \overline{A} \cdot \overline{B} =$

TABLE 4.2 Truth Tables for Various Combinational Operations

Inputs		*Output* OR	NOR	AND	NAND	Exclusive-OR
A	B	$A + B$	$\overline{A + B}$	$A{\cdot}B$	$\overline{A{\cdot}B}$	$A \oplus B$
0	0	0	1	0	1	0
0	1	1	0	0	1	1
1	0	1	0	0	1	1
1	1	1	0	1	0	0

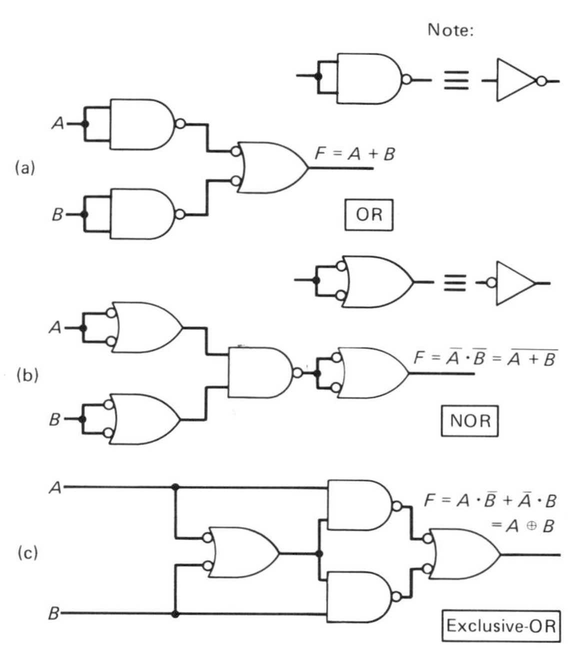

FIGURE 4.2 Use of NAND gates to realize OR, NOR, and Exclusive-OR.

$\overline{A + B}$. By tracing the signals through the functionally drawn logic circuit of Figure 4.2*c*, it's easy to see when input A is high and input B is low (i.e., $A \cdot \bar{B}$), output F is high, or output F is also high when input A is low, and input B is high (i.e., $\bar{A} \cdot B$); therefore, $F = A \cdot \bar{B} + \bar{A} \cdot B$, which is defined as the exclusive-OR function and written $F = A \oplus B$. As an exercise the reader might try to draw the circuits of AND and NAND functions using NOR gates.

With some types of logic, the outputs of gates can be paralleled together giving what is called WIRED LOGIC (Figure 4.3). This is easily achieved using DTL but with TTL only special open collector circuits can be used. A saving in logic gates results

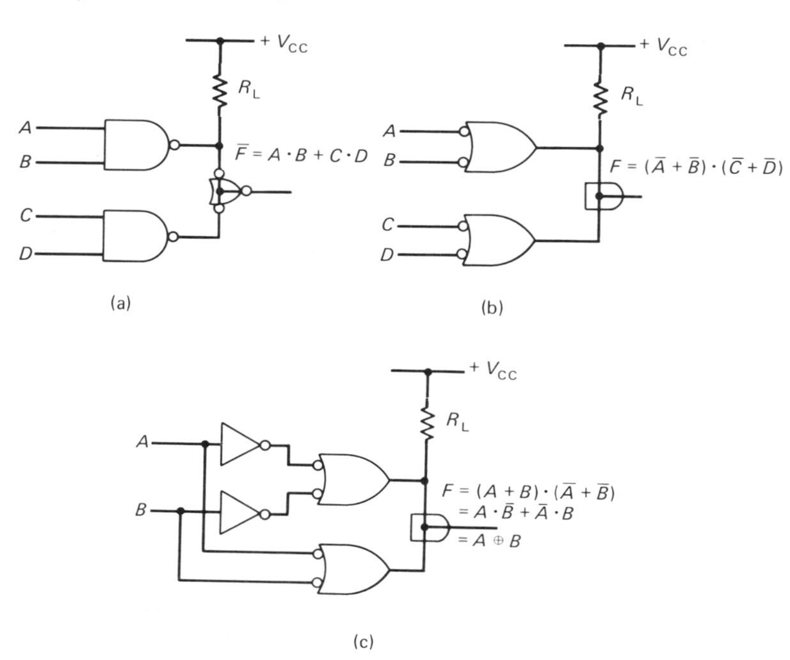

FIGURE 4.3 Wired logic.

from wired logic, and circuits such as the exclusive-OR and comparators can be readily built. Many manufacturers call the circuit of Figure 4.3*a* wired-OR because when inputs A and B are high or inputs C and D are high, the output is low (i.e., $\overline{F}$), so $\overline{F} = A \cdot B + C \cdot D$. However, the function may be considered a wired-AND, which can be shown as follows:

for Figure 4.3*b*

$$F = (\overline{A} + \overline{B}) \cdot (\overline{C} + \overline{D})$$

giving the AND function for the outputs of the two gates.

Using this wired-AND for exclusive-OR, the circuit is as shown in Figure 4.3*c*. The output for this is

$$F = (A + B) \cdot (\overline{A} + \overline{B})$$
$$= A \cdot \overline{A} + A \cdot \overline{B} + \overline{A} \cdot B + \overline{B} \cdot B$$

But $A \cdot \overline{A} = 0$ and $B \cdot \overline{B} = 0$. Therefore

$$F = A \cdot \overline{B} + \overline{A} \cdot B \qquad \text{the exclusive-OR.}$$

The logic circuits discussed thus far are called COMBINATIONAL because a combined set of input conditions are simultaneously required to give a particular output. Another group of logic circuits are called SEQUENTIAL. These are the multivibrators, the counters, shift registers, and memories. The basic difference between the two types of logic circuit is that the SEQUENTIAL circuit possesses a memory, and therefore is able to take into account the previous input states as well as those actually present. Sequential logic circuitry is usually built up from bistables (flip-flops) which can be made by connecting NAND or NOR gates in a particular way. An understanding of all the various bistable types is another essential part of trouble-shooting skill.

A third set of logic circuits is called INTERFACE circuits. All logic units must be able to receive signals from external sources such as transducers, switches, etc., and also to be capable of driving display and indication devices, electromechanical units such as relays or solenoids, or power semiconductors. Interface circuits are used to translate input information into digital logic levels, and to change logic levels into signals capable of driving output devices. Also, within systems, different types of logic may be used, and interfaces are again required between logic; for example, converting logic levels from TTL to CMOS or vice versa.

Being able to diagnose and repair faults in digital units requires a good knowledge of the characteristics of the type of circuits used, and the sort of measuring techniques that yield the quickest results. Before discussing the characteristics, we give the various abbreviations for logic families together with some comments on their present use.

RTL (Resistor-Transistor Logic) This was not manufactured in monolithic IC form. However, discrete circuit blocks are available for industrial use where ruggedness is required and fast speed is not essential.

DCTL (Direct-Coupled Transistor Logic) One of the first types to be made as an IC. However, it suffered from some problems with matching (current hogging) and was soon superseded.

DTL (Diode-Transistor Logic) The first commercially available IC logic family (53/73 series). Now superseded by TTL and CMOS, but still available from many manufacturers.

TTL (Transistor-Transistor Logic) A very successful logic family with a wide range of functions. The 54/74 series is the standard type. Low-power (54L/74L) and high-speed TTL (54H/74H) were also developed. However, it is the later development of Schottky clamped TTL, where the transistors are prevented from saturating, that results in a large improvement in performance. Schottky TTL is available as high-speed (54S/74S) or low power (54LS/74LS).

ECL (Emitter Coupled Logic) A non-saturating type of transistor logic that is extremely fast in operation (10 000 series).

CMOS (Complementary Metal Oxide Logic) Uses n and p channel MOSFETS and has the advantage of low power consumption and very good immunity against noise and interference (4000 B series).

LOCMOS (Locally Oxidized CMOS) Improved performance over CMOS as all outputs are buffered. Same type numbers as CMOS.

PMOS (p-Channel MOS) Used for many LSI devices.

NMOS (n-Channel MOS) Used for many LSI devices.

I^2L (Integrated Injection Logic) A development from DCTL that allows bipolar technology to be used for LSI devices.

SSI means **Small-Scale Integration**, typically below 12 equivalent gates per IC package.

MSI means **Medium-Scale Integration**, typically between 12 and 100 equivalent gates per IC package.

LSI means **Large-Scale Integration**, typically above 100 equivalent gates per IC package.

Summarizing the above information we can say that some of the most commonly used logic IC families at present are

(a) Standard TTL (type 54/74).

(b) CMOS, LOCMOS (type 4000B).

(c) Low-power Schottky TTL (type 54LS/74LS).

(d) Schottky TTL (type 54S/74S).

(e) ECL (type 10 000).

It is on these types that we concentrate in the next few sections of this chapter.

4.2 GENERAL PROPERTIES OF LOGIC GATES

There has to be some common way of specifying and stating the characteristics and performance of the various logic families. These specifications assist in comparisons between families and give essential information for test and service technicians as well as design engineers.

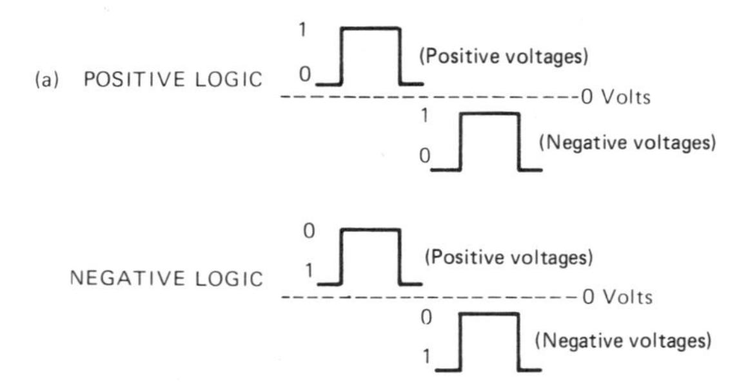

FIGURE 4.4*a* Positive and negative logic.

(a) Logic Convention

For positive logic, the most positive voltage level is called logic 1 and the most negative voltage level is logic 0.

Negative logic is the opposite, so that logic 1 is the most negative level and logic 0 the most positive. This is illustrated in Figure 4.4*a*. Note that, within some logic units, mixed conventions may be used.

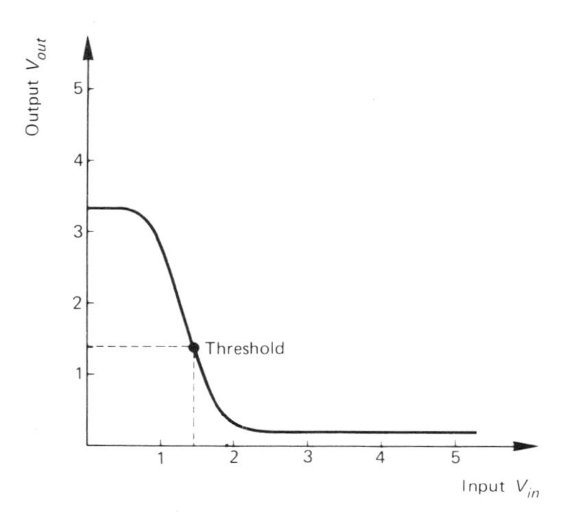

FIGURE 4.4*b* Threshold level (positive logic).

(b) Threshold Level

For any logic gate there will be a point when the level of an input begins to cause a change of state at the output. In Figure 4.4*b*, which shows a transfer characteristic for a TTL gate, if the input level is 1.4 V then the output will be between logic 1 and logic 0 levels, also at 1.4 V. A simple view of threshold is to state that it is the input level that will cause the output to be exactly half-way between logic 1 and logic 0. However, there will be a spread on this value, and manufacturers prefer to quote maximum and minimum input levels which will guarantee that the output will reside in the correct state. For example, for a standard TTL NAND gate

V_{IL} is the maximum input voltage (0) guaranteed to appear as a low state input voltage (800 mV).

V_{IH} is the minimum input voltage (1) guaranteed to appear as a high state input voltage (2.0 V).

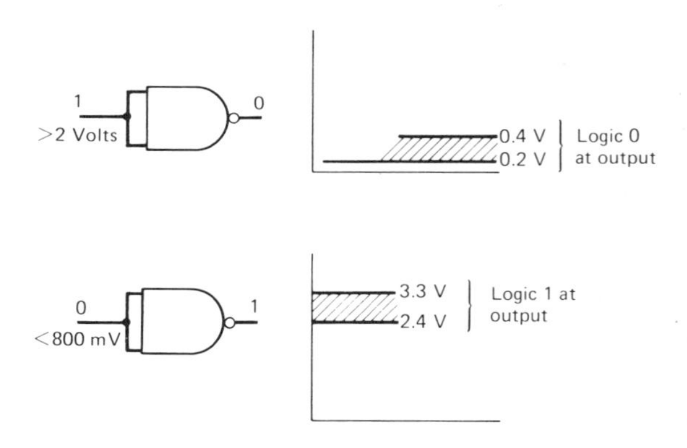

FIGURE 4.4*c* Logic levels at gate outputs.

(c) Logic Levels

For a particular logic type, logic 1 and logic 0 are assigned voltage levels that must be maintained for correct circuit operation (Figure 4.4*c*). For example, in standard TTL these levels at gate outputs are

Logic	Typical Voltage	Minimum	Maximum
1	3.3 V	2.4 V	–
0	0.2 V	–	0.4 V

(d) Noise Immunity or Noise Margin

The level of noise that can be tolerated without falsely triggering gates and introducing errors is of vital importance in any logic system. Noise is any unwanted disturbances on signal leads or power lines and can be generated either externally or internally. Manufacturers usually quote d.c. values of noise margin, giving typical and worst case values.

Taking TTL as an example, the typical noise margin will be the difference between the voltage level from the output of a gate and the threshold of the gate input it is driving as shown in Figure 4.4*d*. Using this criterion, we can see that the best logic 1 or high-state noise immunity is 1.9 V, whereas the logic 0 or low-state noise immunity for TTL is usually taken as 1 V. Worst-case d.c. noise immunity must take into account the minimum and maximum values of output levels and input thresholds. This is shown by Figure 4.4 to be 400 mV, and this is the guaranteed worst-case noise immunity.

(e) Fan-out

This is the term used to describe the capability of a gate output to drive inputs of similar gates. It is really the number of

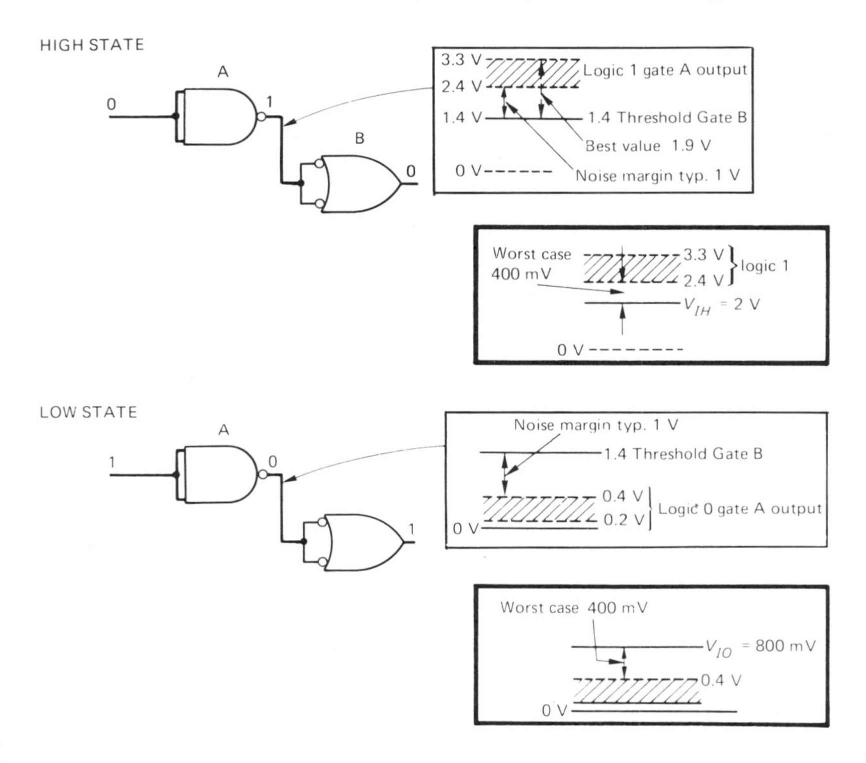

FIGURE 4.4*d* Noise margin (TTL).

inputs that can be driven simultaneously from an output without the output level, whether it be 0 or 1, going out of its specified limits.

(f) Fan-in

This is the number of inputs that can be accommodated on one gate.

(g) Propagation Delay Time

When the input of a gate changes state there will be some finite delay before the output switches. This delay time is made up of several components, such as charge storage effects in transistors, charge carrier transit times through semiconductor elements, and capacitors being charged or discharged. In most bipolar circuits, apart from Schottky TTL and ECL, the charge storage effects predominate. When a transistor is saturated, its base region stores any extra base current in the form of minority carriers (electrons in the p-base region of an npn transistor). When the transistor is switched off, these extra electrons make up a collector current that keeps the transistor conducting for a small time period after its base current supply is cut off. Naturally, rapid speed of operation is essential in many applications. ECL, which biases transistors in the linear region, has a propagation delay of about 2 nanoseconds, whereas for standard TTL, which drives transistors into saturation, the delay is approximately 10 nanoseconds.

(h) Power Consumption

A logic gate dissipates a small amount of power in its static state, that is without signals applied, and also during operation. Since a digital system may be made up of thousands of gates, the power dissipation per gate must be kept to the lowest value

TABLE 4.3 Performance Comparisons of Logic Families

Type of Logic	Propagation Delay Time	Typical Noise Immunity	Fan-out	Power Dissipation per Gate	Usual type of Logic
RTL	100 nsec	Varies, depends on design	5	10 mW	NOR
DTL	25 nsec	Fair 1 V	8	10 mW	NAND/NOR
Standard TTL 54/74	10 nsec	Fair 1 V	10	10 mW	NAND/NOR
Schottky TTL 54S/74S	3 nsec	Fair 1 V	10	19 mW	NAND/NOR
Low-power Schottky 54LS/74LS	9.5 nsec	Fair 1 V	20	2 mW	NAND/NOR
ECL	2 nsec	400 mV	30	40 mW	OR/NOR
CMOS	50 nsec (depends on the fan-out required)	Excellent typically 45% of V_{DD}	100	0.001 mW static 1 mW/MHz	NAND/NOR

possible. This is especially true for units operated from batteries.

Having looked at the various properties of electronic gates, it is useful to compare the typical performances for common logic families. This is shown in Table 4.3.

4.3 OPERATION OF COMMONLY USED LOGIC GATES

It is, of course, possible to trace and repair faulty gates in digital logic systems by simply knowing the logic functions carried out by the various ICs. However, testing and troubleshooting are further assisted by having a good understanding of the actual circuit operation of the ICs. To get this understanding we must take a close look inside the IC itself. Since most logic gates include transistors (bipolar or MOS) being used as switches, a review of basic semiconductor switches follows.

A simple BIPOLAR TRANSISTOR SWITCH, or INVERTER, is shown in Figure 4.5. While the input voltage is at zero, no base current flows into the transistor and it is said to be "cut off." The output voltage, neglecting the effect of the tiny leakage current, is equal to V_{CC}. As the input is taken positive and passes about 0.7 V, base current begins to flow and this is amplified by the transistor. The collector voltage falls, since $V_{out} = V_{CC} - I_C R_C$. Above a certain input voltage, sufficient base current is injected to fully "turn on" the transistor, and the output voltage falls to very nearly zero. In fact the value is called $V_{CE(sat)}$, which for a good switching transistor may be between 50 mV and 100 mV. The transistor is then said to be SATURATED. When a transistor is used as a switch, the input will be either low (logic 0) or high (logic 1), and the transistor must be cut off with a low input and fully turned on (saturated) with a high input.

When the transistor is saturated, any further increase in input voltage will force more base current to flow into the transistor. However, I_C, the collector current, is limited to a value of $(V_{CC} - V_{CE(sat)})/R_C$, so the excess base current has to be stored in the base region of the transistor. What happens is that electrons are attracted from the emitter into the base. Here these electrons are minority charge carriers, since the base region is p-type. If the collector voltage is reasonably high, these electrons, after diffusing across the thin base, would be swept up by the collector; but $V_{CE(sat)}$ is now only 0.1 V or less. Thus, at any instant in a saturated transistor, large

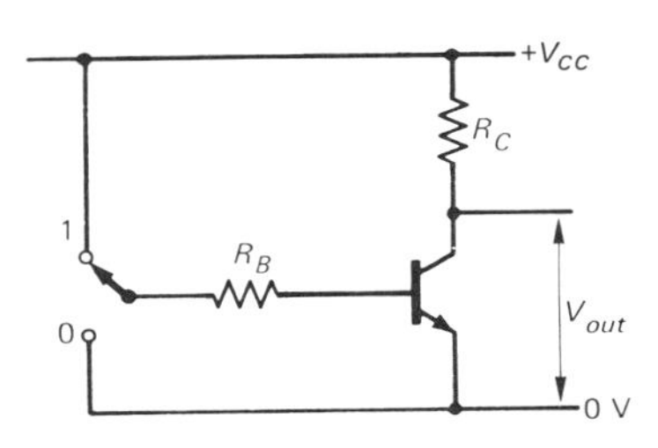

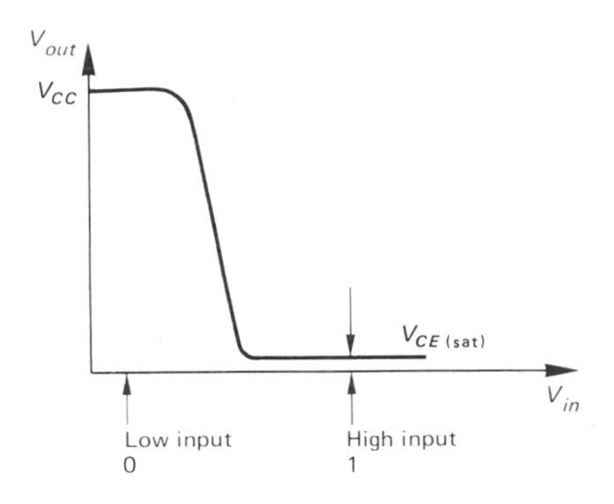

FIGURE 4.5 Transistor inverter. Ratio R_B:R_C usually 10:1 to ensure satisfactory switching.

numbers of minority charge carriers are present in the base region. When the input is returned to a low value, these excess charges in the base must be removed by the collector before the collector current falls and the transistor turns off (see Figure 4.6). The minority charge storage time, even in a small switching transistor, may be several nanoseconds.

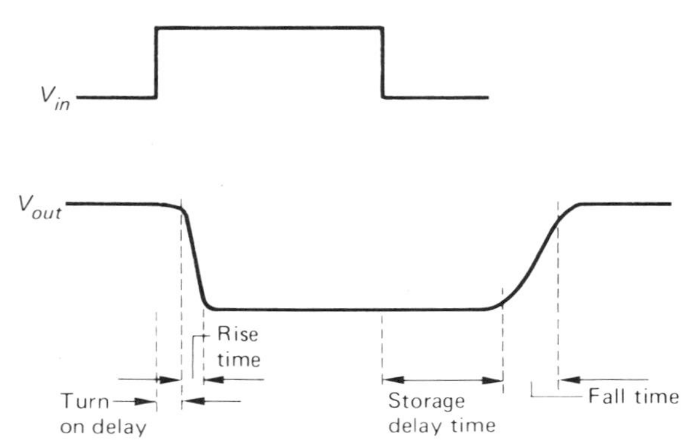

FIGURE 4.6 Effect of minority charge storage.

Several methods have been used to prevent transistors in logic gates from saturating, since its prevention reduces the switching delays. One widely used method is to connect a diode of low forward voltage drop from base to collector (Figure 4.7). A suitable diode can be formed by a metal-to-semiconductor rectifying junction, and this is used in Schottky clamped TTL. The operation is that, as the collector voltage falls to a low value, in response to an increasing input, the diode conducts, and excess input current is diverted through the diode into the collector circuit. In this way, the transistor is prevented from saturating, and switching speeds of 3 nanoseconds are achieved.

As the simple transistor inverter is switched from one state to the other,

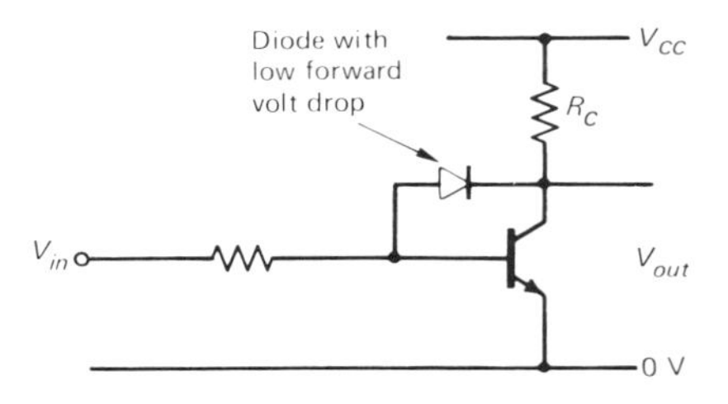

FIGURE 4.7 Nonsaturating circuit.

power will be dissipated by the transistor as shown in Figure 4.8. Note that during the dynamic phase, i.e., the rise or fall of the collector waveform, the transistor dissipation reaches its peak value. When the transistor is ON and in a static condition, then since $V_{CE(sat)}$ is very low, the collector dissipation is also low.

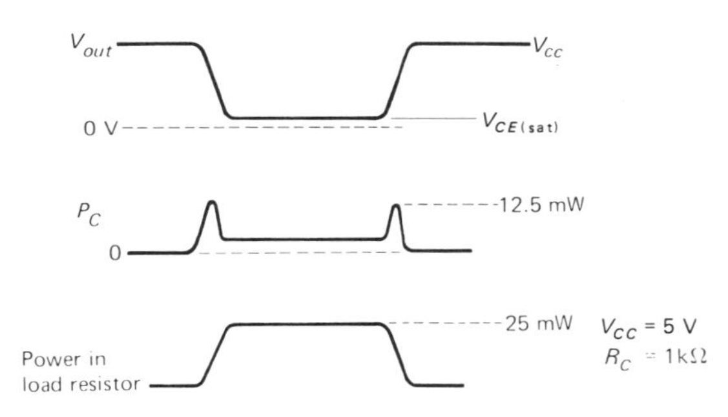

FIGURE 4.8 Power dissipation in simple switch.

A MOSFET SWITCHING CIRCUIT is shown in Figure 4.9. In this type of circuit the load resistor is replaced by a p-channel MOSFET. Both MOSFETs will be enhancement mode types, which means that there is no conducting channel between the source and drain of either device until a suitable bias voltage is applied to the gate. The operation is as follows. If the input voltage is held at zero volts at logic 0, the n-channel MOSFET Q_2 will be OFF, since there is zero voltage bias between its gate and source (NOTE: the substrate is

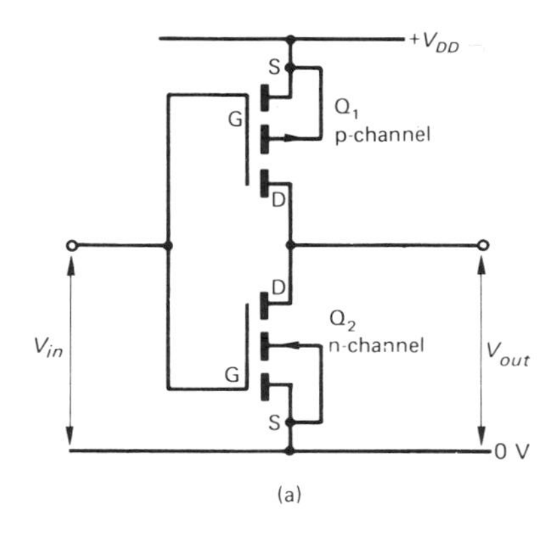

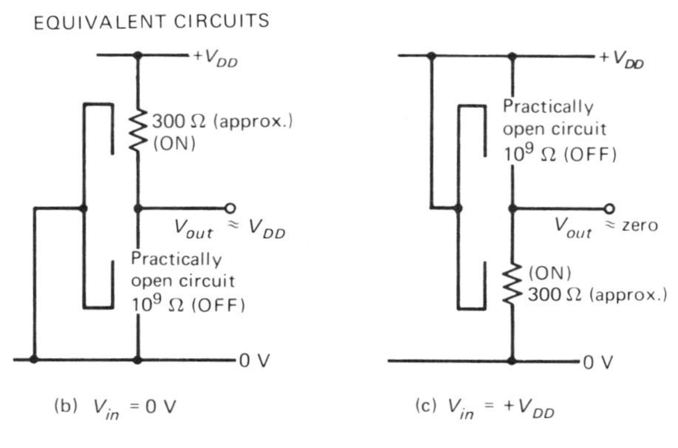

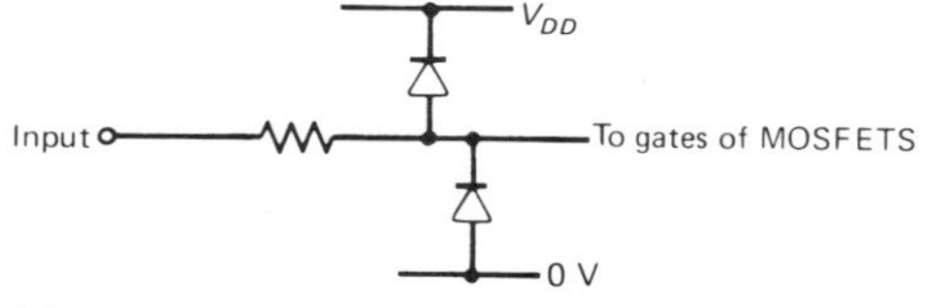

FIGURE 4.9 Complementary MOS inverter.

connected to the source). The resistance between drain and source of Q_2 is then about $10^9\Omega$. However, with the input at zero volts, Q_1 (p-type) will have a conducting channel set up between its source and drain since its gate is 0 V, while its source and substrate are at $+V_{DD}$ volts. The ON resistance of Q_1 is about 300 Ω and the output is therefore high at very nearly V_{DD}. If the input voltage is now taken high to $+V_{DD}$ (logic 1), Q_1 the p-channel MOSFET is turned OFF, effectively disconnecting the output from V_{DD}. At the same time, Q_2 the n-channel MOSFET is biased ON, connecting the output to 0 V via its channel resistance of about 300 Ω. The equivalent circuits for these two possible states are also shown in Figure 4.9. Several very important features of the circuit are

1. Since the gates of the MOSFETs are insulated from the substrate, the input impedance of the circuit is extremely high ($10^{12}\Omega$ or greater). This means that the input is isolated from the output of the inverter and that the fan-out for such circuits at low frequency is practically unlimited.
2. In the static state one MOSFET is on while the other is off. Thus the power consumption is almost negligible, typically 1 nW.
3. The threshold level of the circuit is approximately at the point when the input voltage is midway between 0 V and V_{DD}. This shows that the noise immunity is very good, typically 45% of V_{DD}. Thus with V_{DD} = 15 V, the noise margin is 6.75 V.
4. The power supply can be varied over a wide range of voltages, typically from +3 V up to +15 V and does not need to be highly regulated.

Apart from these advantages to the user, the manufacturer has the further advantages of an extremely simple structure that takes much less area of silicon than its bipolar equivalent and requires fewer process steps, which reduces costs and improves yield and reliability.

Naturally, there are some disadvantages, the most important one being that the very thin insulating oxide between gate and substrate of a MOSFET is easily damaged by voltage spikes in excess of 90 V. Protection diodes have to be included on every input circuit, as shown in Figure 4.9*d* to prevent electrostatic fields during handling or test from punching through the oxide.

Although the power consumption is very low in the static state, when the input is switched from one state to another there will be an instant when both MOSFETs conduct. A current pulse is then taken from the supply. Thus power dissipation will increase with operating frequency and is typically 1 mW/MHz for one CMOS gate.

Each gate input possesses a capacitance of about 5 pF. Therefore, although the fan-out appears almost unlimited, for reasonably fast a.c. operation the fan-out is reduced to 20.

TTL Standard NAND Gate

The circuit of one of the two gates inside the 7420 is shown in Figure 4.10. To ease the understanding of the circuit operation it can be split into three sections:

Q_1: A multi-emitter transistor, performing the AND function on inputs A, B, C, D.

Q_2: a phase splitter.

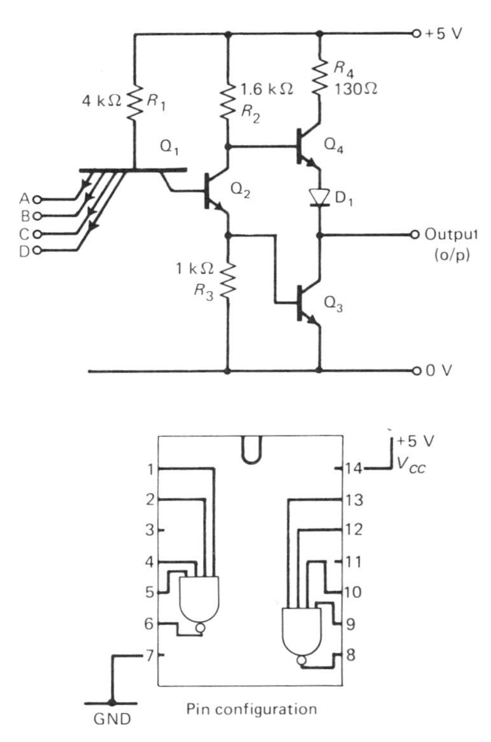

FIGURE 4.10 TTL NAND gate (½7420).

Q_3 and Q_4: an output stage having a low output impedance in both logic states (known as the totem pole output).

Assume that all inputs A, B, C, and D are high. Because the base of Q_1 is connected to $+V_{CC}$ via R_1, all the emitter/base junctions will be reverse-biased. The base/collector junction of Q_1 will be forward-biased and current will flow via R_1 into Q_2 and Q_3, turning them both on. The output of the gate will be at logic 0. Under these conditions, the voltage on Q_1 base will be approximately 2.1 V (i.e., $V_{BE3} + V_{BE2} + V_{BC1}$). Therefore, as long as the input levels on A, B, C, and D are all greater than 2 V, all the emitter/base junctions of Q_1 are reverse-biased. Since Q_2 is saturated, its collector voltage falls to 0.8 V (i.e., $V_{BE3} + V_{CE(sat)2}$) and therefore Q_4 and D_1 are nonconducting because the output is only at approximately 0.2 V. With Q_4 off, there is no connection between the output and V_{CC}. The diode is provided to ensure that Q_4 remains off while Q_3 is on. Q_3 acts as a *current sink* and up to 10 standard inputs of 1.6 mA each can be taken by Q_3 without the output level rising above 0.4 V.

If any one input A, B, C, or D is taken to a logic 0, i.e., 0.2 V, that particular emitter/base junction of Q_1 will be forward biased, and Q_1 base will fall to 0.9 V. This voltage is insufficient to forward-bias Q_2 and Q_3, which then turn off. Q_2 collector potential rises towards V_{CC} and Q_4 turns on, taking the output via D_1 to +3.4 V which is logic 1 level. Note that R_4, a 130 Ω resistor, will limit the output current in the event of an accidental short circuit between output and 0 V.

Effectively the output circuit, called a "totem pole output," acts to give a low output impedance in both logic states. This gives reasonably fast switch on and off since capacitance on the output will be charged and discharged rapidly. Note that, as the circuit switches from one state to another, there is an instant when both Q_3 and Q_4 conduct, taking approximately a 14 mA maximum current pulse from the supply. For this reason the power supply line to TTL ICs must be decoupled to prevent unwanted interference pulses occurring on the power supply leads.

Some of the variations from the standard NAND gate circuit in TTL are shown in Figure 4.11. The 7401, a quad 2 I/P NAND, has no collector load, and is termed "open collector." Several of these gates can be wired in parallel to give the wired-OR facility. The collector load resistor has to be provided externally. With this external resistor at 2 kΩ a reasonable fan-out is still available but, since the circuit has a high logic 1 output impedance, the operating speed is reduced.

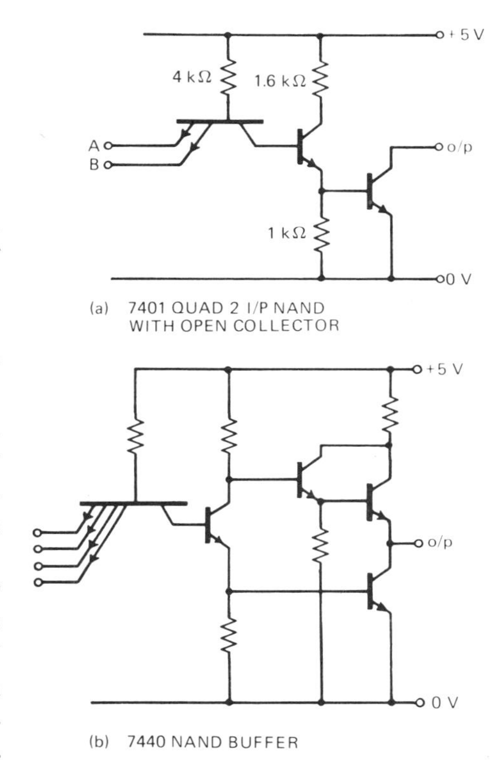

FIGURE 4.11 Variations in TTL gates.

The 7440 is an example of a NAND buffer which is an example of a gate that has higher drive capability. The Darlington emitter follower gives the gate a fan-out of 30.

A NOR gate (7402) is also available as are hundreds of other special circuits such as AND-OR-NOT gates, Schmitt triggers, monostables, bistables, etc. For a complete list consult manufacturer's data books and data sheets.

Schottky TTL

The main limitation in switching speed of standard TTL is the fact that the transistors saturate. Schottky TTL overcomes this by including Schottky barrier diodes between base and collector of the transistors. These metal-to-semiconductor rectifying junctions have a low forward voltage drop (300 mV) so that, when the transistor is turned on, the collector voltage is held at about 400 mV. This is sufficient to prevent saturation of the transistor. The Schottky NAND circuit is shown in Figure 4.12.

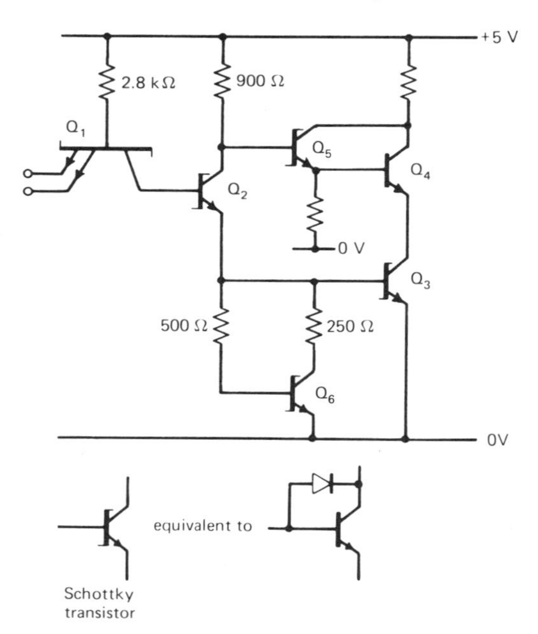

FIGURE 4.12 Schottky TTL NAND gate.

The operation is almost identical to standard TTL except for the additions of Q_5 and Q_6. Q_5 increases output drive in logic 1 state by acting as a Darlington pull-up, and Q_6 is included to decrease still further the switching delays. Q_6, called an "active turn-off" element, operates as a nonlinear load during the switch on and switch off of Q_2 and Q_3. When Q_2 turns off, for example, Q_6 goes rapidly into a high resistance state assisting turn off. Schottky TTL with a propagation delay of approximately 3 nsec and a power dissipation of 19 mW per gate competes with ECL in high-speed logic systems.

The latest important development in TTL is low-power Schottky (54LS/74LS). The NAND gate is shown in Figure 4.13.

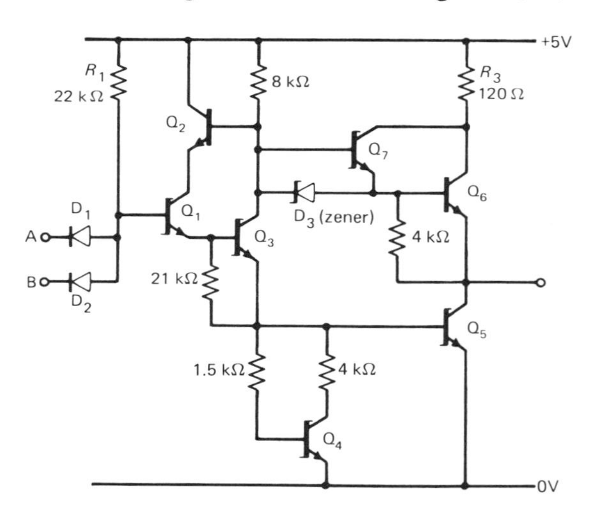

FIGURE 4.13 Low-power Schottky NAND.

This gate has a propagation delay of typically 9.5 nsec, but consumes only 2 mW per gate. It is intended as a replacement for the 54/74 series, since apart from its other improvements in performance, it requires only about one-fifth of the power of standard TTL. The savings on power supply design, and space are important in any large digital system.

The threshold level is 1.5 V, set by the three base/emitter volt drops (Q_1, Q_3, Q_5)

from 0 V minus the input diode drop (D_1 or D_2). Note that the multi-emitter transistor of other TTL circuits is replaced by a simple diode AND circuit. Low-power Schottky TTL is actually an updated DTL circuit. Fast switching speed and high fan-out are achieved by using Schottky barrier diodes across transistors and by clever circuit design. Q_7 is a Darlington pull-up giving fast switching at the output from 0 to 1. Q_4 acts as an active turn off, and Q_2 is an emitter follower supplying the collector current for Q_1. With both inputs going high, Q_2 supplies an initial current pulse into Q_1 to aid switching at the output from high to low. Q_3 and Q_5 turn on rapidly, and the output goes low (400 mV). At this point Q_3 collector goes low (1.1 V), thus reducing the collector current to Q_1 via Q_2 to a lower value during the static state. Diode D_3 also assists by supplying a rapid discharge path for Q_6. R_3, the 120 Ω collector load of Q_6, limits the output current to a safe value of about 30 mA in the event of an output short circuit. The fan-out is typically 20 and the noise margin is better than earlier types of TTL. Table 4.4 shows comparisons.

CMOS Logic

The use of CMOS logic has spread rapidly because of its advantages of low-power consumption, high d.c. fan-out, and excellent noise immunity. It is the obvious choice for systems intended for electrically noisy environments. A typical NOR gate is shown in Figure 4.14*a*, which is an extension of the CMOS inverter switch previously described. Q_1 and Q_2 are p-channel enhancement mode MOSFETs that will only conduct when their gates are negative with respect to their sources, i.e., when both inputs A and B are at 0 V. Q_3 and Q_4 are n-channel enhancement mode MOSFETs that will conduct only when their gates are positive with respect to their sources, i.e., at $+V_{DD}$. The simplest way to understand the operation is by studying the truth table. Note that 0 V = logic 0, $+V_{DD}$ = logic 1. (See table next page.) The gate gives the NOR function, the output being low (0) if either input is high (1).

As stated previously, input protection circuits are included in all CMOS ICs to prevent damage to the thin gate insulations. LOCMOS ICs perform the same functions

TABLE 4.4 Comparison Table for TTL Types (typical values)

Type	Propagation Delay	Fan-out	Worst Case d.c. Noise Margin 1	0	Power Dissipation per Gate
TTL 74	10 nsec	10	0.4 V	0.4 V	10 mW
Schottky 74S	3 nsec	10 (low) 20 (high)	0.7 V	0.3 V	19 mW
Low-power Schottky 74LS	9.5 nsec	20	0.7 V	0.3 V	2 mW

Inputs		States of MOSFETs				Resulting Output
A	B	Q_1	Q_2	Q_3	Q_4	
0 V	0 V	ON	ON	OFF	OFF	High (V_{DD})
0 V	$+V_{DD}$	ON	OFF	OFF	ON	Low (0 V)
$+V_{DD}$	0 V	OFF	ON	ON	OFF	Low (0 V)
$+V_{DD}$	$+V_{DD}$	OFF	OFF	ON	ON	Low (0 V)

as CMOS but have two inverter stages to fully buffer the output. This improves the noise immunity still further because the increased gain gives nearly ideal transfer characteristics.

The circuit of a CMOS NAND gate (Figure 4.14*b*) is the complement to the NOR. Two n-channel MOSFETs are in series between output and 0 V while the two p-channel MOSFETs are in parallel with $+V_{DD}$. It is left to the reader to work out a truth table to explain operation.

One final point on CMOS concerns its noise immunity. Thus far our discussion has shown that the noise margin, that is the difference between logic states and threshold for CMOS, is very high, typically 4.5 V with V_{DD} = 10 V. However, stating the noise margin in this way does not take into account the effect of any current-injected noise. Such noise can occur from signals on adjacent leads switching large signals and capacitively coupling current into the connecting leads or p.c.b. traces on the output of a CMOS gate. Since the output impedance of a CMOS gate is relatively high (300 Ω typically), it only requires a few milliamps of noise-injected current to pull the output into the threshold region. Thus, although CMOS has much greater voltage

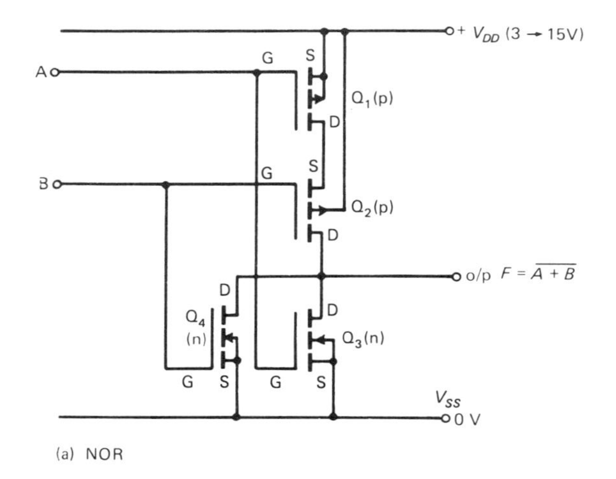

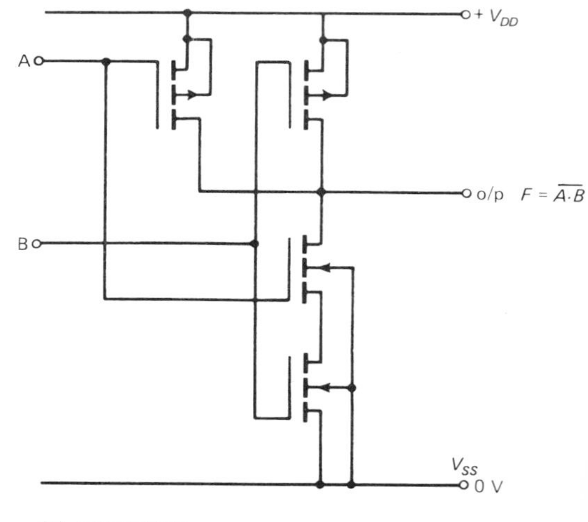

FIGURE 4.14 Basic CMOS gates.

noise immunity than TTL, it is more susceptible to current-injected noise; care must be taken during layout to avoid such noise generation.

ECL Logic

Emitter coupled logic, sometimes called current mode logic (CML), is the fastest available to date, with a propagation delay time of 1 to 2 nsec. A standard circuit in the 10 000 series is the OR/NOR gate in Figure 4.15. The operation of the gate depends on the fact that Q_1, Q_2, and Q_3 form a differential switch, and are not allowed to saturate. A reference voltage of −1.29 V is developed in the IC and applied to the bias input of the differential amplifier (Q_3 base). If the two inputs are at logic 0 (−1.75 V), then the current through R_3 is supplied by Q_3. This is because its base voltage at −1.29 V is more positive than the base voltages of Q_1 and Q_2. There will be a voltage drop of 0.85 V across R_2, and the OR output via emitter follower Q_5 will be −1.75 V (logic 0). The emitter voltage of Q_6, the NOR output, will be at −0.9 V (logic 1).

If a logic 1 level (−0.9 V) is applied to either A or B, Q_1 or Q_2 will conduct diverting current from Q_3. A voltage of 0.85 V will be dropped across R_1, the NOR output will be −1.75 V (0), and the OR output −0.9 V (1).

The noise margin of ECL is low (400 mV) but the fan-out, because of the relatively high input impedance of the input transistors and low output impedance of the emitter followers, is relatively high (30). Another advantage of the circuit is that power supply noise generation is virtually eliminated, since the current taken from the supply remains almost constant even when switching takes place.

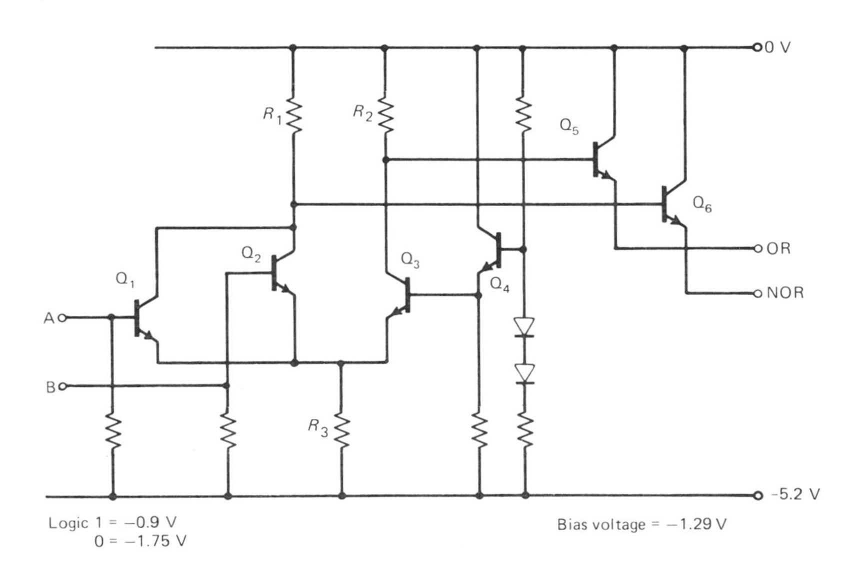

FIGURE 4.15 ECL logic gate.

4.4 BISTABLE CIRCUITS, COUNTERS, AND REGISTERS

Bistables, or flip-flops, are circuits that can be triggered into either of two stable states. Since a great portion of most digital systems are sequential in nature, it follows that a good understanding of the various bistable types and their operation is vital. Confusion can arise over the variety of types, namely R–S, clocked R–S, T, D, and JK; but by starting with the simplest, the R–S latch, it is relatively easy to build up an understanding of the more complex types.

The **R–S latch** can be made using two cross-coupled transistor switches or two cross-coupled gates as shown in Figure 4.16. It follows that, if one output is low, the other must be high. These two output pins are called Q and $\overline{Q}$. The two inputs are known as Set (S) and Reset (R). The Set input, if it is taken to logic 0 momentarily, will force the Q output to assume logic 1 state, and Q will then remain at logic 1 until a Reset input is applied. Note that the other output $\overline{Q}$ will always be in the opposite state to Q as long as only one input, that is either S or R, is made 0 at any one time. The output states will be indeterminate, Q and $\overline{Q}$ both logic 1, if both S and R are made logic 0 simultaneously.

The R–S latch is actually a memory circuit and this action can also be described by truth table (Table 4.5). As well as taking the states of the inputs (R and S) into account, the table must include the state of the Q output *before* the input signals are applied; this is written as Q_n. The state of the Q output after the application of an

TABLE 4.5 Full Truth Table for R–S Flip-Flop (using NAND gates)

Inputs R	S	Previous State of Q Q_n	Final State of Q Q_{n+1}
0	0	0	Indeterminate output
0	0	1	
0	1	0	0 } 0 on R always results in a 0 at Q.
0	1	1	0
1	0	0	1 } 0 on S always results in a 1 at Q.
1	0	1	1
1	1	0	Q_n } No change of state.
1	1	1	Q_n

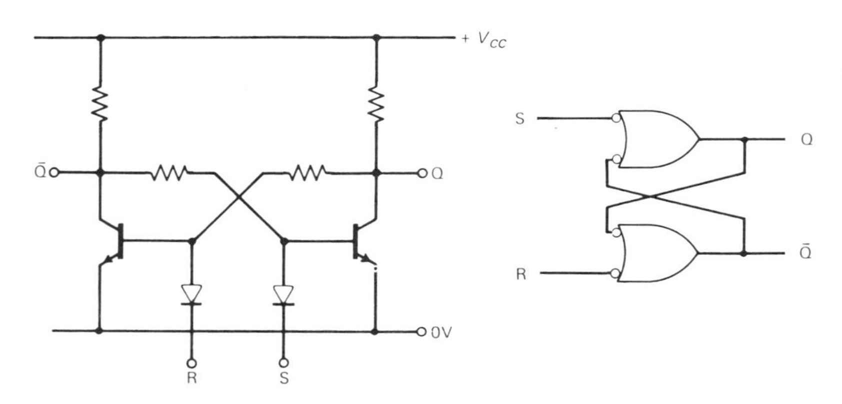

FIGURE 4.16 R-S bistable.

input is written as Q_{n+1}, which is the final or resultant state of the flip-flop.

If an R–S flip-flop is made by cross-coupling two NOR gates then the output level that will force a change of state will be logic 1. This is because a 1 on any input of a NOR gate will force the output to go to 0.

The truth table for the R–S bistable using NOR gates is

R	S	Q_{n+1}	
0	0	Q_n	no change of state.
0	1	1	1 on S forces Q to assume logic 1.
1	0	0	1 on R forces Q to assume logic 0.
1	1	Indeterminate.	

Note that here the full truth table has not been written—this is the case in most data sheets.

With both of these simple flip-flops a change of state at the output occurs a few nanoseconds after the input data changes. This action is called *asynchronous*. If a **clock input** is added as in Figure 4.17 SYNCHRONOUS operation is achieved, since data on the inputs can only be transferred to set or reset the bistable when the clock signal is high. Synchronous operation is important, since it is useful to control the operation of a complete digital system from a central clock pulse generator and also to avoid any buildup of delays within counters or shift registers. The method of

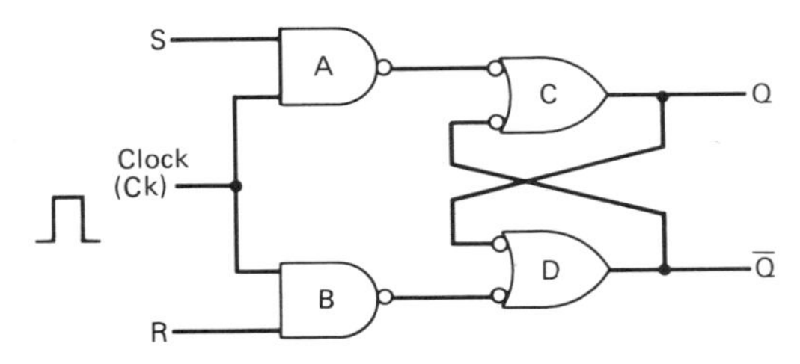

FIGURE 4.17 Clocked R-S bistable.

connecting the block in Figure 4.17 is said to be a positive gated latch, since gates A and B will be open to S or R data when the clock goes high. Most modern flip-flops are arranged so that input data is only transferred, during the edge of the clock pulse, data being "locked-out" or inhibited after the positive clock edge. This type of bistable is called an edge-triggered flip-flop and thus prevents any change in input data during the clock pulse width from affecting the output state of the circuit.

The **D bistable** or data latch shown in Figure 4.18 is an example of a "clocked"

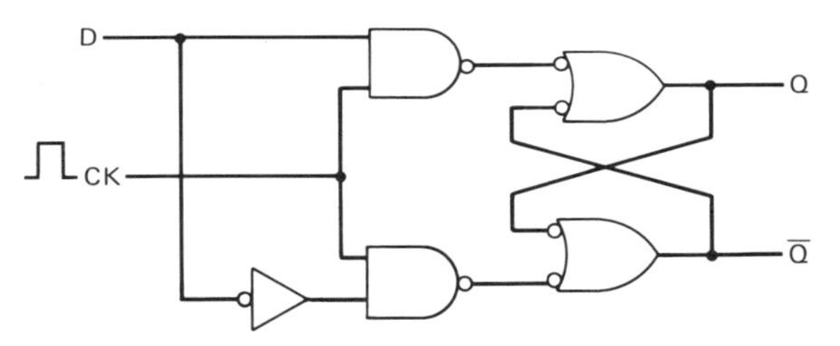

FIGURE 4.18 D bistable.

IC Flip-Flop. It is useful for temporary data storage. The state of the D input is transferred to the Q output when the clock goes high. When the clock goes low, the Q output retains this state. Edge-triggered versions are available. The truth table is shown in Table 4.6.

TABLE 4.6 Truth Table for D Bistable in Figure 4.18

Clock	D	Q_n	Q_{n+1}
⎍	0	0	0
⎍	0	1	0
⎍	1	0	1
⎍	1	1	1

(Note that the output always follows the same state as the input after the clock pulse.)

A bistable that was often used in discrete form is the **T bistable** or toggle type (Figure 4.19*a*). The circuit has a pulse-steering circuit from the outputs to force the negative edge of the T input pulse to that gate input that will cause a change of state. The output thus changes state on every negative edge of the T input. The circuit therefore divides by 2. Since IC bistables such as the D and JK can easily be wired to divide by 2, the T is not often used. An example of a ÷2 using a positive edge-triggered D bistable is shown also in Figure 4.19*b*.

One of the most versatile and widely used bistables is the **JK bistable**, shown in its simplest form in Figure 4.20*a*. It has an advantage in that it cannot be forced into an indeterminate state by presenting identical inputs. The truth table, for a narrow positive clock pulse, is shown in Table 4.7.

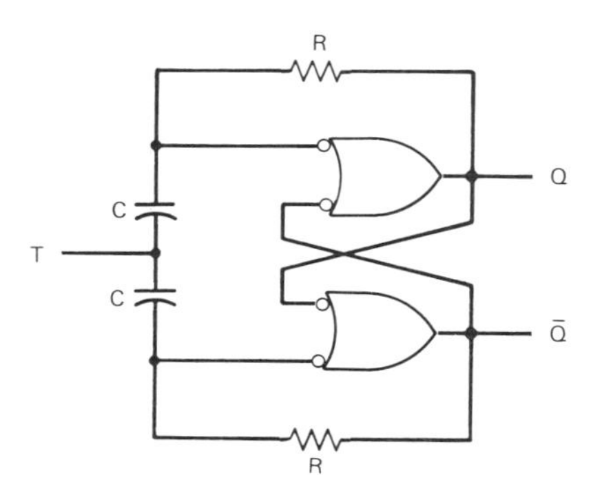

FIGURE 4.19*a* T bistable.

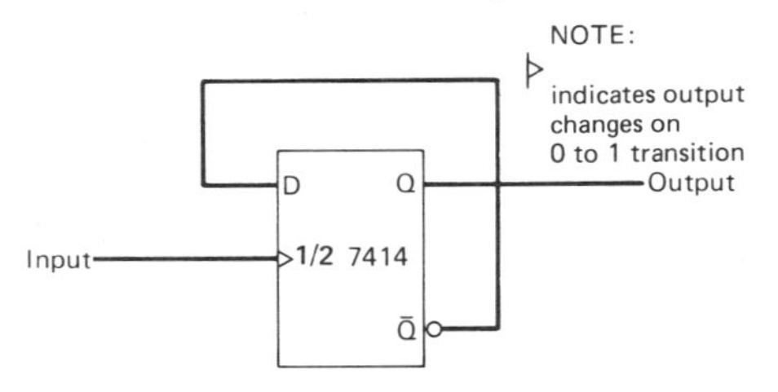

FIGURE 4.19*b* Using a D type positive edge-triggered flip-flop as a ÷ 2.

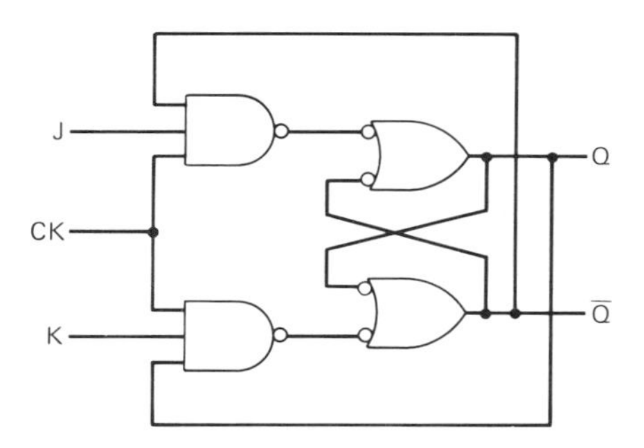

FIGURE 4.20*a* Basic JK bistable.

TABLE 4.7 Truth Table for JK Bistable

J	K	Q_n	Q_{n+1}	
0	0	0	0	Output remains in previous
0	0	1	1	state (Q_n).
0	1	0	0	When J = 0, K = 1,
0	1	1	0	output goes to 0.
1	0	0	1	When J = 1, K = 0,
1	0	1	1	output goes to 1.
1	1	0	1	When J = K = 1 output
1	1	1	0	always complements ($\bar{Q}_n$).

The JK bistable achieves this operation by virtue of its additional feedback from Q to K gate and $\bar{Q}$ to J gate. Only one input gate can be enabled at any one time. The basic circuit, because of the feedback, can suffer from timing problems, usually referred to as "race hazards." For example, if the positive clock pulse width is too wide, the outputs may change state several times during the positive clock pulse. Problems like these are virtually eliminated by the use of **masterslave circuits** (Figure 4.20*b*). As the clock pulse goes high at point A on the clock pulse input waveform,

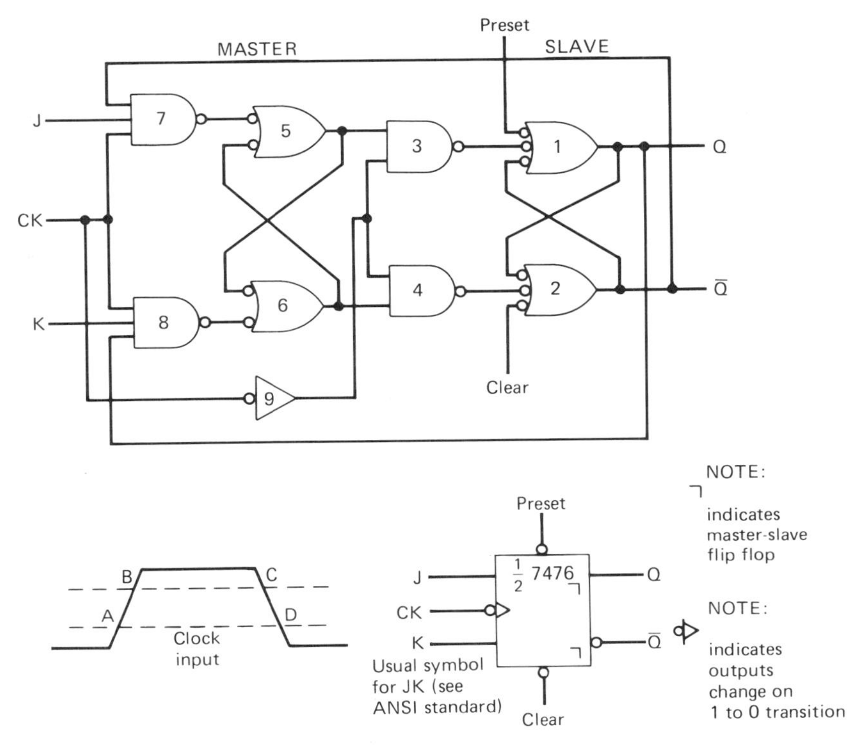

FIGURE 4.20*b* Master-slave JK bistable.

gates 3 and 4 close, isolating the slave from the master. At point B, gates 7 and 8 are opened, allowing the J and K input data to change the state of the master. As the clock goes low at point C, gates 7 and 8 close, disconnecting inputs from the master. Then finally at point D, gates 3 and 4 open, allowing the master to change the state of the slave. Thus the outputs change state on the trailing edge of the clock pulse. From this discussion it should be clear that master-slave flip-flops are pulse-triggered flip-flops that trigger on the trailing edge of the clock pulse.

Naturally flip-flops, such as the master-slave JK, are not drawn in full on a circuit; instead, a logic symbol as shown is used. Preset and clear inputs are shown with a bubble, since a logic 0 (low) is required on the Preset to force the Q to assume logic 1, and a 0 is required on the Clear to force the Q to assume logic 0. Note that these two inputs override the clock and are therefore asynchronous. They are important in counters, dividers, shift registers, since they allow the states of each flip-flop to be set or cleared.

Multiple JK master-slave flip-flops can be wired to form asynchronous (ripple through) or synchronous binary counters as shown in Figure 4.21. Both circuits divide by 16 and have a count sequence of pure binary. The synchronous counter appears to be more complex but has the advantage

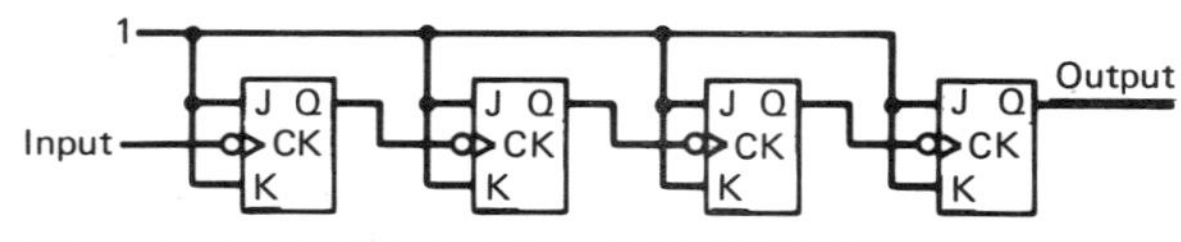

(a) ASYNCHRONOUS (RIPPLE THROUGH) ÷ 16

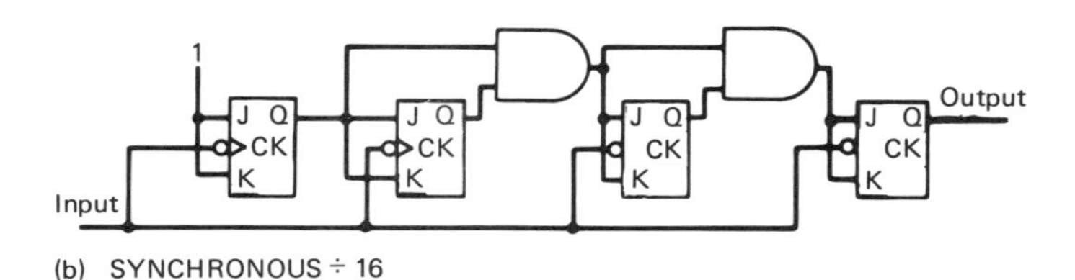

(b) SYNCHRONOUS ÷ 16

(c) ASYNCHRONOUS DECADE COUNTER

(d) TWISTED RING OR JOHNSON COUNTER (Modulo 8)

FIGURE 4.21 Counting circuits.

of a lower total delay. Also in Figure 4.21 are shown examples of dividers and counters of numbers other than binary. Troubleshooting examples are given later in this chapter for simple counting circuits.

The trend is for manufacturers to create many JK and D flip-flops, counters, and shift registers inside one IC package. Several types exist in TTL, ECL, and CMOS:

7490A	TTL asynchronous decade counter
7493A	TTL 4-bit binary counter
74192/193	TTL up/down decade counter
4017B	CMOS decade counter-divider
4020B	CMOS 14-stage binary counter
4018B	CMOS presettable divide-by-n counter

A **shift register** is a device used for the temporary storage of digital information

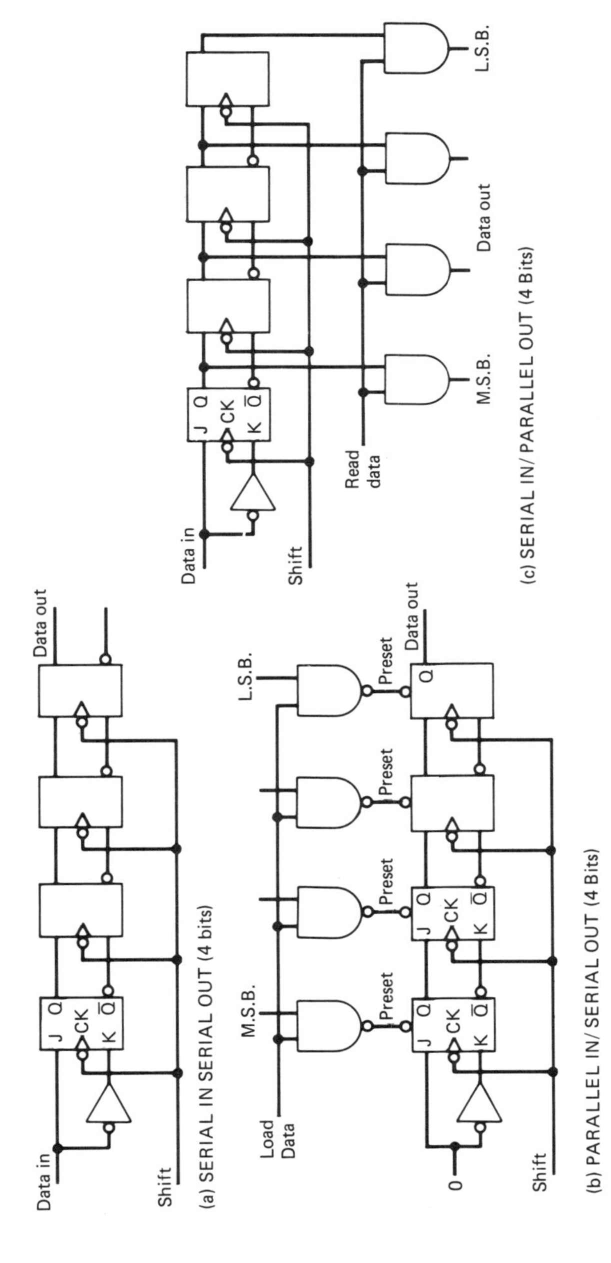

(a) SERIAL IN SERIAL OUT (4 bits)

(b) PARALLEL IN/SERIAL OUT (4 Bits)

(c) SERIAL IN/PARALLEL OUT (4 Bits)

FIGURE 4.22 Basic shift registers.

which can then be moved (shifted) at a later time. Shift registers can be readily constructed using JK flip-flops to take the form of

(a) Serial in/Serial out.

(b) Parallel in/Serial out.

(c) Serial in/Parallel out.

as shown in Figure 4.22.

Data stored in the Shift Register is loaded either in series with shift pulses or in parallel by setting the flip-flops. The data can be shifted to the right one place with every shift pulse.

Large-number shift registers (serial in/serial out) are made in MOS and are the basis of recirculating memories. A bistable can be formed using MOS devices (Figure 4.23). If the S input is taken high (1), Q_5 conducts taking $\bar{Q}$ low. This turns off Q_2, forcing Q to assume logic 1. Similarly if the R input is taken high (1), Q_6 conducts and Q assumes logic 0 state. Such a bistable forms the basic element for static MOS shift registers such as the 2 bit shift register

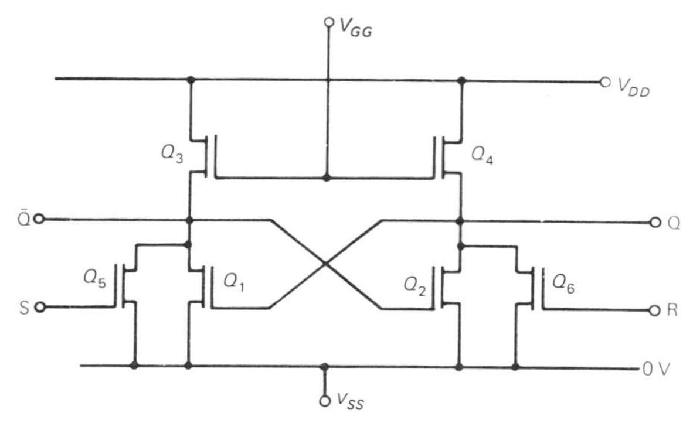

FIGURE 4.23 MOS bistable.

shown in Figure 4.24. Q_2, Q_5 and Q_7, Q_{10} form the two bistables and Q_3, Q_4 and Q_8, Q_9 are cross-coupling elements. These cross-coupling elements are switched off and on by $\overline{\text{clock 1}}$ and $\overline{\text{clock 2}}$ signals. Q_1 and Q_6 are data transfer switches. The phase relationship between the three clock waveforms is important. To shift data, the clock line is taken high, turning on Q_1 and Q_6, and at the same time the cross-coupling elements are switched off by $\overline{\text{clock 1}}$ and $\overline{\text{clock 2}}$ going low. Input data from Q_1 to Q_2 is stored by the gate capacitance of Q_2, and similarly data from bistable A is stored

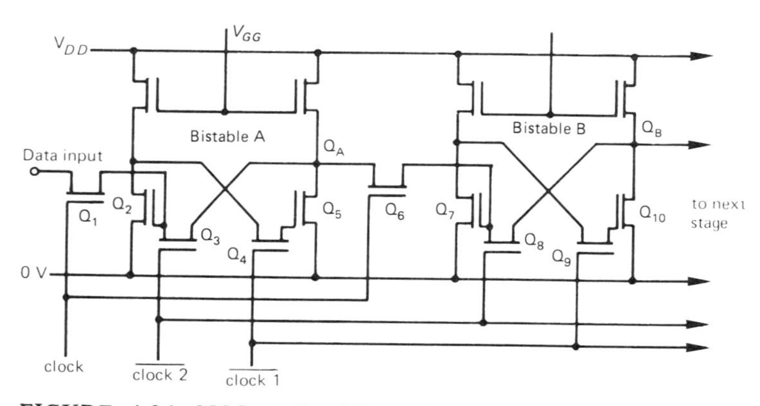

FIGURE 4.24 MOS static shift register (2 bits shown).

by the gate capacitance of Q_7. When the clock goes low, Q_1 and Q_6 turn off, $\overline{\text{clock 1}}$ goes high first to switch Q_4, Q_9. This forces Q_5 and Q_{10} to assume a new state. After a short delay $\overline{\text{clock 2}}$ also goes high, turning on Q_3 and Q_8. Note that, while the clock pulse is not applied, the bistables remain in their set states. Thus some power is always consumed. Shifting of information only takes place when the clock waveforms are applied.

Dynamic MOS shift registers (Figure 4.25) are simpler in structure and they operate to switch load devices on and off with clock pulses. Much less power is consumed from the supply, but the clock signal must be continually applied or the data stored will be lost. A two-phase (ϕ_1 and ϕ_2) clock is required. When ϕ_1 switches

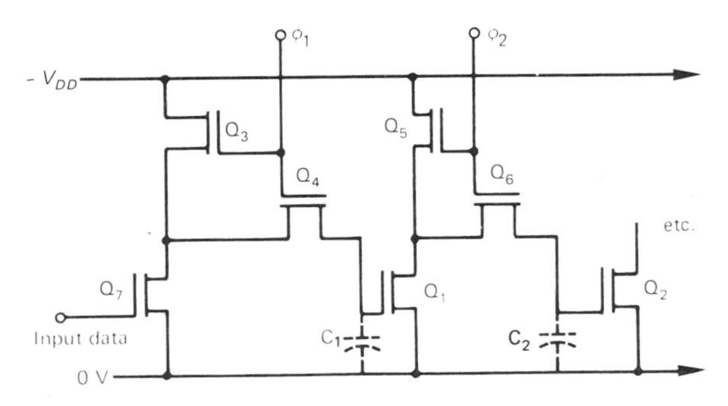

FIGURE 4.25 Dynamic MOS shift register (1 bit).

low, ϕ_2 switches high. Q_3, Q_4 turn off and Q_5, Q_6 turn on. The level at Q_1 drain is now transferred to Q_2 gate. One complete cycle of ϕ_1 and ϕ_2 clock is required to shift the data one stage. On ϕ_1, Q_3 and Q_4 turn on while Q_5 and Q_6 are off. Data applied will be transferred from Q_7 to Q_1 to be stored on the gate capacitance of Q_1.

The two-phase clock signals must not be allowed to overlap, since correct storage and transfer of data would not take place.

4.5 TROUBLESHOOTING DIGITAL LOGIC CIRCUITS

Before looking at digital troubleshooting techniques in detail a few obvious but sometimes easily overlooked points are well worth remembering. As in the servicing of any instrument or system, first make certain that

1. An up-to-date service manual with circuits, layout diagrams, and specifications is available.
2. The necessary tools and test instruments and spare parts are available.

Then proceed to define the fault as accurately as possible. This means carrying out a functional test and noting all the fault symptoms. This is necessary to ensure that there really is a fault and not an operator error. In addition:

3. Be aware of the types of logic ICs used in the equipment. In particular, know the expected logic levels and the specification for power supply voltages (typical and maximum values).
4. Avoid the use of large test probes; these can easily short out IC pins and cause more faults.
5. Never remove or insert an IC while power is applied. The current surges set up may possibly cause permanent damage.

6. Do not apply test signals while the the power is off.
7. Always check power supply voltages at the actual IC pins, not between board connections or on the p.c.b. traces. An open circuit power line to the IC may be missed in this way.

The actual process of diagnosing faults is by sequentially operating gates and ICs within the system and then comparing the resulting outputs with those that should be present. This can be done dynamically by applying suitable test signals and checking the resulting operations with a wide bandwidth oscilloscope. The CRO bandwidth must be at least 10 MHz, and its triggering has to be good otherwise some pulse information may be missed. Testing in this way will assist in narrowing down the search for a fault to a small area of the overall system. Then static checking has to be used, i.e., one gate or IC function at a time. This may necessitate switching off or slowing down any system clock generator. At this stage there are several test aids; some are very sophisticated and therefore costly, and can be used to speed up troubleshooting. The basic aids are IC test clips, probes, and "pulsers."

However, before looking at these let us consider some of the faults and their symptoms associated with digital logic circuits. ICs themselves are fairly sturdy and reliable devices; so, many faults in systems are caused by dry joints, breaks in p.c.b. traces, or shorts between traces on the p.c.b. A thorough visual inspection of the p.c.b. is therefore a good idea, checking for trace or print faults, slightly burned resistors or solder between traces. A useful aid for this is a small magnifying glass.

For most logic ICs, apart from the special case of TRI-STATE outputs, the output should be either

(a) Low at logic 0 (for TTL, 0.8 V maximum).

(b) High at logic 1 (for TTL, 2 V minimum).

(c) Pulses switching back and forth between 1 and 0.

The type of output naturally depends on the conditions at the IC inputs.

What is not allowed (apart from TRI-STATE) is a static output level that is not a 0 or 1. For TTL the output must not be greater than 0.8 V or less than 2 V.

The sort of faults that can occur at gate outputs are:

1. "Stuck" at 0, i.e., for TTL always less than 0.8 V irrespective of input states.
2. "Stuck" at 1, i.e., for TTL always above 2 V irrespective of input states.
3. High impedance state output, i.e., output not 0 or 1.

Such fault conditions for a single gate are illustrated in Figure 4.26.

Within a system a gate will have its inputs supplied by other gate outputs and its output may itself be driving several other gate inputs. Therefore, a fault may seem to be caused by a particular gate, but the same symptoms may result from a bad trace on the p.c.b. that is connected to

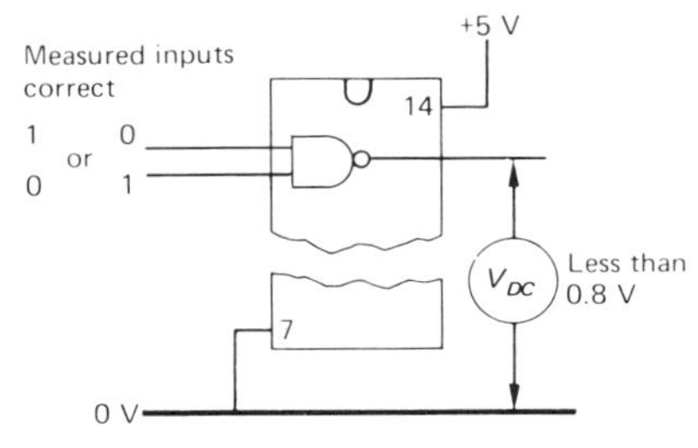

(a) Output stuck at 0
Output should be logic 1
Possible faults:
internal transistor
short circuit
or +5 V power line
open either internally or
externally

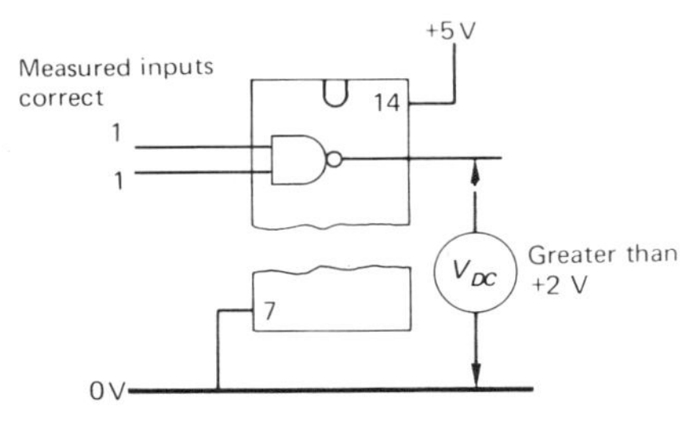

(b) Output stuck at 1
With logic 1s on inputs, output
should be less than 0.8 V
Possible faults:
Internal transistor
open circuit
or 0 V line open circuit
either internally or
externally

FIGURE 4.26 Possible fault conditions in a single gate.

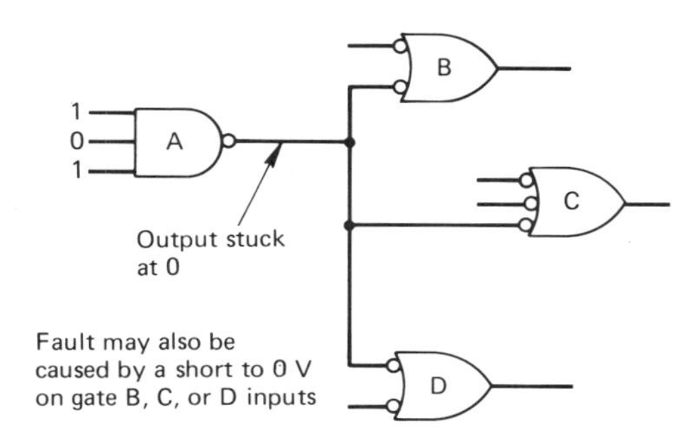

FIGURE 4.27

the inputs of driven gates. For example, consider Figure 4.27, in which gate A has its output permanently stuck at 0. Having checked that the correct inputs will not force a change of state, i.e., taking one input down to 0, we might then assume that the fault lies with gate A. But *this may not be true* since a short circuit to 0 V of any input on gates B, C, or D will also hold the output of A at 0 V.

Finding the actual IC or track (path) that is causing the short can, of course, be rather difficult in a complex system. One good method (Figure 4.28) is to switch off the system power supplies and inject a standard a.c. signal (500 kHz) via two test clips as shown. The current flows along the track that is shorted and sets up a magnetic field, which can be followed by a current sensing probe. Once the probe passes the short its output falls to zero. The output can either be an audio tone or an LED display. Power line shorts can also be located using a similar technique, but a d.c. current source must be used because any decoupling capacitors will act as a short to an a.c. test signal. A standard d.c. current of about 10 mA is injected as shown in Figure 4.28*b* from a constant current source. Two probes are used to measure the voltage drop along the track caused by the flow of the test current. Naturally this is only tens of microvolts per centimeter. The voltage displayed on a meter will increase uniformly as the right-hand probe is moved along. When the short is reached, no further increase in voltage takes place. Methods like these, for finding shorts on power and signal lines, are incorporated in commercially available fault tracers.

Other standard test and service aids are:

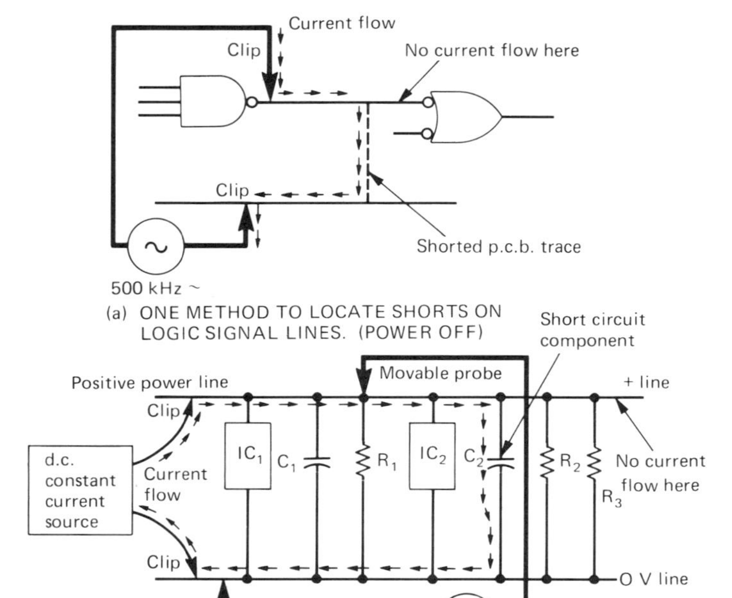

(a) ONE METHOD TO LOCATE SHORTS ON LOGIC SIGNAL LINES. (POWER OFF)

(b) LOCATING SHORTS ON POWER LINES. (POWER OFF)

FIGURE 4.28

(a) **IC clips** These snap onto an IC and bring out the pins for easy testing.

(b) **Logic state monitors** Devices that clip onto the IC and display the static or dynamic state of each IC pin. They find their own power from the IC being tested. An LED indicates the presence of a 1 (LED on) or 0 (LED off) for each pin.

(c) **Logic probes** Probably one of the most useful aids. The usual type has two LED display indicators: a green LED for logic 0 and a red LED for logic 1.

These units can be quite versatile, for example:

Display	Meaning of Input
Green ON	Below logic 0 threshold
Red ON	Above logic 1 threshold
Both flash	Pulses (>100 msec) at a frequency of less than approx. 100 Hz.
Both dim	Wide pulses or square waves above 100 Hz.

Therefore,

Green flashes	Narrow positive-going pulses <100 Hz.
Red flashes	Narrow negative-going pulses <100 Hz.
Green dim	Narrow positive-going pulses >100 Hz.
Red dim	Narrow negative-going pulses >100 Hz.

The logic probe being described also allows the user to detect a single asynchronous pulse as narrow as 50 ns.

(d) **Logic pulser** A unit used to force a change of state at an input junction allowing outputs of gates to be checked with a logic probe. The pulser has a tri-state output, and until operated its output remains in a high impedance state. The pulser output pin is held against the appropriate IC input and the push button is operated. A narrow output pulse of a few microseconds is delivered. The output of the pulser can sink or source up to 500 mA of current, which is sufficient to override any other IC outputs on the junction. The narrow pulse width prevents damage to any ICs. A single pulse, group of pulses, or a continuous pulse train can usually be selected.

All the maintenance aids described are not always available. However, it is worth remembering that an analog multirange meter will give much information on IC outputs. Logic 0 and 1 levels can be checked, and a square wave output will give a meter indication about halfway between these levels.

The following exercises are intended to assist in understanding troubleshooting of digital logic circuits.

4.6 EXERCISE: CAR SEAT BELT ALARM (FIGURE 4.29)

This is an example of combinational logic used to operate an alarm in a car if the driver and/or the passenger seats are occupied and the seat belts are not fastened when the car is started. The signals that indicate the presence of the driver and perhaps a passenger can be taken from pressure-operated switches set to logic 1 when a seat is occupied.

When the car ignition is switched on and the gears are engaged, if either of the

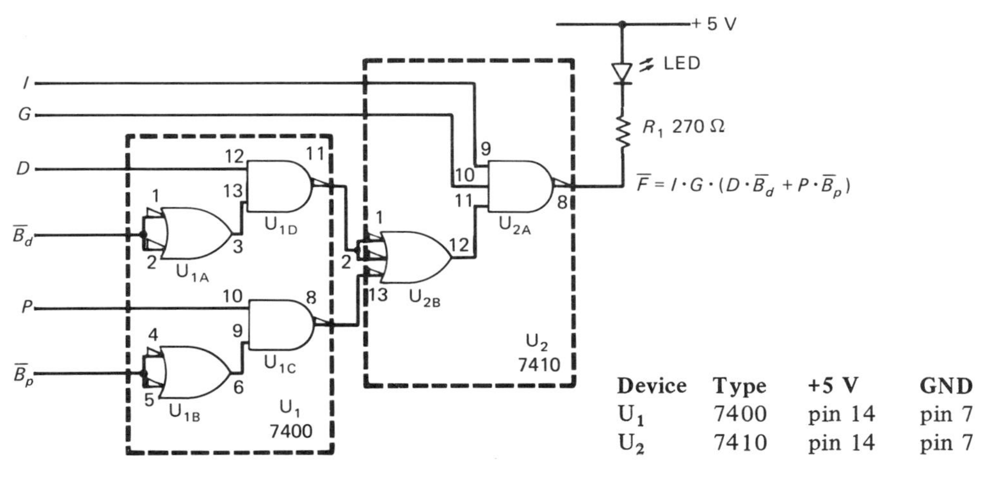

Device	Type	+5 V	GND
U_1	7400	pin 14	pin 7
U_2	7410	pin 14	pin 7

FIGURE 4.29 Car seat belt alarm.

seats is occupied and its belt is not fastened, then the alarm will be operated. In this example an LED is used but an audible alarm could easily be added.

If $\overline{F}$ = alarm signal on:
G = gears engaged
I = ignition on
D = driver's seat occupied
P = passenger's seat occupied
$\overline{B}_d$ = driver's belt unfastened
$\overline{B}_p$ = passenger's belt unfastened

then the Boolean expression that clearly describes the system is

$$\overline{F}, L = I \cdot G \cdot (D \cdot \overline{B}_d + P \cdot \overline{B}_p), L$$

Note that the circuit uses all NAND gates to achieve the desired function. If there is a high on the output of U_{2B} then all three inputs to U_{2A} will go high when the car is started. This will cause the output of U_{2A} to go low (i.e., output $\overline{F}$ will go low), forward-biasing the LED. A current-limiting resistor must always be used with an LED because the forward voltage drop of a red LED is only about 1.6 V at 20 mA. Note that a high on the output of U_{2B} must mean a seat is occupied without the associated seat belt being fastened. It is common practice on functionally drawn logic diagrams to use a bar over an output to signify an active low output (i.e., $\overline{F}$). The practice may also be applied to inputs (i.e., $\overline{B}_d$ and $\overline{B}_p$) as shown in Figure 4.29.

QUESTIONS

1. The possible states that can cause a 1 or high output at U_{2B} can be listed in tabular form. Complete a table to show this.
2. A fault exists such that the alarm light is on continuously, irrespective of any input states. What is the most likely fault?
3. What would be the symptoms for the following faults:
   (a) R_1 open circuit.
   (b) P.C.B. is printed open circuit from U_{1C} output.
   (c) U_{1A} output failed stuck at 0.
4. The alarm lights up when the car is started and a passenger is present and his seat belt fastened. Correct operation results with the driver only. List the possible faults and write a test sequence that could be used to find the fault.
5. In the example NAND gates only have been used. Draw the logic diagram for the expression

$$\overline{F}, L = I \cdot G \cdot (D \cdot \overline{B}_d + P \cdot \overline{B}_p), L$$

   using AND, OR, and NOT gates.
6. What would be the effect of an open circuit + 5 V line to U_1 (the 7400)?

4.7 EXERCISE: CONTACT BOUNCE ELIMINATOR (FIGURE 4.30)

Apart from the mercury-wetted types, nearly all contacts on mechanical switches and relays suffer from an effect known as

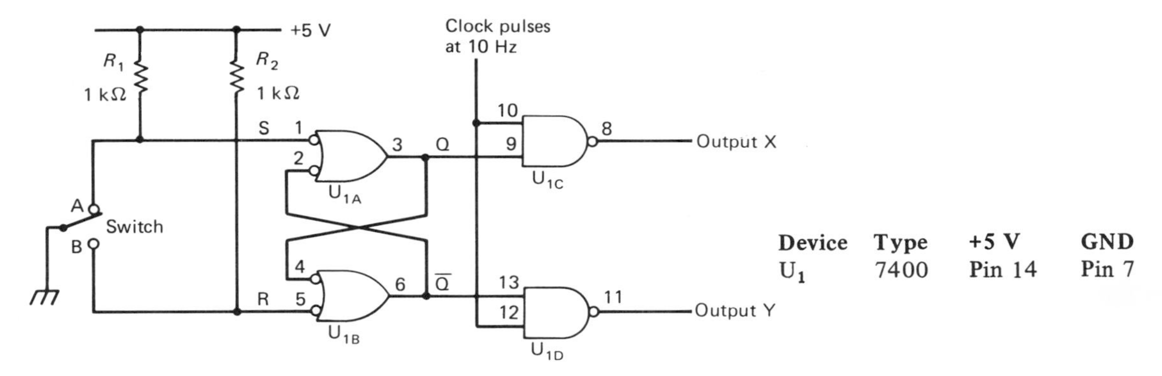

Device	Type	+5 V	GND
U_1	7400	Pin 14	Pin 7

FIGURE 4.30 Contact bounce eliminator using 7400.

"bounce." When the switch is operated, the contacts make and break several times before finally setting down in the closed position. This bounce time may last several milliseconds and, if the switch is used to operate logic circuits, all of these vibrations will be recorded as input pulses and cause false triggering. One common method of eliminating the effects of bounce is to use an R–S flip-flop circuit. In this case a quad 2 Input/Positive (I/P) NAND gate IC (7400) has been used.

With the switch in position A, the set input will be low (0) and the reset input will be high (1). The Q output will be held high (logic 1) and U_{1C} will be open. Since $\overline{Q}$ is low (logic 0), U_{1D} will be closed. When the switch is operated, the contact first breaks from A but the flip-flop cannot change its state until the contact is at B, forcing the reset input to go low. The first time the contact makes at B, the flip-flop will be reset and any switch bounce will be ignored. The circuit uses the ability of a flip-flop to be set or reset by a momentarily applied low signal as indicated by the functional form of the circuit that shows bubbles on the inputs of U_{1A} and U_{1B}. Bounce-less switch circuits are used in the control of nearly all logic systems.

QUESTIONS

1. The circuit fails such that no pulses under any conditions can be obtained from either pin 8 or pin 11. State the possible cause of failure.

2. The circuit fails with the following symptoms. State the probable cause of failure in each case.

Switch Position	Pin 3	Pin 6	Pin 8	Pin 11
Fault (i)				
A	1	Pulses	1	Pulses
B	Pulses	1	Pulses	1
Fault (ii)				
A	0	1	1	Pulses
B				
Fault (iii)				
A	1	1	Pulses	Pulses
B	0	1	1	Pulses
Fault (iv)				
A	0	1	1	Pulses
B	0	1	1	Pulses

3. Redesign the circuit using Quad 2 I/P NOR gates.

4.8 EXERCISE: ALARM CIRCUIT USING CMOS LOGIC (FIGURE 4.31)

An alarm circuit is often required that will operate when a switch contact is momentarily made, and which will continue to give an alarm output until a clear signal is given. Quite complicated circuits can be constructed but the basic principles are illustrated by this circuit. By using CMOS logic the possibility of false triggering by noise is virtually eliminated and the current drain on the battery is very low. One IC (U_1), a Quad 2 I/P NAND gate type 4011B is used. Two of the gates (U_{1C} and U_{1D}) are wired as an S–R flip-flop. The Q output (gate U_{1C}) will be high if a trip signal is given and low when the clear button is pressed. The other two gates are wired to form a low-frequency astable oscillator. This oscillator will be inhibited from operating while the level from Q is low. When a trip signal is received, the bistable is immediately set, the Q output switches to high, and the oscillator runs causing the LED to flash on and off.

CMOS logic gates are ideal for low-frequency oscillator circuits, since their very high input impedance allows high-value timing resistors to be used. With the inhibit line low, (i.e., after the clear switch has been pressed) U_{1A} pin 3 is high and U_{1B} pin 4 is low causing C_1 to change via R_3 toward $+V_{DD}$ (9V). Since point x goes high, U_{1A} pin 1 goes high but the oscillator is inhibited from oscillating. The inhibit line goes high after the trip switch is pressed, and this starts the oscillator oscillating. When the inhibit line goes high, U_{1A} pin 3 goes low and U_{1B} pin 4 goes

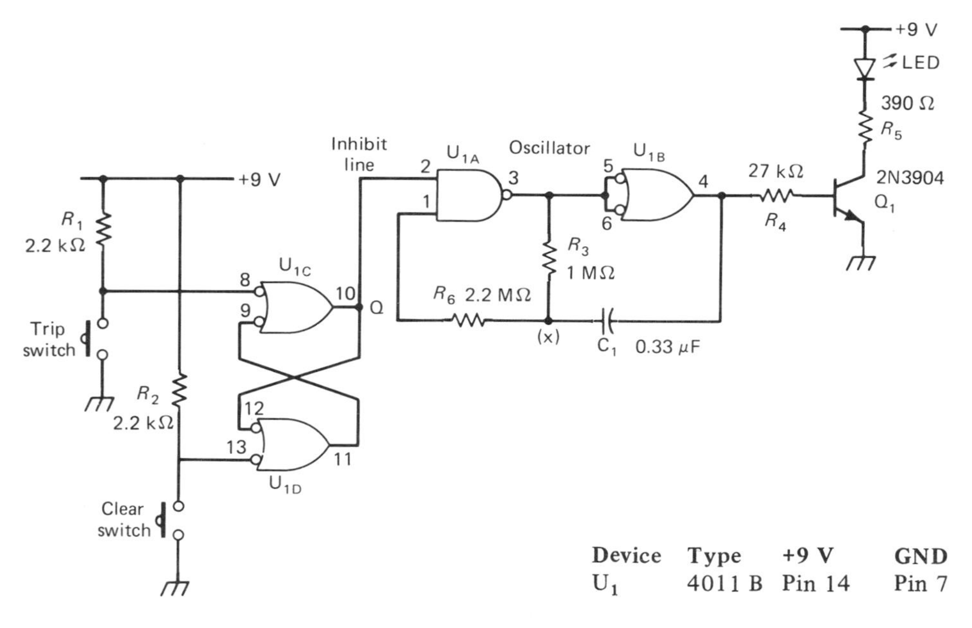

Device	Type	+9 V	GND
U_1	4011 B	Pin 14	Pin 7

FIGURE 4.31 Alarm circuit.

high causing: (1) C_1 to transfer a positive going step to U_{1A} pin 1; and (2) C_1 to discharge. When the level at point x crosses the lower threshold of U_{1A} pin 1, the output at U_{1A} pin 3 switches high. U_{1B}'s output goes low and a negative-going step is transmitted via C_1 to U_{1A}. C_1 now charges pin 1 via R_3 toward $+V_{DD}$ (9 V). Again when point x crosses the upper threshold point, the gates switch and the cycle continues. A transistor switch is used to drive the light-emitting diode. Note that a LED will switch on, giving off visible light when it is forward biased by about 1.6 V. The forward current should be typically 20 mA. A current-limiting resistor is essential to prevent the LED from being overdriven.

QUESTIONS

1. Explain how the astable circuit is inhibited from oscillating by a logic 0 (a low) on pin 2 of U_{1A}.
2. Sketch the waveform that you would expect to measure at point x after an alarm trip signal is received.
3. For the following faults the alarm fails to operate and the LED remains off. Deduce the component that is faulty and the type of fault.

	Pin 10	Pin 1	Pin 3	Pin 4	Q_1 Base
Fault (i)	High	Low	High	Low	0 V
Fault (ii)	High	Varying	Varying	Varying	0 V

4. How could the LED and the transistor be quickly checked to verify their operation?
5. A fault occurs so that the LED is permanently on and does not flash after a trip signal is received, but it can be turned off by pressing the clear button. What is the faulty component?
6. State the symptoms that would exist for the following faults:
   (a) A short circuit between pins 6 and 7.
   (b) An open circuit between pin 7 and ground.
7. A modification is required so that additional audible alarm tone (approx. 400 Hz) is provided. Sketch a circuit that will do this.

4.9 EXERCISE: TELEPHONE TONE GENERATOR (FIGURE 4.32)

In this circuit three TTL ICs are used to generate the four telephone tones:

Dial tone	25 Hz
Number unobtainable	400 Hz
Number busy	400 Hz switching at 1.6 Hz
Ringing tone	400 Hz modulated at 25 Hz switching at 1.6 Hz and 0.4 Hz.

The three square waves of 400 Hz, 25 Hz, and 1.6 Hz are generated by three Schmitt trigger gates wired with positive feedback. One IC U_1, a Quad NAND Schmitt (74132), is used. Because of the hysteresis present in a Schmitt circuit,

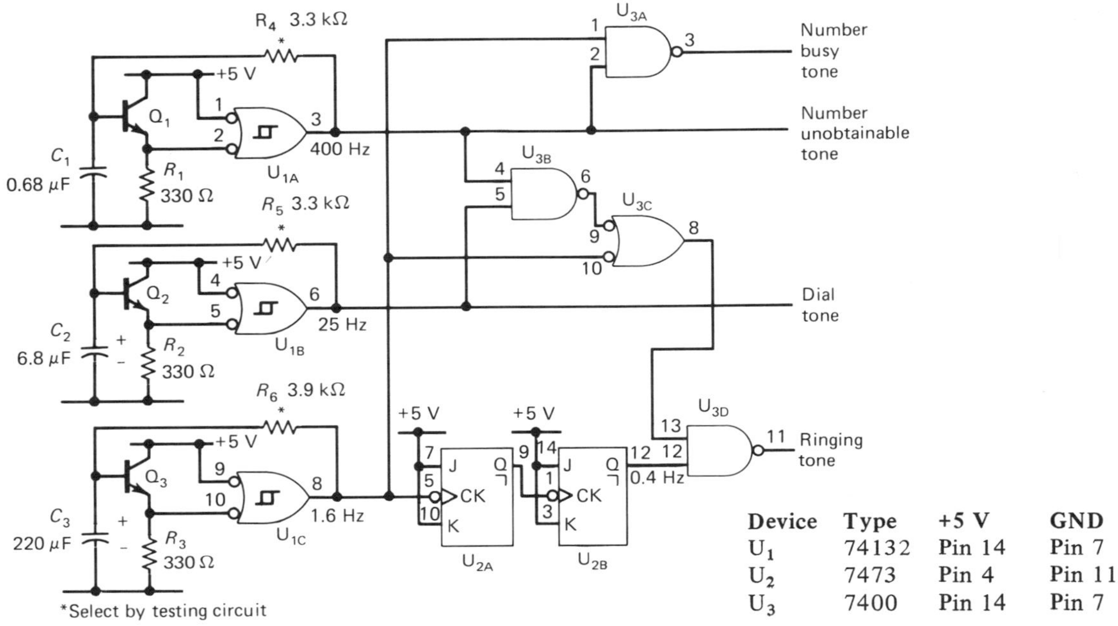

Device	Type	+5 V	GND
U_1	74132	Pin 14	Pin 7
U_2	7473	Pin 4	Pin 11
U_3	7400	Pin 14	Pin 7

FIGURE 4.32 Telephone tone generator.

square wave oscillations will occur if an RC network is connected from the gate output to its input. When the output goes high, the capacitor is charged via the resistor until the voltage across the capacitor exceeds the upper trip point of the gate. The gate output then switches low and the capacitor is discharged. Because of the hysteresis effect, the gate cannot switch high again until the voltage across the capacitor falls below the lower trip point. With a TTL Schmitt, the upper threshold is typically 1.7 V and the lower threshold is 0.9 V, giving a hysteresis of about 1.2 V. Unfortunately, the value of the feedback resistor is limited to a maximum of about 390 Ω; this is because of the relatively high input current supplied by a TTL gate. To get low-frequency oscillators an emitter follower circuit can be used, as in the example, to allow higher values of timing resistor. The values of R_4, R_5, and R_6 can be selected to give the most realistic sound by testing the circuit.

The 1.6 Hz oscillator output is divided down by a 7473 dual JK bistable wired to divide by 4. This gives the 0.4 Hz square wave.

The oscillator outputs are combined in the following way by the 2 I/P NAND gates of a 7400 IC. The Number-busy tone is 400 Hz switched by the 1.6 Hz waveform; the Number-unobtainable tone is 400 Hz unmodulated; the Dial tone is 25 Hz; and the Ringing tone is achieved by first modulating the 400 Hz with the 25 Hz, then switching this with 1.6 Hz, and finally gating it with the 0.4 Hz waveform. A circuit such as this can be used as part of a demonstration telephone system. The outputs can be easily checked by a small loudspeaker via an emitter follower circuit.

Suppose a fault occurred such that no Number busy or Ringing tones were available although the Dial and Number-unobtainable tones were correct. The signal common to this first two is the 1.6 Hz oscillator. If this failed with its output stuck at 0, the symptoms above would result. If on the other hand the 1.6 Hz oscillator failed with its output high, then the Number-busy tone would be identical with the Number-unobtainable, the Dial tone would be correct, but no Ringing tone would be present.

QUESTIONS

1. (a) Explain the operation of the divide-by-4 circuit. Why do pins 14, 3, 7, and 10 have to be connected to +5 V?

   (b) Sketch the waveforms of each output.

2. What would be the symptoms for the following faults?

   (a) Q_1 base emitter open circuit.

   (b) C_2 short circuit.

   (c) R_3 open circuit.

3. State the component faults that could cause the following symptoms:

   (a) No Ringing tone, but a Number-busy tone present on pin 13 of the U_{3D}.

   (b) No Number-busy tone available. Rest of outputs correct.

   (c) Dial tone only.

   (d) Number busy tone 400 Hz. Rest of outputs correct.

4.10 EXERCISE: SINGLE DECADE COUNTER WITH DECODER AND DISPLAY (FIGURE 4.33)

Many counting circuits consist of counters that count up in the 8421 Binary Coded Decimal (BCD) code. A typical example of this is the 7490A, which consists of four master-slave JK flip-flops giving a ÷2

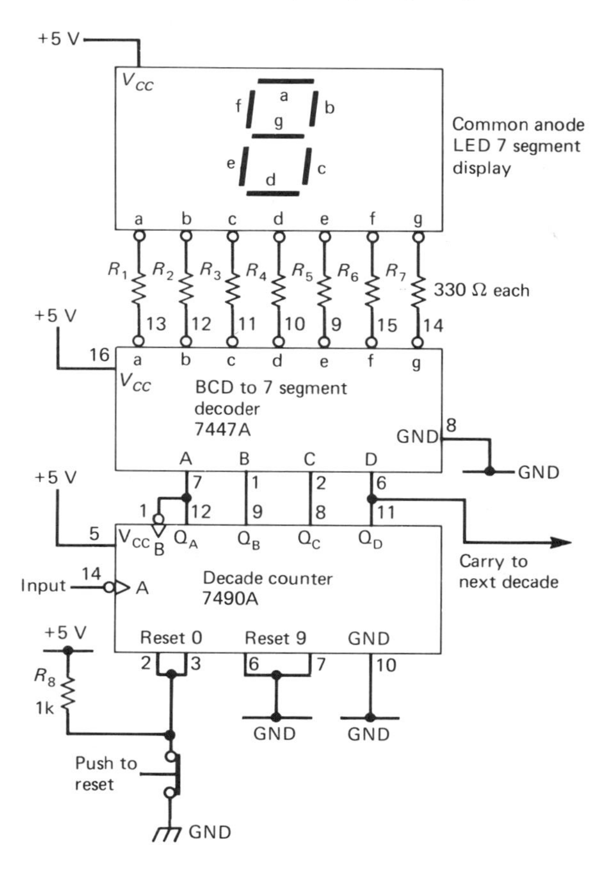

FIGURE 4.33 Single decade counter with decoder and display.

followed by a ÷5 arrangement. By connecting the output of the ÷2 to the input of the ÷5, a decade counter of ÷10 is achieved. Since master-slave flip-flops are used, this counter operates on the negative edge of the clock input. In order to be able to display the number of input pulses the counter has received, the state of the counter has to be decoded. Several decoder circuits are available such as the 7445, which converts the four-line BCD information into decimal (0 to 9) to drive a cold cathode numerical indicator tube. However, in this example an LED seven-segment display is used and therefore a BCD to seven-segment decoder driver (7447A) is required.

Before discussing the decoder circuit, let us first consider the seven-segment display. This consists of eight light emitting diodes, seven arranged so that the numerals 0 to 9 can be displayed with the eighth LED a decimal point. The unit used is a common anode type, so that each LED segment will be forward biased and will therefore emit light when its appropriate pin is taken to ground via a current-limiting resistor. A current-limiting resistor is essential, since the forward volt drop across an LED is typically 1.6 V. It is easy to see that, to display the numeral 3, for example, requires segments a, b, c, d, and g to be on. The truth table that follows will show the arrangement for the other numerals.

Many different types of LED 7-segment displays are available, some with common cathodes which require positive voltage for switch-on, others with built-in decoders and even counters. The 7447A decoder has to take the BCD information from the four flip-flops in the 7490A and convert the states of these flip-flops into low signals, hence, the bubbles on the outputs of the 7447A, to drive the appropriate display segments. For example, suppose seven input pulses have been received by the counter; the states of the four flip-flops Q_D, Q_C, Q_B, Q_A will be 0, 1, 1, 1 respectively. The decoder output must be low for segments a, b, and c and high for all others so that 7 is displayed. The full truth table for the counter and decoder is shown below.

The display will naturally change continually while pulses are applied to the 7490A input. For larger counting systems,

Count input	*Outputs from* 7490 A D C B A	*Outputs from decoder* a b c d e f g	*Resulting display* 0	*Count input*	*Outputs from* 7490 A D C B A	*Outputs from decoder* a b c d e f g	*Resulting display* 0
0	0 0 0 0	0 0 0 0 0 0 1	0	5	0 1 0 1	0 1 0 0 1 0 0	5
1	0 0 0 1	1 0 0 1 1 1 1	1	6	0 1 1 0	1 1 0 0 0 0 0	6
2	0 0 1 0	0 0 1 0 0 1 0	2	7	0 1 1 1	0 0 0 1 1 1 1	7
3	0 0 1 1	0 0 0 0 1 1 0	3	8	1 0 0 0	0 0 0 0 0 0 0	8
4	0 1 0 0	1 0 0 1 1 0 0	4	9	1 0 0 1	0 0 0 1 1 0 0	9

like those used in frequency counters, a temporary memory or latch (7475) is used between the counter stage and the decoder. Control logic then samples the states of the counters, and this information is held in the temporary memory so that the display is stationary. The simple circuit given in this example could be used as a revolution counter and can be extended to give more decades, say units, tens, and hundreds by taking the carry output and feeding it to the next stage and so on.

Systems such as these require several ICs. Using ICs that contain a latch/decoder/driver (4511B) or a counter/latch/decoder/driver (74143) in a single package reduces the IC count. A recent Ferranti circuit, the ZN1040E, contains a four decade counter with latches and a decoder plus other gating circuits in one 28-pin DIP. As in other similar systems, only one decoder circuit is used in the ZN1040E because the display is multiplexed. An internal oscillator in the IC, running at a fairly high frequency, switches the decoder to each of the four decade counters in turn, and at the same time only one seven-segment display is energized. A multiplexed system saves power and components.

QUESTIONS

1. What signal is required at the counter's input?
2. A fault exists such that segment d will not light. The LED is not faulty. Deduce the probable fault.
3. Although input pulses are present, the display remains at 0. What is the most likely fault?
4. The display counts 0,1,0,1 while input pulses are present. What is the fault? How could the ICs be quickly checked?
5. What would be the symptoms for the following faults?
   (a) R_7 open circuit.
   (b) Pin 8 to ground open circuit on the 7447A.
   (c) An open circuit, pins 6,7 to ground on the 7490A.
6. Following reset the circuit counts in the sequence 2,3,2,3,6,7,6,7. State the fault.
7. Design a simple circuit that would detect slowly moving objects and produce a pulse suitable for driving the counter.

4.11 EXERCISE: TWO-PHASE CLOCK PULSE GENERATOR (FIGURE 4.34)

In many logic systems fairly stable square wave or pulse generating circuits are required to provide synchronizing or strobing signals throughout the system. In MOS shift registers, clock signals have to be provided that do not overlap. This circuit, using a crystal to give good frequency stability, produces outputs ϕ_1 and ϕ_2 which are negative-going and do not overlap. A suitable TTL-to-MOS interface is also shown that converts the TTL logic level into pulses going from +4 V to −22 V.

Two gates of a 7400 IC (U_1) are connected with the crystal to provide the 1 MHz square wave oscillation. The crystal

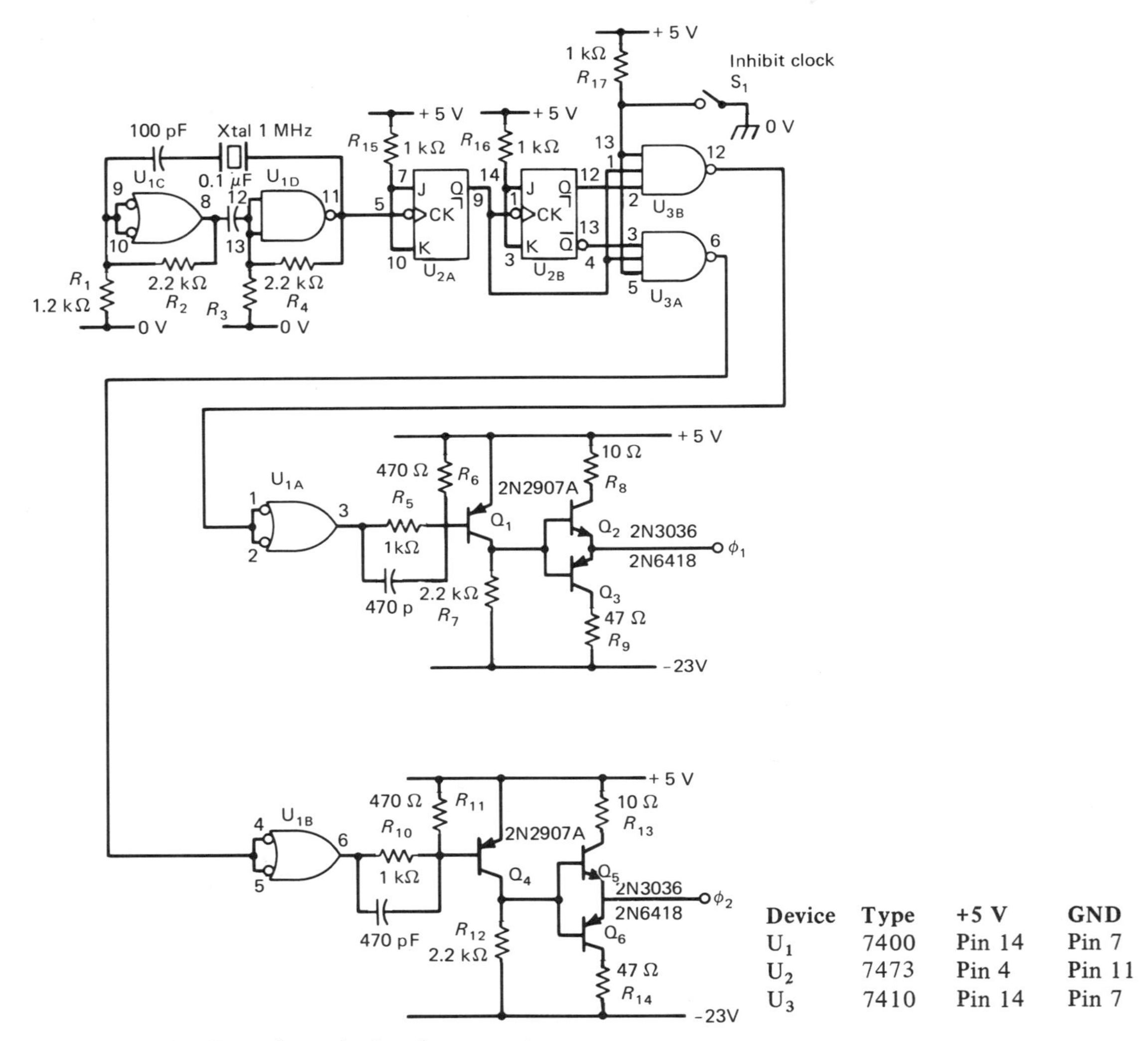

Device	Type	+5 V	GND
U_1	7400	Pin 14	Pin 7
U_2	7473	Pin 4	Pin 11
U_3	7410	Pin 14	Pin 7

FIGURE 4.34*a* Two-phase clock pulse generator.

gives positive feedback from the output of U_{1D} to the input of U_{1C}. This 1 MHz signal is divided by 2 by FF U_{2A} and the 500 kHz square waves are fed to FF U_{2B} and inputs of U_{3B} and U_{3A}. The output of U_{3B} can only go low when both Q outputs of U_{2A} and U_{2B} are high. Similarly the output of U_{3A} can only go low when Q of U_{2A} and $\overline{Q}$ of U_{2B} are both high. Since Q and $\overline{Q}$ outputs of U_{2B} can never be identical the outputs of U_{3B} and U_{3A} cannot overlap. This can be seen from the timing diagrams. Since U_{3B} and U_{3A} are triple-input NAND gates, the clock pulses can be inhibited by connecting an input of each to ground.

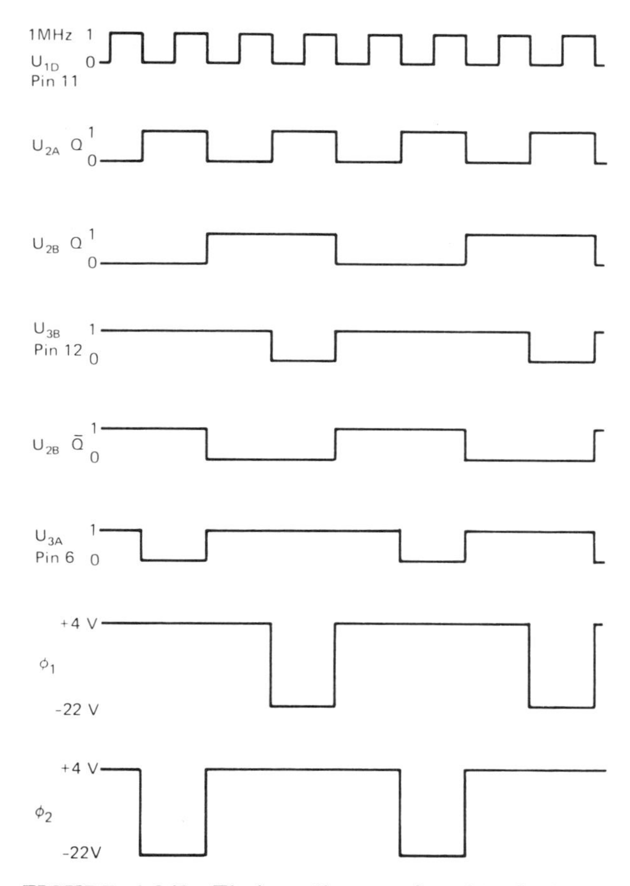

FIGURE 4.34*b* Timing diagram for 2ϕ clock pulse generator

The pulses from the outputs of U_{3B} and U_{3A} are inverted by NAND gates U_{1A} and U_{1B} and then are fed to the TTL to MOS interface circuits. When the output of U_{1A} is low, Q_1 a pnp transistor is forward-biased and its collector voltage rises to very nearly +5 V. Q_2 and Q_3 are connected as a complementary emitter follower, so Q_2 switches on and Q_3 is off. The ϕ_1 clock will be at about +4.0 V. When, however, U_{1A}'s output goes high, Q_1 turns off. This turns off Q_2 and Q_3 conducts, taking the ϕ_1 clock signal to about −22 V. The action of the circuit formed by Q_4, Q_5, and Q_6 is identical.

QUESTIONS

1. What is the frequency and pulse width of the ϕ_1 and ϕ_2 output signals?
2. Why is a complementary emitter follower used?

3. If the "inhibit clock" switch is closed, what will be the voltage level at ϕ_1 and ϕ_2 outputs?
4. Suppose ϕ_1 failed high (+4 V) while ϕ_2 continued to operate correctly. How could the operation of the suspect interface circuit be quickly checked?
5. What would be the symptoms for the following faults:
   (a) An open circuit from U_{2B} pin 13 to pin 3 of U_{3A}.
   (b) R_{10} open circuit.
   (c) R_{17} open circuit.
6. A fault is suspected on the ϕ_1 interface circuit, the inhibit clock switch is closed, and the following voltage readings with respect to 0 V are taken with a 20 kΩ/V multimeter. Determine, with reasons, the faulty component or components.

	Q_1 Base	Q_1 Collector	Q_2 Emitter
Fault A	5 V	-22.9 V	-22.2 V
Fault B	4.3 V	4.3 V	+3.6 V
Fault C	3.3 V	-22.9 V	-22.2 V
Fault D	4.3 V	4.85 V	0 V

4.12 EXERCISE: DIGITAL RAMP GENERATORS (FIGURE 4.35)

One of the advantages of digital circuits is that they can be used to generate special waveforms, and they are especially useful for generating very low frequency signals. Ramp-type waveforms are, of course, used in many instruments and measurements, the requirements usually being good linearity and rapid flyback. A digital method commonly used is to count the pulses from a stable oscillator with, say, an 8-bit counter and then convert the states of the counter into a ramp with a digital to analog converter (DAC). Two simple circuits are shown in Figure 4.35. The first is based on a 7-bit CMOS counter 4024B, which is driven from a CMOS oscillator. Seven output lines are available from the 4024B, which divide the input frequency by 2, 4, 8, 16, 32, 64, and 128. These outputs are fed direct to an R-2R ladder network. This type of resistor network is generally used in DACs. As the counter counts up, a linearly rising waveform made up of 128 discrete steps will be generated. On the 128th pulse the counter resets and the ramp returns to zero. The cycle then repeats itself.

The second circuit using TTL gives a ramp output of only 16 steps, and is therefore easier to follow. An oscillator based on a Schmitt trigger produces pulses to advance a 4-bit binary counter (7493A). This counter divides the input frequency by 2, 4, 8, and 16 so that a staircase waveform of 16 steps appears at the output of the R-2R ladder. The oscillator runs at about 2 Hz so the staircase waveform can be displayed on a chart recorder.

Using the logic levels direct from the counters will not give perfectly even steps. This is because the levels may vary slightly from one output line to another. A complete digital-to-analog converter made up of buffer amplifiers, ladder switches, and an R-2R ladder network would be required for a precision ramp. The ladder first switches a complementary pair of transistors for each bit, and then switches a preci-

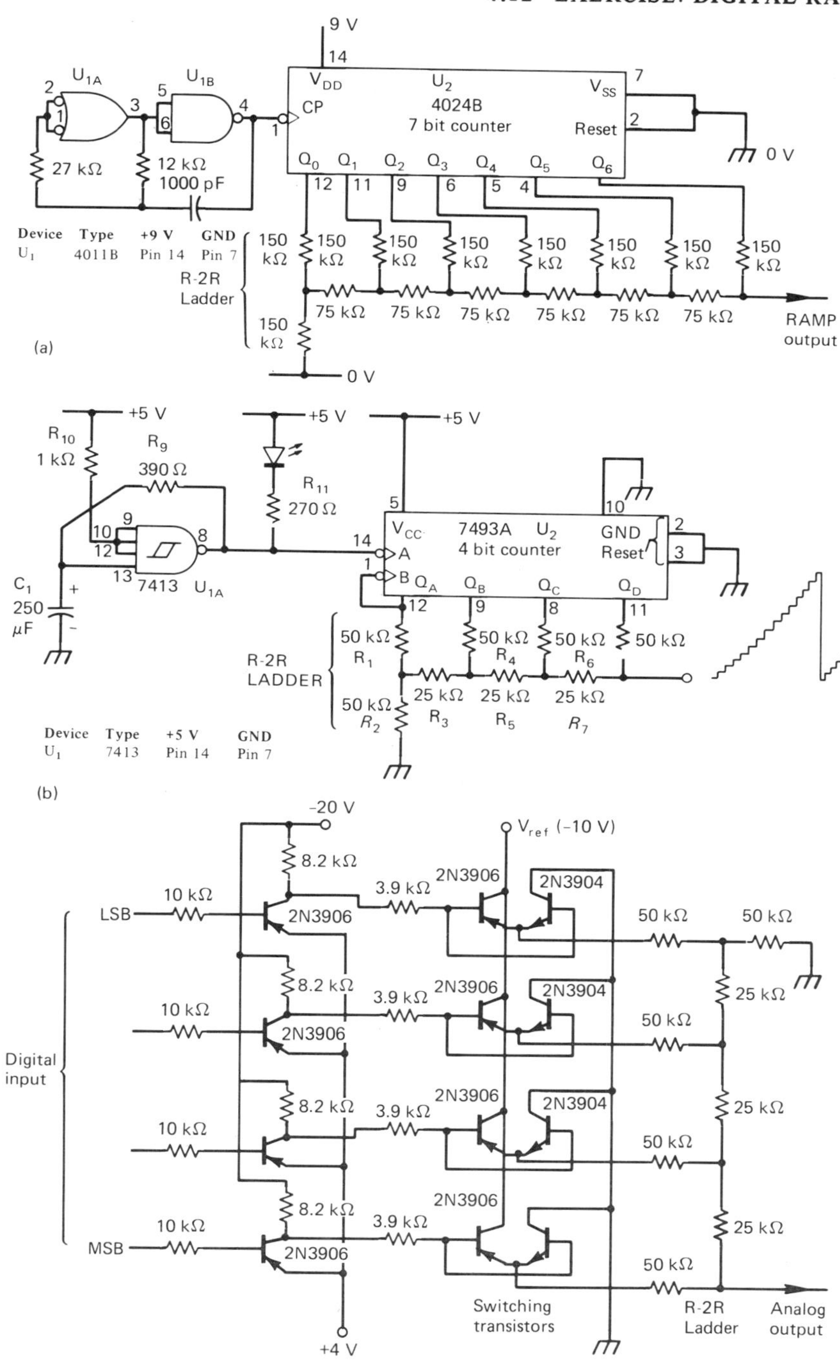

FIGURE 4.35 Digital ramp generators.

sion reference voltage level to the R-2R network as shown in Figure 4.35*c*.

QUESTIONS

1. Determine the frequency of the ramp signals of both circuits.
2. These questions refer to the TTL circuit.
   (a) A fault exists causing the output frequency to double in value and the staircase to have only 8 steps (see Figure 4.36*b*). Which resistor in the R-2R ladder has failed?
   (b) What would be the effect of an open circuit of pin 2 or pin 3 of the 7493A to ground?
   (c) A resistor open circuit causes the output to be a square wave at the same frequency as the ramp. State the resistor that has failed.

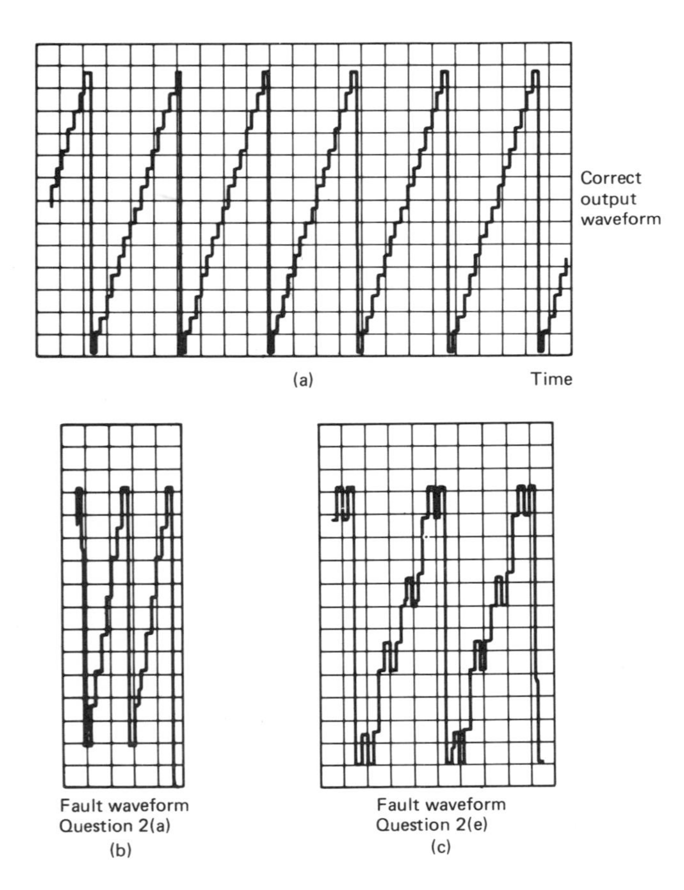

FIGURE 4.36 Output waveforms (chart recorder traces).

(d) Sketch the resulting output waveform for R_5 open circuit.

(e) The output waveform is as shown in Figure 4.36*c*. What is the fault?

4.13 EXERCISE: PROGRAMMABLE DIVIDER (FIGURE 4.37)

A divider circuit consists of a number of bistables with feedback to modify the

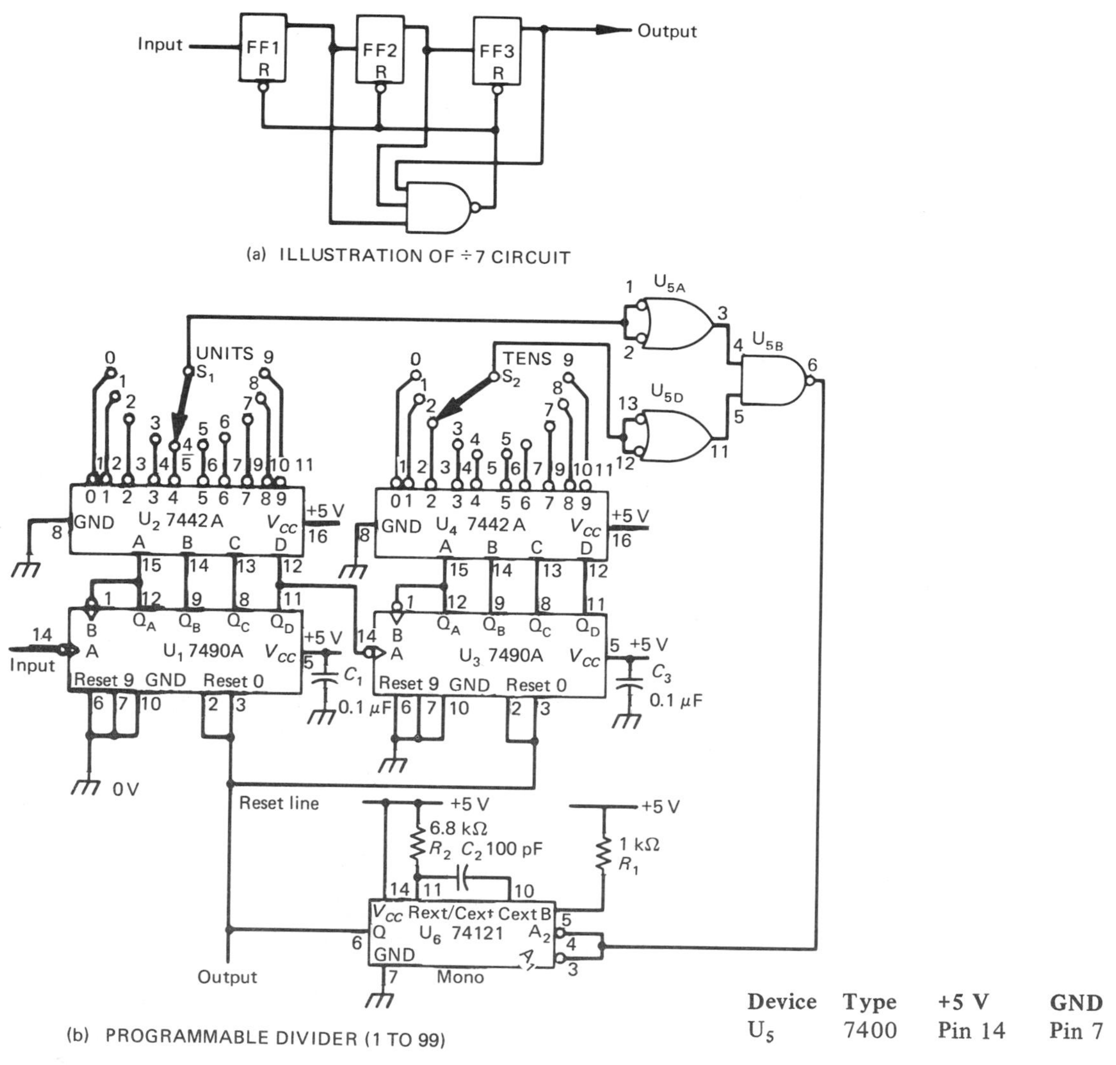

(a) ILLUSTRATION OF ÷7 CIRCUIT

(b) PROGRAMMABLE DIVIDER (1 TO 99)

Device	Type	+5 V	GND
U_5	7400	Pin 14	Pin 7

FIGURE 4.37 Programmable divider.

count sequence. For example, using three bistables an input frequency would be divided by eight. Suppose the circuit has to be divided by 7. One simple way to achieve this is to have a gate circuit that looks at the outputs of the bistables and causes them to be reset when the seventh pulse is received. The state of the three bistables on the seventh pulse will be momentarily 111. At this instant the NAND gate output goes to 0 and resets all three bistables. A pulse about 20 nsec wide will appear at the output for every seven input pulses (see Figure 4.37*a*). A monostable could be added to increase the pulse width.

In this example (Figure 4.37*b*) the input frequency can be divided by any number from 1 to 99. Two 7490A decade counters are used and these are both reset when the appropriate division is reached. The BCD outputs from the 7490As are decoded by the 7442A BCD to decimal decoders, one output line from a decoder going low at any one time. Two switches are used to select the units and tens division number, for example, 24 as shown in Figure 4.37*b*. As the input is applied, and the counters advance, after 20 input pulses the 2 output line of U_4 goes low. Following another 4 input pulses, the 4 output line of U_2 also goes low. Two logic 1s (highs) appear on the inputs of NAND gate U_5B and a negative step is applied to the monostable. The Q output of the monostable will be a short-duration positive pulse which will reset both decade counters. Thus one output pulse will appear for every 24 input pulses. By changing the setting of two switches, division by any number up to 99 is possible.

QUESTIONS

1. Which components determine the output pulse width and what is the output pulse width? (Look up 74121 data.)

2. If the output pulse width is increased to 10 μsec what effect would this have on the circuit?

3. What would be the symptoms for the following faults?
   (a) A reset line open circuit to U_3.
   (b) S_1 wiper open circuit.
   (c) Gate U_5A output stuck at 1 (a high).
   (d) S_2 wiper open circuit.

4. Which components or print faults on the p.c.b. could cause the following symptoms:
   (a) Circuit only produces output pulses with S_2 set to positions 8 or 9. The division is incorrect.
   (b) Circuit will divide only with S_2 in position 0.
   (c) The division of the circuit is incorrect. Output divides correctly for S_1 positions as seen by temporarily grounding pin 12 of U_5D. For S_2 positions, the division for "tens" is as follows with S_1 set to 0.

S_2 Position	Actual Division
1,2,4,5,8,9	No division
3	10
6	40
7	50

5. Modify the circuit so that division up to 999 is possible.

CHAPTER 5

Circuits Using Linear Integrated Circuits

5.1 INTRODUCTION

The word *linear* is used to describe the class of circuits and ICs that respond mainly to analog rather than digital signals. An analog signal is one that is variable and can, therefore, take any value between some defined limits. A good example of an analog system would be the amplification of the small voltage generated by a thermocouple to a level sufficient to give an indication of temperature on a 1 mA meter movement. A linear IC, in this case an op-amp, is used as shown in Figure 5.1 to increase the thermocouple output voltage. As the temperature being measured by the thermocouple varies, so a small change in thermocouple voltage takes place; this is an analog signal. The amplifier, operated in its linear region, increases the thermocouple's voltage by a fixed gain factor dependent on the ratio of the feedback resistors. The meter

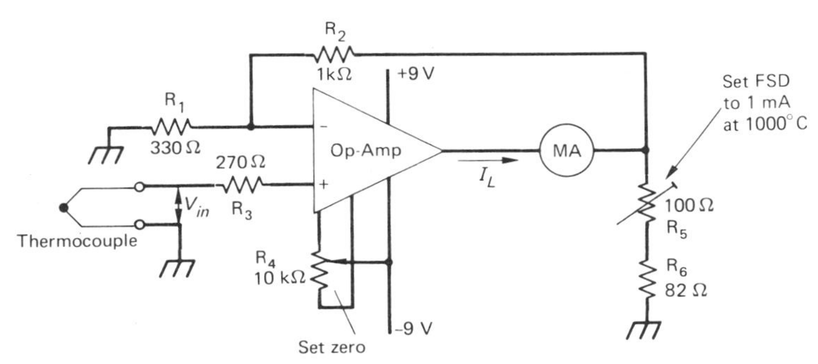

FIGURE 5.1 Typical analog system: thermocouple amplifier

$$\frac{I_L}{V_{in}} \cong \frac{R_1 + R_2}{R_1} \cdot \frac{1}{R_5 + R_6}$$

indication can then be calibrated in terms of temperature.

However, linear type ICs do not have to be operated in their linear region, and op-amps can be used, for example, to produce square wave oscillations. Some devices labeled "linear" combine a mixture of linear and digital type circuits; timers such as the 555, 556, and ZN1034E are good examples of this, and their outputs switch between high and low states. Thus the actual dividing line between linears and digitals is rather blurred. Most manufacturers list the following types of circuits under the general heading of linear:

Operational amplifiers (op-amps) and comparators.

Video and pulse amplifiers.

Audio and radio frequency amplifiers.

Regulators.

Phase locked loops (PLL).

Timers.

Multipliers.

Analog-to-digital converters.

Waveform generators.

Naturally the field is very wide so we can only look at some of the more popular types of circuits. The most commonly used linear IC is the operational amplifier, of which a large number of different types are available. These useful devices can be wired up, with a few other components, to give amplifiers, oscillators, monostables, active filters, and a wide variety of other applications. We start by looking briefly at the basic principles of the op-amp.

5.2 BASIC PRINCIPLES OF OP-AMPS

An IC op-amp is essentially a d.c. coupled differential amplifier with very high gain. The symbol in Figure 5.2 shows that two input terminals are provided, one called the noninverting input marked +, and the other the inverting input marked −. The open loop voltage gain A_{vol} is typically 100 dB (100 000 in voltage ratio) so that only a small differential input is required to cause

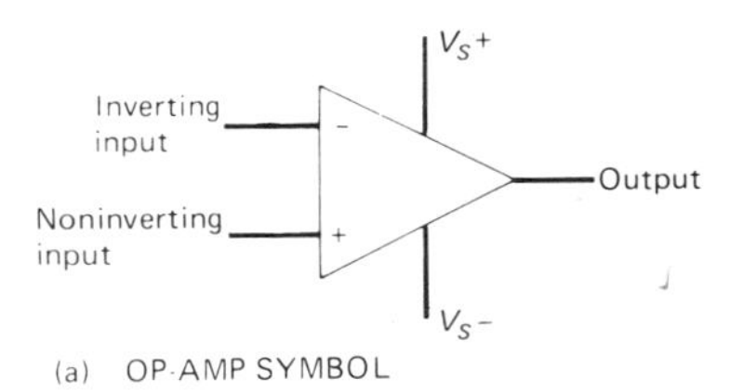

(a) OP-AMP SYMBOL

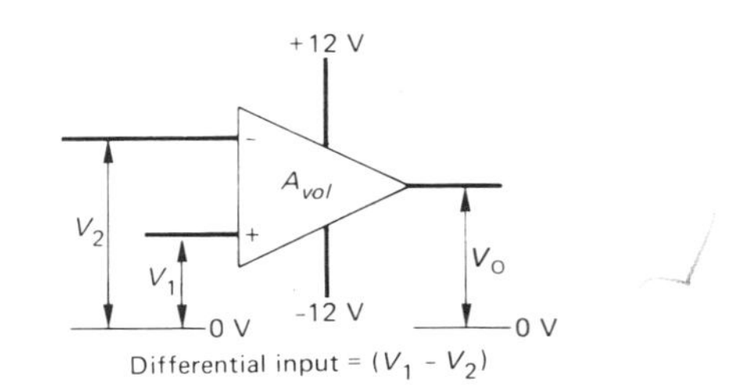

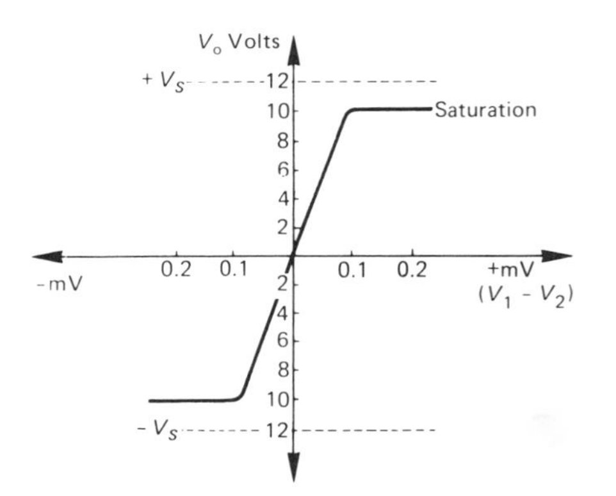

(b) TYPICAL TRANSFER CHARACTERISTIC

FIGURE 5.2 Op-amp symbol and transfer characteristics

a large output change.* By differential is meant a signal that causes a difference of fractions of a millivolt between the two input connections. For example, if the inverting input is held at zero volts and the noninverting input level made +0.1 mV, then the output will go positive to nearly +10 V. If the noninverting input level is then made −0.1 mV, the output will go negative to nearly −10 V. In a similar way if the noninverting input is held at zero volts and the inverting input level is made +0.1 mV, the output will go to −10 V. The amplifier responds to the difference in voltage between the two input leads, and when this is zero the output should be nearly zero. Thus the op-amp must be provided with both a positive and negative supply voltage so that the output can swing about zero.

A typical transfer characteristic is shown in Figure 5.2*b*. This shows that, when $(V_1 - V_2)$ is positive, the output goes positive. The output will saturate if $(V_1 - V_2)$ exceeds about +0.1 mV. Similarly when $(V_1 - V_2)$ is negative, the output goes negative. The characteristic has been drawn passing through zero at the point when $V_1 = V_2$. In practice some "offset" always occurs, and a potentiometer must be added to trim out or "null" any such offset voltage. This is discussed later.

The typical op-amp has a differential input stage which is supplied from a constant-current source, a second stage of amplification and d.c. level shifting, and finally a complementary class B type output stage. The actual circuit diagrams shown in manufacturer's data sheets do appear rather complicated because of the d.c. coupling, use of transistors as resistors, and necessary protection circuits, but the operation as described above is fairly straightforward and depends to a great extent on the differential input stage. With any d.c. coupled amplifier the DRIFT of the output signal must be kept to a low value. Drift is defined as any change in output voltage when the input is short-circuited or otherwise held at zero. Two of the major causes of drift are temperature changes causing the V_{BE} of transistors to be altered by about −2 mV per °C, and power supply voltage changes. By using a differential input stage in which two transistors are connected together in a balanced arrangement, drift caused by temperature and power supply variations can be minimized. The great advantage of the differential arrangement is that, if signals of the same polarity are applied to the two inputs, then they effectively cancel each other out and the resulting output is very small. Signals such as these are called "common mode" and one measure of the quality of an op-amp is its COMMON MODE REJECTION RATIO (CMRR):

$$\text{CMRR} = \frac{\text{Differential gain}}{\text{Common mode gain}}$$

If the temperature changes, the V_{BE} of both input transistors change together, giving common mode input signals. With an IC, both input transistors and the associated components are diffused into the same piece of silicon and can, therefore, be closely matched. An op-amp with a high CMRR can be used to measure a small differential signal that accompanies a large common mode signal as in the case of

---

*A_{vol} in dB = $20 \log_{10} \dfrac{V_o}{V_1 - V_2}$

electrocardiogram signals from two electrodes attached to the body to measure heart contracts. The differential signal from the electrodes has an amplitude of about 1 mV; however, both electrodes may contain a common-mode signal of about 0.1 V at the power-line frequency. An op-amp with a high CMRR detects and amplifies the differential signal and rejects the common mode signal.

In any linear application the op-amp is wired with an external feedback network to give a stable gain. For an amplifier system the GAIN WITH NEGATIVE FEEDBACK is given by the formula:

$$A_c = \frac{A_o}{1 + A_o\beta}$$

where A_c is the closed loop gain, i.e., gain with feedback applied

A_o is open loop gain i.e., $\frac{V_o}{V_1 - V_2}$ in Figure 5.2*b*

β is the fractional gain of the feedback network

$A_o\beta$ is the loop gain.

Since with an op-amp A_o is typically 100 000, the loop gain $A_o\beta$ is usually much greater than unity. In this case the formula reduces to:

$$A_c \approx \frac{A_o}{A_o\beta} \approx \frac{1}{\beta}$$

(a) INVERTING AMPLIFIER

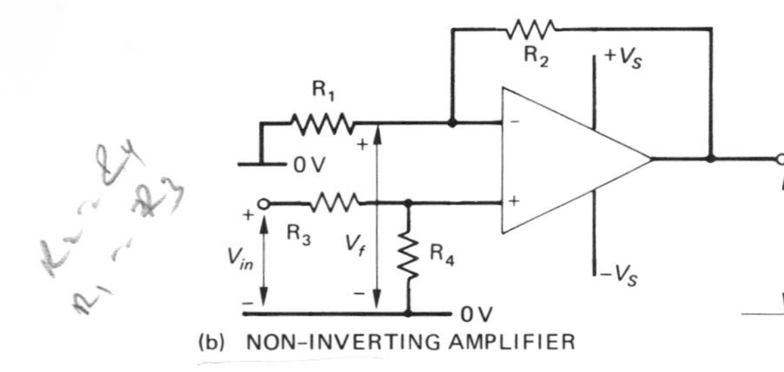

(b) NON-INVERTING AMPLIFIER

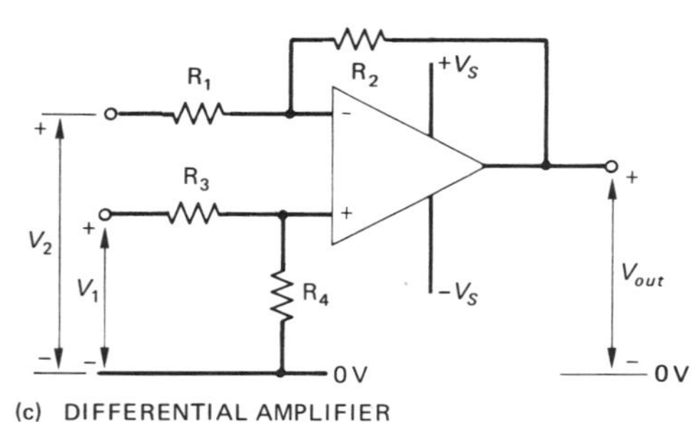

(c) DIFFERENTIAL AMPLIFIER

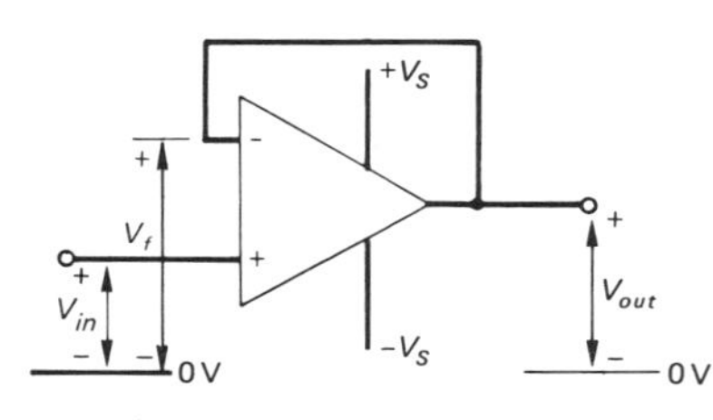

(d) VOLTAGE FOLLOWER

FIGURE 5.3 Methods of applying negative feedback to an op-amp

(a) Voltage gain $\approx -R_2/R_1$
Input impedance = R_1

(b) Voltage gain $\approx \frac{R_1 + R_2}{R_1}$

Input impedance = $R_{in}\frac{A_o}{A_c}$

(c) Voltage gain $\approx \frac{R_2}{R_1}(V_1 - V_2)$
Usually, $R_1 = R_3$, $R_2 = R_4$

(d) Voltage gain is unity. Very high input impedance. Very low output impedance.

This shows that the closed loop gain is then solely dependent on the component values of the feedback loop and, since these can be made close tolerance resistors, the gain of amplifier systems can be accurately set. The ways of applying negative feedback are shown in Figure 5.3.

In the inverting amplifier (Figure 5.3*a*), the current flowing into terminal x via R_1 is approximately V_{in}/R_1. This is because the effective voltage change between the inverting and noninverting terminals is very small, since A_o is very large. The bias current, I_B, required to flow into both the inverting and noninverting inputs of an IC op-amp is generally less than 1 μA; therefore, the current flowing in R_2 must be nearly equal to that flowing in R_1. R_3 in this application is generally chosen to be equal to the value of the parallel combination of R_1 and R_2, and the voltage drop across R_3 will be very small (i.e., for R_3 = 1 kΩ, $V_{R3} = I_B R_3 = 1 \times 10^{-6} \times 1 \times 10^3$ = 1 mV). Node s at the inverting input in Figure 5.3*a* is often called the summing junction. Since the voltage across R_3 is very small and the voltage between the op-amp input terminals is also very small, the voltage at the summing junction with respect to ground may be approximated as zero.

$$V_{in}/R_1 \approx -V_{out}/R_2$$

$$\frac{V_{out}}{V_{in}} \approx -\frac{R_2}{R_1}$$

so,

$$\text{Voltage gain } A_c = -R_2/R_1$$

In circuit *b* a noninverting amplifier is shown. A fraction of the output signal is fed back to the inverting input terminal. This opposes the input signal on the noninverting input. Recall that V_{R3} is very small so V_{in} is approximately the voltage at the noninverting input.

So,

$$V_f = V_{out} \cdot \frac{R_1}{R_1 + R_2}$$

$$V_{out} = A_o(V_{in} - V_f) = A_o\left(V_{in} - V_{out} \cdot \frac{R_1}{R_1 + R_2}\right)$$

$$\frac{V_{out}}{A_o} = V_{in} - V_{out} \cdot \frac{R_1}{R_1 + R_2}$$

Since A_o is very large, $V_{out}/A_o \approx 0$ and can be neglected.

$$V_{in} \approx V_{out} \frac{R_1}{R_1 + R_2}$$

$$\text{Voltage gain} \approx \frac{R_1 + R_2}{R_1}$$

Note that with this circuit the input impedance is very large and the output resistance is very low.

For the differential input type of amplifier the closed loop gain will be R_2/R_1. This type of circuit would be used to amplify the signals from, say, a bridge circuit.

The voltage follower (Figure 5.3*d*) is a useful impedance buffer. It has 100% negative feedback, since the output is connected back directly to the inverting input making $V_f = V_{out}$

$$V_{out} = A_o(V_{in} - V_f)$$

$$V_{out}/A_o = V_{in} - V_f = V_{in} - V_{out}$$

However, since A_o is typically 100 000, $V_{out}/A_o \approx 0$ and can be neglected, then

$$V_{out} = V_{in}$$

In other words, the output follows the input. The main advantage in this circuit is that the input impedance is very high ($> 100\ \mathrm{M\Omega}$) while the output impedance is very low, approaching less than an ohm.

Op-amp Performance Characteristics

Having looked at feedback methods let us now consider the main performance characteristics of an op-amp. These are:

(a) **Open loop voltage gain A_{vol}** The low-frequency differential gain without any feedback applied.

(b) **Input resistance R_{in}** The resistance looking directly into the input terminals under open loop conditions. Typical values for bipolar ICs are 1 MΩ while FET input stages may be greater than $10^{12}\ \Omega$.

(c) **Input offset voltage** With the inputs both grounded the output of an op-amp should ideally be zero. However, because of slight mismatches of the bias voltages in the input circuits, an offset voltage occurs. A typical value of this differential input offset is about 1 mV. Most modern op-amps are provided with the means for nulling this offset.

(d) **CMRR** (common mode rejection ratio) The ratio of differential to common mode gain, i.e., the ability of the amplifier to reject common mode signals.

(e) **Supply voltage rejection ratio** The ability of the amplifier to reject variations in supply voltage.

(f) **Slew rate** If a sudden step input is applied to an op-amp, its output will not be able to respond immediately. Instead, its output will move to the new value at a uniform rate. This is *slew rate limiting*, which is, in effect, the maximum rate of change of voltage at the output of the device. Typical figures vary from 1 volt/μsec (741) to 35 volt/μsec (Signetic NE531). (See Figure 5.4.)

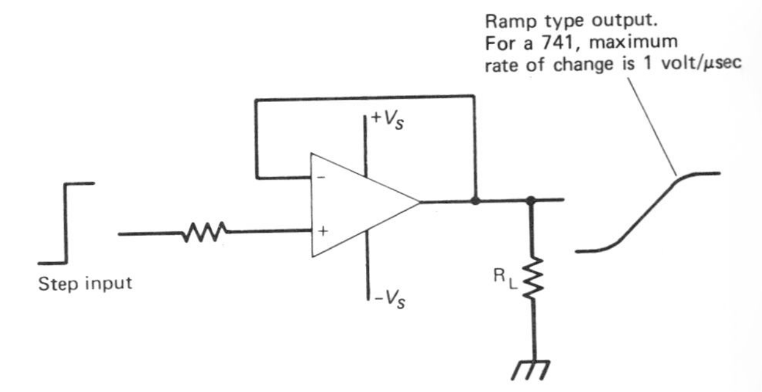

FIGURE 5.4 Op-amp slew-rate limiting: the response to a sudden change at the input cannot be immediate.

(g) **Full-power bandwidth** The maximum signal frequency at which the full voltage output swing can be obtained.

(h) **Output voltage swing** The peak output swing, referred to zero, that can be obtained.

TABLE 5.1 Op-amp Parameters and Characteristics

Device	741	NE531	709	FET Input NE536
Supply voltage range	3 V to 18 V	5 V to 22 V	9 V to 18 V	6 V to 22 V
Max. differential input voltage	30 V	15 V	5 V	30 V
Output short circuit duration	Indefinite	Indefinite	5 sec	Indefinite
Open loop voltage gain A_{vol}	106 dB	96 dB	93 dB	100 dB
Input resistance	2 MΩ	20 MΩ	250 kΩ	10^{14} Ω
Differential input offset voltage	1 mV	2 mV	2 mV	30 mV
CMRR	90 dB	100 dB	90 dB	80 dB
Slew rate	1 V/μsec	35 V/μsec	12 V/μsec	6 V/μsec
Full-power bandwidth	10 kHz	500 kHz	–	100 kHz
Output voltage swing	13 V	15 V	14 V	10 V

Table 5.1 lists the important parameters and characteristics for a few of the more common op-amps.

The 709 op-amp was one of the first linear IC devices made available. Note that it is not short-circuit proof. Another drawback to the early IC op-amps is the possibility of "latch-up" caused by overdriving, the amplifier's output remaining in a saturated state. The 709 also requires external components to give frequency compensation and to prevent the amplifier from bursting into unwanted oscillation. Most of these problems have been overcome in the design of later IC op-amps. The 741 and NE531 are short-circuit protected and are provided with offset voltage null capability and have no latch-up problems.

The gain/frequency characteristic of op-amps is another important factor in any design. The large signal open loop gain A_{vol} is quoted for d.c. or very low frequency operation. As the signal frequency is increased, the open loop gain falls in value. The frequency response for a 741 op-amp is shown in Figure 5.5. Note that at 10 kHz the open loop gain has dropped to 40 dB (100 as a voltage ratio) and at 100 kHz it is down to 20 dB. The 741 has internal frequency compensation components to prevent unwanted oscillation, and these cause the gain to fall off in this way. If a wider power bandwidth is required, a Motorola MC1741S or Silicon General SG741S which has a full power bandwidth of 200 kHz might be used. Of course, when

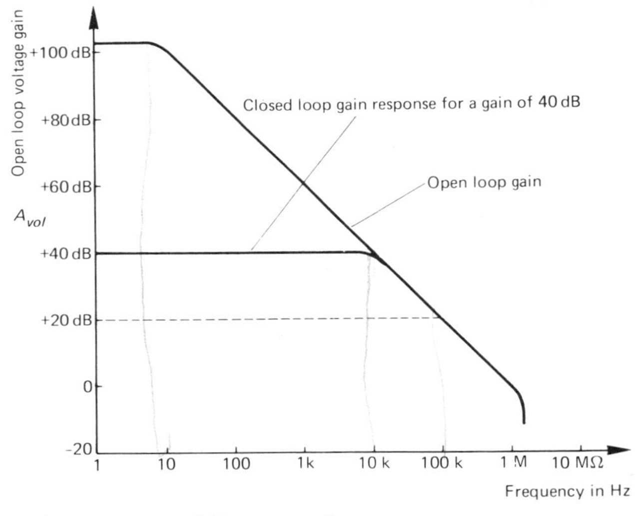

FIGURE 5.5 741 op-amp frequency response.

negative feedback is applied the frequency response is flat, but the value of the closed loop will be contained within the open-loop response. If a closed loop gain of 100 (40 dB) is required, then the response will be as shown in Figure 5.5.

A test circuit for obtaining the transfer characteristic of an op-amp is shown in Figure 5.6. The oscilloscope must have both its X and Y channels d.c. coupled. The test signal is a sawtooth waveform of about 1 V peak-to-peak at 20 Hz, and this is applied to the noninverting input via potential divider R_1 and R_2. The signal level on the noninverting input is then approximately 1 mV peak-to-peak. This is sufficient to drive the amplifier from negative to positive saturation. If a large offset exists in the op-amp, the test signal amplitude may have to be increased. Before measurement the oscilloscope input leads must be grounded and the trace centered on the screen to give a reference point. The transfer characteristic will be displayed as the sawtooth test signal moves from a negative through zero to a positive value. The open loop voltage gain A_{vol} will be the slope of the characteristic, and an indication of input offset voltage can also be obtained. Note that, if a suitable slow sawtooth source is available, an XY plotter can also be used for this measurement.

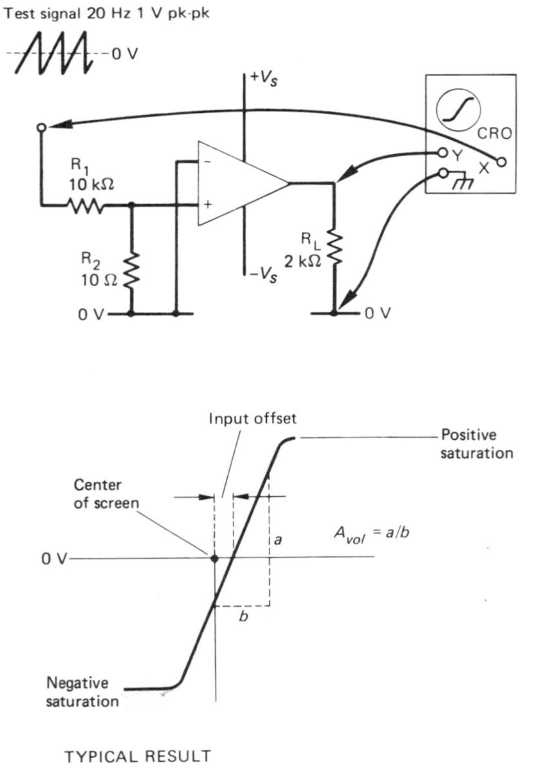

Test signal 20 Hz 1 V pk-pk

0 V

$+V_s$

CRO

Y

X

R_1 10 kΩ

R_2 10 Ω

0 V

$-V_s$

R_L 2 kΩ

0 V

Input offset

Positive saturation

Center of screen

$A_{vol} = a/b$

0 V

a

b

Negative saturation

TYPICAL RESULT

FIGURE 5.6 Test circuit for transfer characteristics.

5.3 TROUBLESHOOTING OP-AMP CIRCUITS

Most of the precautions stated for digital IC components, such as not plugging in devices with the power still on and avoiding the use of large test probes, apply equally to linear circuits such as the op-amp and need not be repeated. Instead, consider some simple circuit arrangements and some possible faults and their symptoms.

Figure 5.7*a* shows first a 741 op-amp used as the error amplifier in a simple voltage regulator circuit. Actually the 741 is being used as a voltage comparator feeding a series control element Q_1. The circuit gives a regulated output voltage of about 10 V from an unregulated 15 V supply. The noninverting input of the 741 is fixed at the reference voltage given by the 5.6 V zener diode. The output of the amplifier

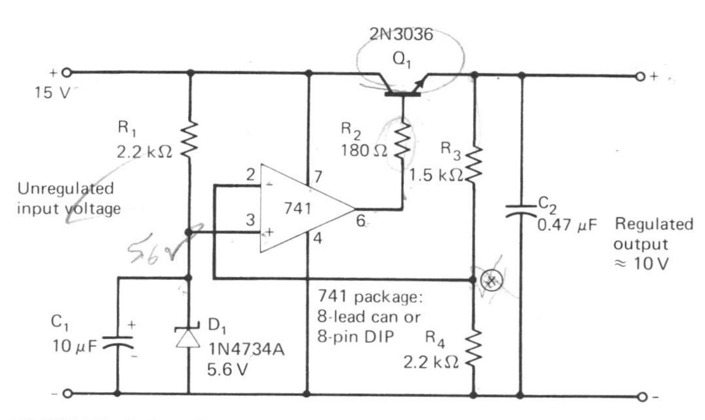

FIGURE 5.7*a* Op-amp used as the error amplifier in a simple voltage regulator circuit.

will assume a voltage level that will cause the difference between the zener voltage and point x to be as small as possible. If the voltage at point x goes low, just below the reference voltage, the 741 output will rise positive, driving the series element Q_1 harder into conduction, and forcing the output voltage to rise and correcting point x back to be more nearly equal to the reference voltage. Since the voltage at x is about 5.6 V, the total voltage across R_3 and R_4 will be about 10 V. Note that there is no current limit with this simple circuit and if the output is accidentally short circuited, the 2N3036 will burn out.

It is, of course, possible for many different faults to occur inside the IC itself, such as internal shorts or open circuits. What is required during troubleshooting is to rapidly track down the fault to either the IC or one of the other components. Note that external faults such as open circuit or shorted p.c.b. traces or shorted IC pins are also possible, and these may suggest that the IC itself is faulty. Since ICs are now highly reliable devices, it is more likely that a fault is caused by other components. In any case, it is wise to check these possibilities before unsoldering the IC as this may be a lengthy job and could well cause damage.

Take for example, a fault that gives the following voltage readings on the IC pins. An ordinary d.c. voltmeter with 20 kΩ/V resistance is used.

741 pin no.	2	3	7	6	Output of power supply
Meter reading (MR)	5.7 V	8.2 V	15.3 V	14.4 V	13.8 V

These measurements are made between the actual IC pins and 0 V and show that there must be an internal open circuit on the inverting input of the 741 because pin 3 is much higher in voltage than pin 2, i.e., the differential input voltage is typically less than 0.1 mV. The output of the 741 has been driven positive, and an excessive differential input exists that should drive the output toward zero volts.

The reverse symptoms would occur if

the internal noninverting input lead became open circuit. The meter readings would then be as follows:

741 pin no.	2	3	7	6
MR	5.7 V	1.75 V	15.3 V	3.4 V

Suppose no fault existed with the 741 but the following measurements were taken

741 pin no.	2	3	7	6
MR	5.7 V	0 V	0 V	0 V

In this case, since pin 7 is at 0 V, the fault must be an open circuit supply lead to the IC from the 15 V unregulated line. The p.c.b. trace or wiring would have to be checked. If the IC had an internal power supply open circuit, similar voltage readings would result except that pin 7 would be at 15.3 V.

The second example (Figure 5.7*b*) shows a 741 used as an a.c. amplifier with a voltage gain of about 30. Negative feedback is applied from the output via R_2, C_2, and R_3. Actually, for a.c. signals the reactance of C_2 is small compared with R_3 so the circuit gain is

$$A_c = \frac{R_2 + R_3}{R_3} \approx 30$$

The purpose of C_2 is to provide the circuit with a high value of input impedance over its useful frequency range. It is a technique called BOOT-STRAPPING. Imagine the input going positive; since the amplifier is noninverting the output also goes positive and a signal level of almost the same magnitude as the input appears across R_3. The a.c. current flowing in R_1 is thus negligible which, therefore, increases the effective input impedance of the amplifier to a very high value. A d.c. feedback path is provided by R_2, which gives the circuit good d.c. stability. Thus the d.c. level at pin 6 should be approximately zero.

Suppose a short circuit occurred between pins 2 and 3. This could be caused by a small solder bridge between traces on the p.c.b. The effect would be to give almost zero a.c. gain and, since no offset control is provided, the d.c. level at pin 6 could take any value between ±7 V.

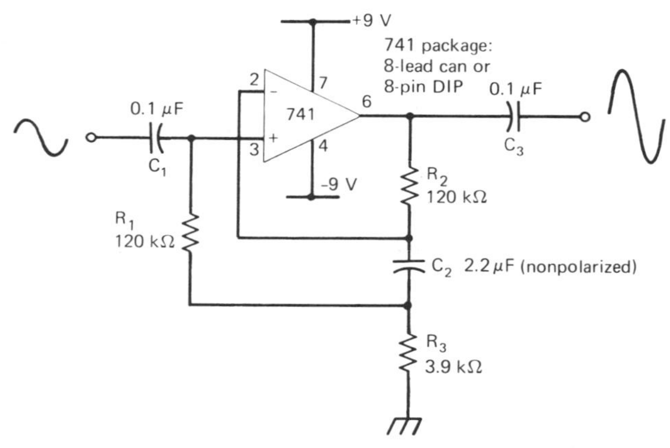

FIGURE 5.7*b* A.C. amplifier with a gain of 30.

If C_2 became open circuit, the symptoms would be that the a.c. gain would fall to about unity, since nearly all the output would be fed back via R_2 to the inverting terminal. In addition, the circuit's input impedance would fall to equal R_1 (120 kΩ). If you have built the circuit, try out various other faults.

Later in this chapter there are several other examples of circuits using standard IC op-amps.

5.4 TIMER ICs

A fairly large number of monolithic timer circuits are now available, but probably the best known are the 555, the 556, and the ZN1034E.

Timing circuits are those that will provide an output change of state after a predetermined time interval. This is, of course, the action of a monostable multivibrator. Discrete circuits can be easily designed to give time delays from a few microseconds up to a few seconds, but for very long delays mechanical devices generally had to be used. The 555 timer IC, first made available in 1972, allows the user to set up quite accurate delays or oscillations from microseconds up to several minutes, while the ZN1034E can be set to give time delays of up to several months.

The basic operation of the 555 can be understood by referring to Figure 5.8. For **monostable operation** the external timing component R_A and C are wired as shown. Without a trigger pulse applied, the $\overline{Q}$ output of the flip-flop is high, forcing the discharge transistor to be on and holding the output low. The three internal 5 kΩ resistors R_1, R_2, and R_3 form a voltage divider chain so that a voltage of $\frac{2}{3}V_{CC}$ appears on the inverting input of comparator 1 and a voltage of $\frac{1}{3}V_{CC}$ on the non-inverting input of comparator 2. The trigger input is connected via an external resistor to V_{CC} so the output of comparator 2 is low. The outputs of the two comparators control the state of the internal flip-flop. With no trigger pulse applied, the $\overline{Q}$ output will be high and this forces the internal discharge transistor to conduct. Pin 7 will be at almost zero volts, and the capacitor C will be prevented from charging. At the same time the output will be low.

When a negative-going trigger pulse is applied, the output of comparator 2 goes high momentarily and sets the flip-flop. $\overline{Q}$ output goes low, the discharge transistor turns off, and the output switches high to V_{CC}. The external timing capacitor C can now charge via R_A so the voltage across it rises exponentially towards V_{CC}. When this voltage just exceeds $\frac{2}{3}V_{CC}$, the output of comparator 1 goes high which resets the internal flip-flop. The discharge transistor conducts and rapidly discharges the timing capacitor, and at the same time the output switches to zero. The width of the output pulse t_{pw} is equal to the time taken for the external capacitor to charge from zero to $\frac{2}{3}V_{CC}$.

$$t_{pw} = 1.1\,CR_A$$

R_A can have a value from 1 kΩ up to (1.3 V_{CC}) MΩ. In other words if a 10 V supply source is used, R_A can be 1 kΩ minimum or a maximum of 13 MΩ. In practice, medium values of R_A from 50 kΩ to 1 MΩ are used, since these tend to give the best results.

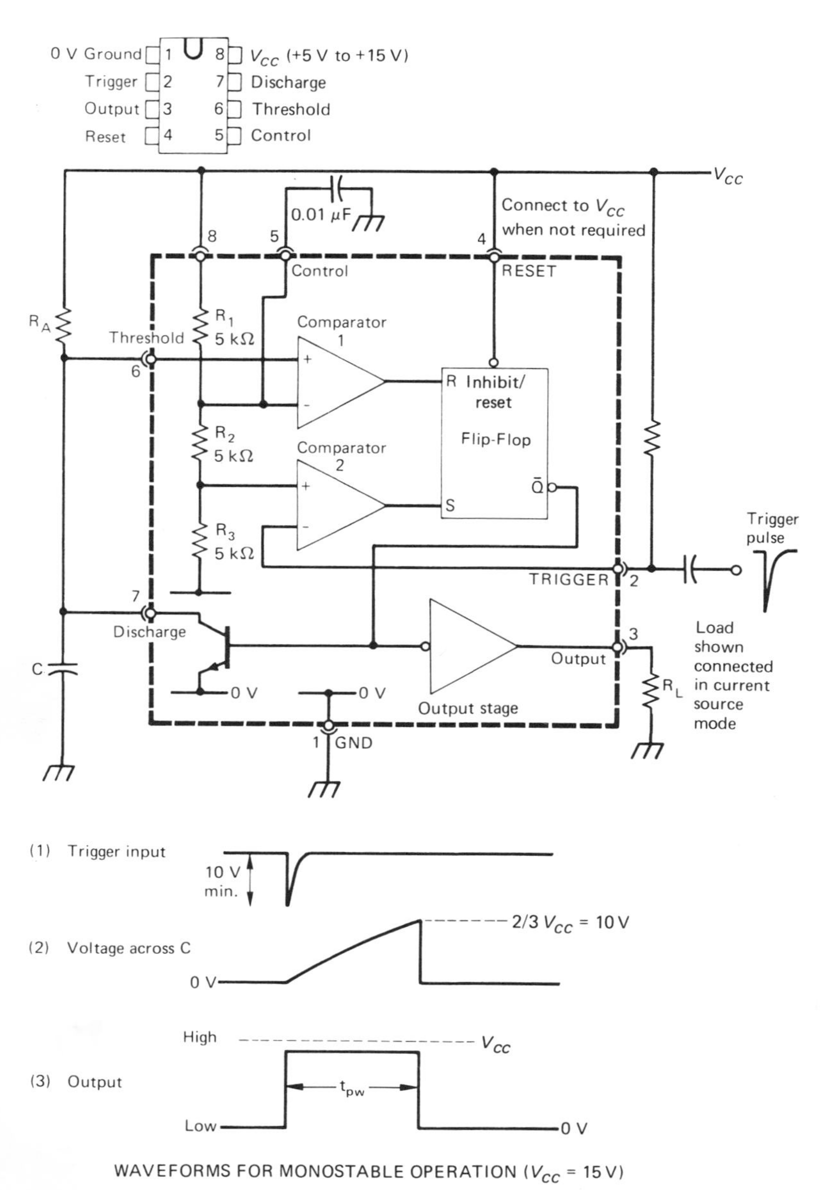

FIGURE 5.8 The 555 timer.

The 555 output switches between almost zero (0.4 V) to about one volt below V_{CC} with rise and fall times of 100 nsec. The load can be connected either from the output to ground or from the output to V_{CC}. The first connection is known as the current source mode and the second as the current sink. In both situations up to 200 mA of load current can be accommodated.

Two other input pins are provided. Pin 4, the *reset* terminal, can be used to interrupt the timing and reset the output by application of a negative-going pulse. Pin 5, called the *control*, can be used to modify or modulate the time delay. A voltage applied to pin 5 overrides the d.c. level set up by the internal resistors. In normal timing applications when no modulation is required, pin 5 is usually taken to ground via a 0.01 μF capacitor. This prevents any pick up of noise affecting the timing.

One of the important points about the 555 is that the timing is relatively independent of supply voltage changes. This is because the three internal resistors fix the ratio of the threshold and trigger levels at $\frac{2}{3}V_{CC}$ and $\frac{1}{3}V_{CC}$. A typical change of time delay with supply voltage is 0.1% per volt. In addition, the temperature stability of the microcircuit is excellent at 50 ppm per °C. Thus the accuracy and stability of the timing delay depends to a great extent on the quality of the external timing components R_A and C. Electrolytic capacitors may have to be used for long-time delays, but the leakage current must be reasonably low. Also, because the tolerance of electrolytic capacitors is wide (typically −20% +50%) part of the timing resistor may have to be a preset to enable the delay to be fairly accurately set.

An example of a 555 used as a simple 10-second timer is shown in Figure 5.9. Pressing the start button momentarily takes pin 2, the trigger input, down to 0 V. The output will switch high and the LED will come on. C_1 then charges from 0 V toward $+V_{CC}$. After 10 sec the voltage across C_1 reaches 6 V ($\frac{2}{3}V_{CC}$) and the 555 is reset, the output switching back to the low state.

Suppose the circuit failed with the symptoms that the output always remained low and could not be triggered on. A list of the possible faults giving this symptom are

(a) Open circuit power supply leads to the IC.

(b) Failure of the trigger circuit, i.e., an open circuit switch contact or open circuit connection to pin 2 of the IC.

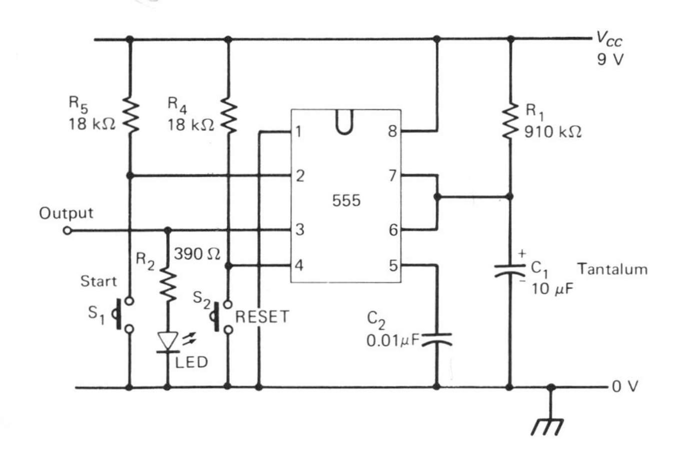

FIGURE 5.9 10-second timer using a 555.

(c) An internal fault in the 555 IC.

(d) An open circuit from pin 3 to the load.

If C_1, the timing capacitor, or its connections, were open circuit, the time delay would be very short but pressing the start switch would cause the output to go high, and it would remain high for the length of time the start switch was held in.

To locate the fault the following sequence of checks should be made:

1. Check for power supply voltage using a voltmeter at the IC between pins 8 and 1.
2. Investigate the trigger circuit. Pressing the start switch should cause pin 2 to fall from a positive value to 0 V. Since the trigger of the 555 is very sensitive, connecting a meter lead to pin 2 may cause the timer to turn on. This alone would be an indication that the start switch circuit was faulty and not the IC.
3. Check the output between the IC pins 3 and 1.
4. Check that pin 4 the reset is positive (V_{CC}) and that pin 5 is $\frac{2}{3} V_{CC}$.

A fault such as R_1 open circuit would result in the output, once triggered, remaining high. This is because C_1 will no longer have a charging path to V_{CC}. With this fault the circuit would be reset by pressing S_2. Similar symptoms would occur if the p.c.b. trace or wiring from C_1 to pins 6 and 7 became open, except that the voltage across C_1 would rise positively. Note that if voltage measurements are made across C_1 or at pins 6 and 7, a high impedance meter must be used.

The 555 can also be wired up to operate as an **astable multivibrator**, the output being a train of positive pulses with the width and frequency determined by external timing components (Figure 5.10*a*). Pins 2 and 6 are connected together, which

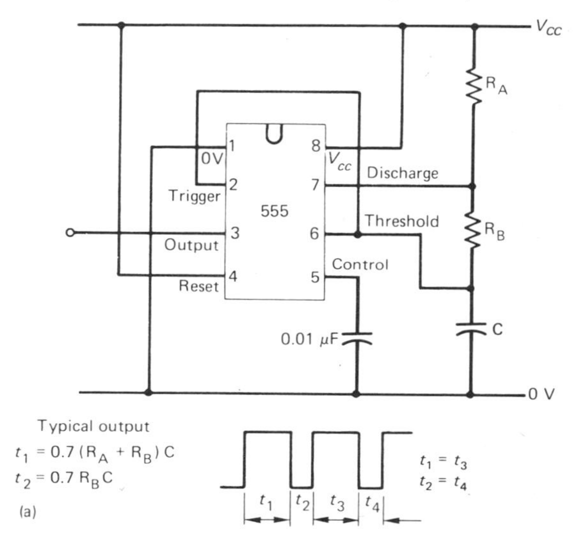

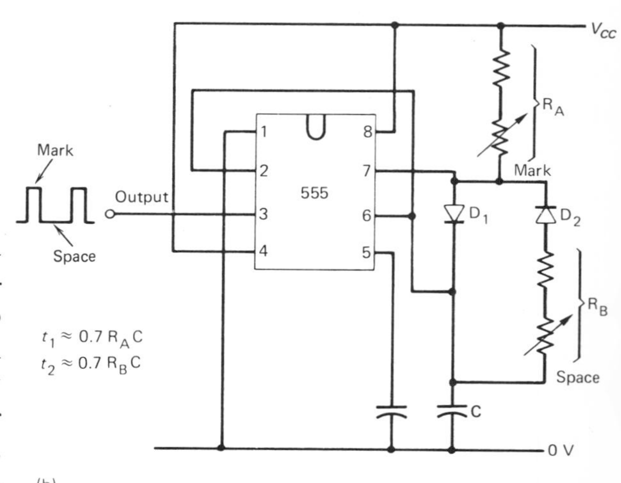

FIGURE 5.10 555 used as an astable.

allows the capacitor to charge and discharge between the threshold and trigger levels. At switch-on, C charges via R_A and R_B toward V_{CC}. When the voltage across C reaches $\frac{2}{3} V_{CC}$, the output changes state and C is discharged via R_B toward 0 V. When the voltage across C falls to $\frac{1}{3} V_{CC}$, the circuit again changes state, the internal discharge transistor turns off, and C charges via R_A and R_B toward V_{CC}. Thus a continuous train of pulses appears at the output. To get an almost symmetrical output waveform, R_B should be made very large with respect to R_A, say 50 times, and then the frequency f will be primarily determined by R_B and C.

$$f \approx \frac{1}{1.4\, R_B C} \quad \text{when } R_B \gg R_A$$

To achieve some control over the output waveform, that is the ratio of mark-to-space of the output waveform, a circuit like that shown in Figure 5.10*b* is used. Here diode D_1 conducts when C is charging, and D_2 conducts when C is discharging. The ratio of mark-to-(mark + space) is called the duty cycle (DS) of a waveform.

Gated oscillators can be created by using pin 4, the reset, as a control. While pin 4 is held at 0 V, the oscillator is inhibited, but if a positive voltage is applied then the circuit is allowed to oscillate. A suitable control signal can be the output of another 555 wired as a monostable. A later exercise shows this.

Many other useful circuits can be made by using the 555 or the 556 timers such as temperature- or light-controlled oscillators. The 556 is in effect two 555 devices within one common package.

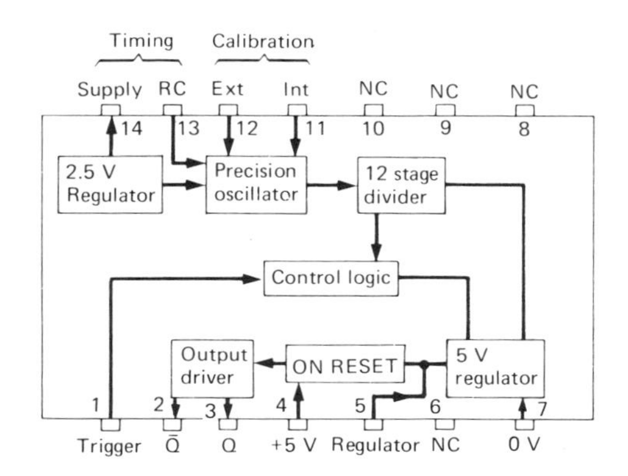

FIGURE 5.11*a* The ZN1034E timer.

The other popular IC timer, the ZN1034E, is shown in Figure 5.11*a*. This has an internal oscillator, the frequency of which can be controlled by an external resistor and capacitor. The oscillator output is connected inside the IC to a 12-stage divider so the oscillator has to complete (4096 – 1) cycles before the timing period ends. Thus very long time delays can be achieved by using only medium values of timing components. The time delays of weeks or months, if required, can be repeated with high accuracy as long as good-quality components are used for the oscillator timing.

The total timing period t_p is related to R_t and C_t by the formula

$$t_p = 4095\, K C_t R_t$$

where K is a constant determined by the value of the calibration resistance. An internal calibration resistor of 100 kΩ is provided by connecting pins 11 and 12 and this gives $K = 0.668$. This method is recommended for best temperature stability. Thus, if $C_t = 100\ \mu F$ and $R_t = 2.2\ M\Omega$, and $K = 0.668$, then

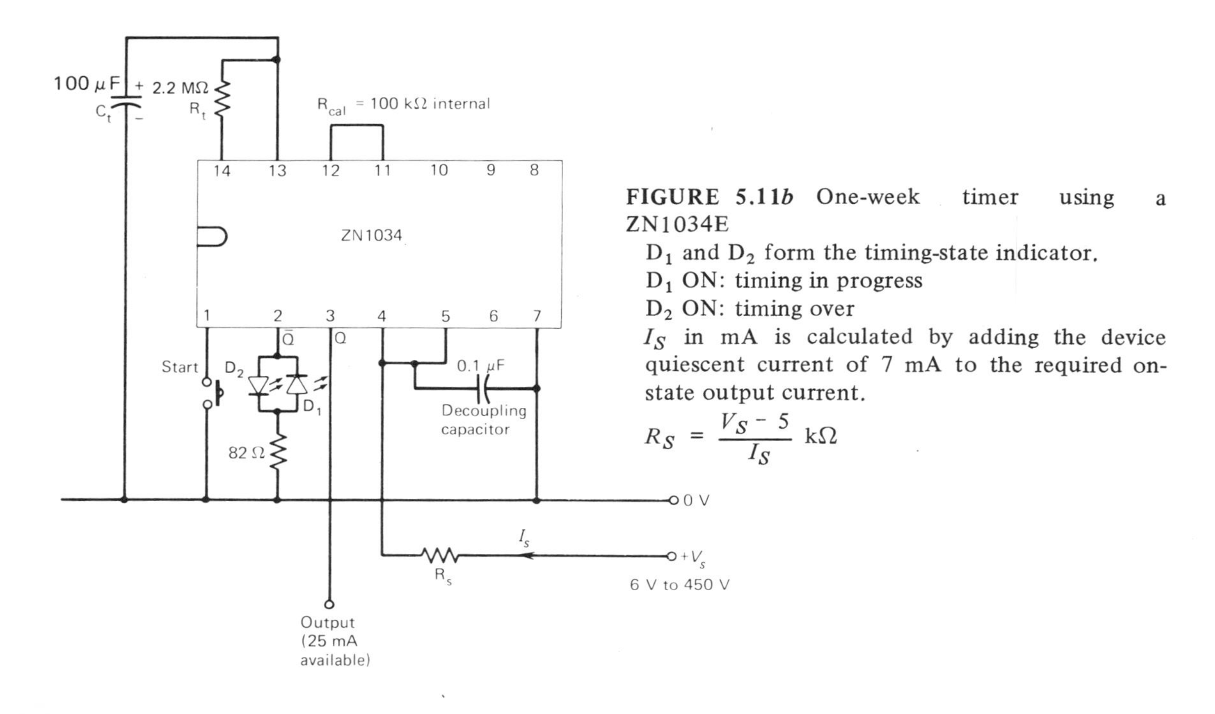

FIGURE 5.11*b* One-week timer using a ZN1034E

D_1 and D_2 form the timing-state indicator.
D_1 ON: timing in progress
D_2 ON: timing over
I_S in mA is calculated by adding the device quiescent current of 7 mA to the required on-state output current.

$$R_S = \frac{V_S - 5}{I_S} \text{ k}\Omega$$

$$t_p = 7 \text{ days}$$

$$\text{Oscillator time period} = \frac{7 \text{ days}}{2735}$$

A circuit for this is shown in Figure 5.11*b*. Two complementary outputs are provided: $\bar{Q}$ on pin 2 is normally at a low voltage but rises to about +3 V at the end of the timing period; Q on pin 3 will be high at 3 V after the start button is pressed and will fall to less than 0.4 V (low) at the end of the timing period. The outputs can each sink or source up to 25 mA and can, therefore, be used to drive relays via a transistor switch, or thyristors, or small signal lamps. The calibration can be checked by measuring the time period of the oscillator on pin 13. However, a high impedance meter must be used with an impedance of at least 10 R_t, otherwise the oscillator period will be changed. For this circuit the oscillator will have a periodic time equal to the total time period, 7 days, divided by 4095 K, i.e.

5.5 OTHER LINEAR ICs

Some of the other linear ICs mentioned in the introduction such as regulators and analog-to-digital convertors will be discussed in other chapters. The IC regulators, for example, will appear in Chapter 6, which is concerned with troubleshooting power supplies.

One type worth consideration at the moment is the **phased locked loop** or PLL IC. This is basically (Figure 5.12) a feedback control system made up of a phase detector, a low pass filter, and a voltage controlled oscillator. The VCO is an oscillator whose frequency will vary from its free running value when a d.c. voltage is

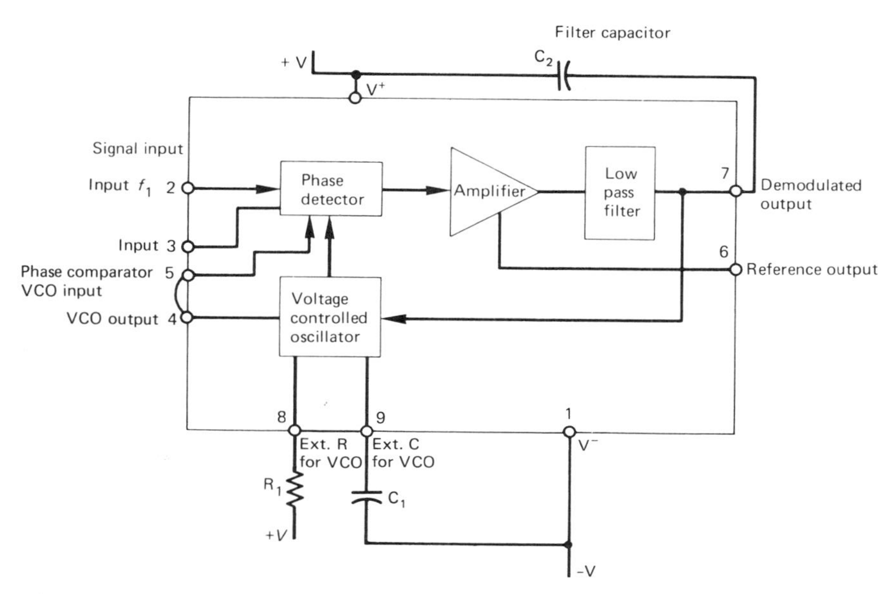

FIGURE 5.12 Basic phase locked loop.

applied. The analysis of a PLL is beyond the scope of this book but the operation is fairly straightforward. With no input signal applied, the output voltage will be zero and the VCO will free run at a frequency determined by the external components R_1C_1.

When an input signal of frequency f_1 is applied, the phase comparator circuit compares the phase and frequency of the incoming signal with that of the VCO. An error voltage is generated that is proportional to the difference between these two frequencies. This error is amplified and is filtered by a low pass filter to appear at the output as a low frequency signal. This is fed back to the input of the VCO and forces the VCO to alter its frequency so that the difference or error signal reduces. If the input frequency f_1 is sufficiently close to f_O then the VCO will synchronize its operation to the incoming signal, in other words it locks onto the input frequency. Once synchronized the VCO frequency is almost identical to the input frequency except for a small phase difference. This small phase difference is essential so that a d.c. output is produced that keeps the VCO frequency equal to the input. If the input frequency or phase changes slightly, the d.c. output will follow or track this change.

A PLL can, therefore, be used as an FM demodulator, or for FM telemetry, and for FSK receivers. FSK stands for frequency shift keying and is a method used for transmitting data using frequency modulation of a carrier. Logic 0 level will be one frequency, say, 1700 Hz, while logic 1 will be represented by a frequency of 1300 Hz. At the transmitter the logic levels are applied to a VCO to force the output to shift in frequency. The receiver will be a PLL that locks to the input frequency and then produces a d.c. level shift at its output as

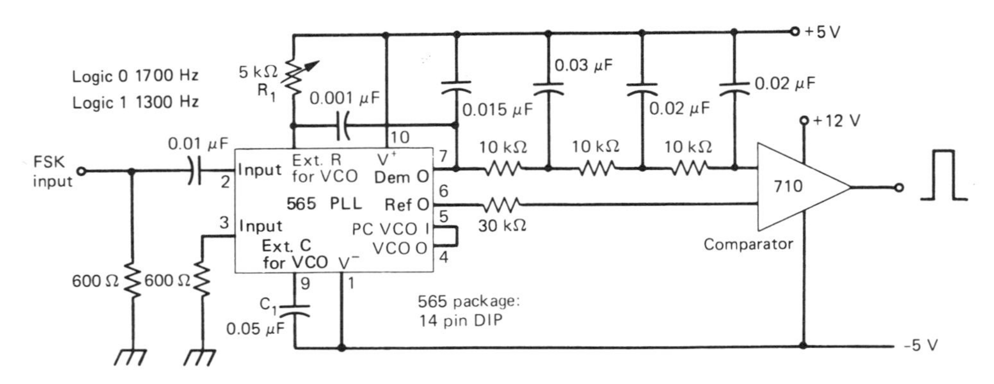

FIGURE 5.13 FSK receiver/decoder.

the received frequency shifts. Such a system has the advantage of being less affected by noise and interference.

A typical FSK receiver using a 565 IC PLL is shown in Figure 5.13. This is intended to receive and to decode FSK signals of 1700 Hz and 1300 Hz. The output of the PLL IC, which will be a voltage level dependent on the input frequency, is passed through a three-stage RC filter which removes the carrier frequency. A 710 comparator IC gives a high-state output for a 1300 Hz signal and a low-state output for a signal of 1700 Hz. The signaling rate, that is, the rate at which changes between the two frequencies is made, is typically 150 Hz max.

5.6 EXERCISE: AUDIO-FREQUENCY AMPLIFIER USING A 741 OP-AMP (FIGURE 5.14)

This audio amplifier uses a 741 connected in the noninverting mode to drive an 8 Ω loudspeaker via a complementary class B output stage. The frequency range is from about 15 Hz to 15 kHz, and the power output is about 3.5 W.

An input signal applied via C_1 to pin 3 of the 741 will give an output at pin 6 of the same polarity. This is coupled via an emitter follower Q_1 to the bases of the two output transistors Q_3 and Q_2. A portion of the output signal is fed back to the inverting input of the 741 via the potential divider R_3 and R_2. These two resistors determine the overall voltage gain of the circuit. Naturally these feedback resistors also provide the essential d.c. coupling so that test point B is held at, or very near, zero volts.

On the positive half-cycle of the input, Q_3 conducts and Q_2 is off; and on the negative half cycle, Q_2 conducts and Q_3 is off. Power is, therefore, supplied to the loudspeaker on positive half-cycles via Q_3 and on negative half-cycles via Q_2. These two output transistors should be a matched pair, and both must be mounted, via insulating washers, on to a heatsink. Diodes D_1 and D_2 are included to assist in eliminating crossover distortion by setting up a small amount of forward bias for Q_3 and Q_2.

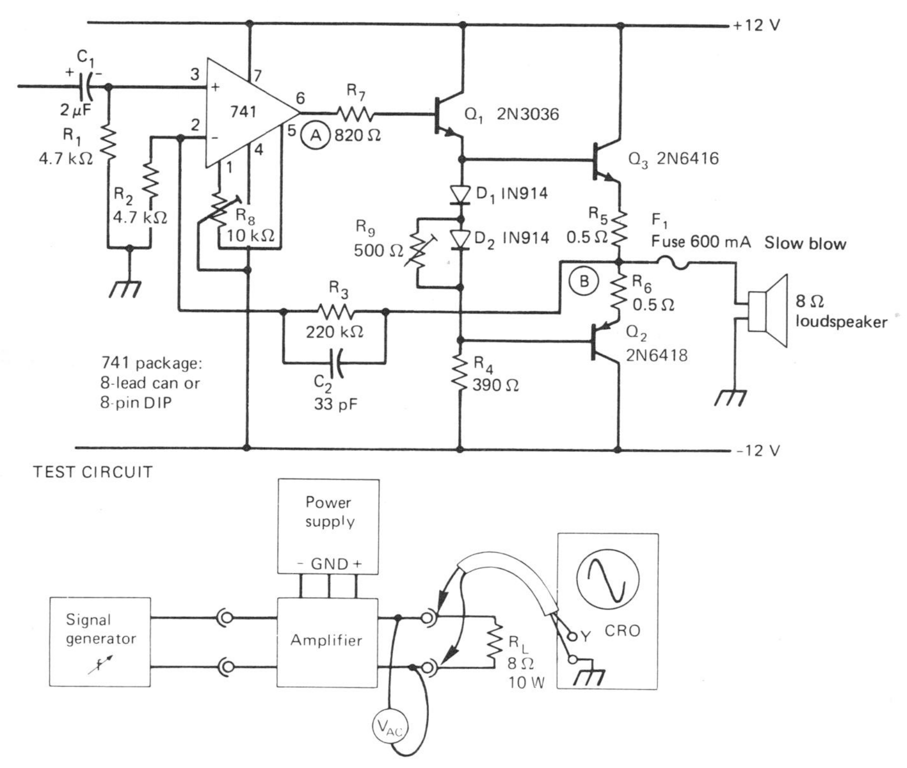

FIGURE 5.14 AF amplifier using a 741.

Any differential input offset voltage of the 741 will be amplified and appear at point B as a few hundred millivolts either positive or negative. This would set up an undesirable d.c. current through the speaker. R_8 is included as an offset null control to eliminate this. To set up the amplifier, and for testing, the speaker should be disconnected, and an 8 Ω 10 W wire-wound resistor should be connected as the load; R_9 should be set to zero. With a voltmeter connected from point B to 0 V, R_9 should be adjusted until the d.c. voltage at point B is zero. Measurement of frequency response, power output and sensitivity can be made with an oscilloscope and an a.c. voltmeter as shown. An input signal of 50 mV peak to peak at 1 kHz should give an output across the 8 Ω load of approximately 2.3 V peak to peak. R_9 can then be adjusted to give minimum cross-over distortion. The bandwidth can be found by first reducing the signal generator frequency until the output drops to 0.707 of the mid-frequency (1 kHz) value. This frequency is the low-frequency 3 dB point. The high-frequency 3 dB point can then be found by increasing the signal generator frequency until the output voltage again falls by 3 dB. During this test the input signal amplitude must be held constant.

The maximum available power output

can be approximately determined by increasing the input signal amplitude while monitoring the output waveform with the oscilloscope. The maximum rms a.c. voltage across the load with negligible distortion can be used to give power output as follows:

$$\text{Power output} = \frac{V_{rms}^2}{R_L} \quad \text{where } R_L = 8\ \Omega$$

Q_2 and Q_3 can both be damaged if excessive current is allowed to flow through them. This could occur if, for example, Q_1 became short circuit. The power supply lines should be fitted with current limits of about 1 A so that the maximum power dissipation for Q_3 and Q_2 is not exceeded.

QUESTIONS

1. Calculate the voltage gain of the circuit.
2. Explain what is meant by the sensitivity of an audio amplifier and state the value of the sensitivity for the amplifier in this example.
3. Which components determine
   (a) the low-frequency 3 dB point?
   (b) the high-frequency 3 dB point?
4. Under quiescent (no-signal) conditions what will be
   (a) The d.c. voltage from pin 6 of of the IC to ground?
   (b) The d.c. voltage from Q_2 base to ground?
   (c) The current flowing through Q_1?
   (d) The power dissipation of Q_1?
5. The amplifier develops a fault with the following symptoms:

   F1 blown
   Voltage TP(B) to ground −11.3 V
   Voltage TP(A) to ground +10.2 V
   No transistors are overheating

   State, with a supporting reason, the possible faulty component (or components) and the type of fault.
6. The amplifier develops a fault such that the gain becomes very low. The output voltage is almost identical with the input. State, with a supporting reason, the possible faulty component (or components) and the type of fault.
7. What would be the symptoms and effects for the following faults:
   (a) Q_3 base emitter open circuit.
   (b) R_3 open circuit.
   (c) An internal open circuit of the output of the 741 IC.
   (d) R_9 open circuit.

5.7 EXERCISE: SQUARE WAVE GENERATOR USING A 741 OP-AMP (FIGURE 5.15)

Because of their very high values of open loop gain, and the fact that differential inputs are available, op-amps can be used to create square wave or pulse generators. The basic circuit of an astable multivibrator is

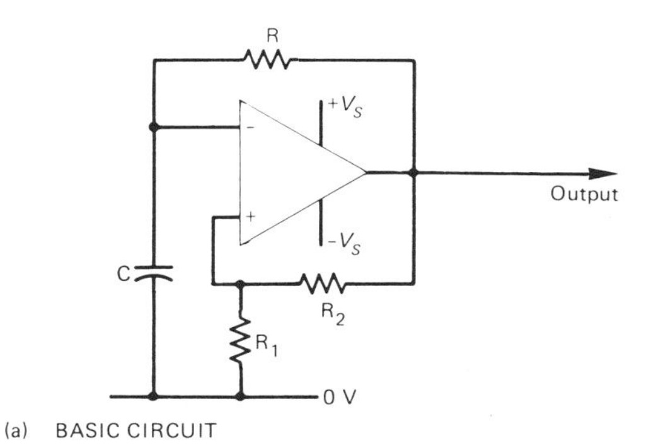

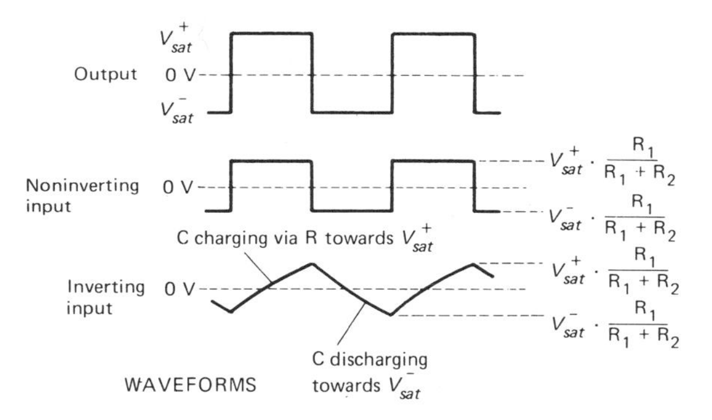

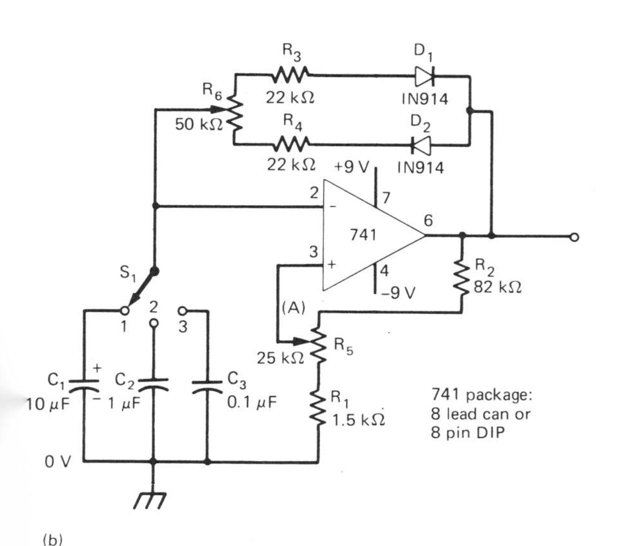

FIGURE 5.15 Square wave generator using a 741.

shown in Figure 5.15*a*. When power is applied, C will be uncharged so the op-amp output will saturate at its positive level (V_{sat}^+). A portion of this output voltage is fed back via R_2 and R_1 to the noninverting input. The voltage on the noninverting input will be, in fact,

$$V^+ = V_{sat}^+ \cdot \frac{R_1}{R_1 + R_2}$$

While the voltage on the inverting terminal is less than V^+, the output must remain at the positive saturated level. However, C charges via R causing the voltage at inverting terminal to rise. When this voltage just exceeds the level on the noninverting terminal, the op-amp output switches to its negative saturated level V_{sat}^-. The voltage on the noninverting terminal now reverses in polarity to

$$V^+ = V_{sat}^- \cdot \frac{R_1}{R_1 + R_2}$$

Now the capacitor discharges via R towards V_{sat}^- until the voltages on the two input terminals are again equal causing the op-amp output to switch to its positive value again. So the circuit produces square wave oscillations, with waveforms as shown. Note that both feedback paths control the frequency, since the *RC* time constant will determine the charge and discharge rate, while the potential divider R_2, R_1 determines the switching points. The frequency is given by the formula

$$f = \frac{1}{2RC \log_e \left(1 + 2\frac{R_1}{R_2}\right)}$$

The circuit in Figure 5.15*b* shows the modification to give variation in frequency and mark-to-space ratio. S_1 switches in different values of capacitor for range control of frequency while R_5 acts as a variable control for the frequency. With R_5 set to point A, for example, a greater portion of the output is fed back to the noninverting input. This means that the selected capacitor has to charge and discharge over a larger amplitude. The frequency will, therefore, be at its minimum value. The frequency ranges are approximately as follows:

S_1 Position	Frequency
1	2 Hz to 20 Hz
2	20 Hz to 200 Hz
3	200 Hz to 2 kHz

The mark-to-space ratio can be varied by R_6. When the output is positive, D_2 conducts to charge the selected capacitor via R_4 and a portion of R_6. When the output switches negative, D_1 conducts to discharge the capacitor via R_3 and the other part of R_6. In this way the setting of R_6 only affects the relationship between the positive and negative portions of the waveform and alters the frequency only slightly.

QUESTIONS

1. What is the range of mark-to-space control with R_6? How could this be increased to, say, 10:1?
2. State the approximate values for (*a*) output amplitude and (*b*) rise and fall times.
3. What would be the effect of the following component faults:
   (a) R_5 open circuit.
   (b) C_2 short circuit.
   (c) R_5 wiper open circuit.
4. For the following cases, state, with a supporting reason, the component (or components) and type of fault, that would cause the symptoms:
   (a) Variation of R_6 causes large changes in frequency on all ranges, but only small changes in mark-to-space ratio.
   (b) The frequency goes high on all ranges.
   (c) No output obtainable. The d.c. voltage at pin 6 is approximately +8 V and the d.c. voltage on pin 3 is positive and can be changed by R_5. The voltage on pin 2 is zero.

5.8 EXERCISE: TIMER UNIT USING 555s (FIGURE 5.16)

This unit uses three 555 timer ICs (or one 556 and one 555) to enable time delays from 20 seconds to about 240 seconds to be set on a calibrated dial. At the end of the time delay, an audible warning signal is given that lasts for approximately 1 sec. The operation is quite straightforward. U_1 forms the basic timing element. It is connected as a monostable and is triggered into a high output state when the start button is pressed. C_1, a tantalum electrolytic capacitor, will then charge via the

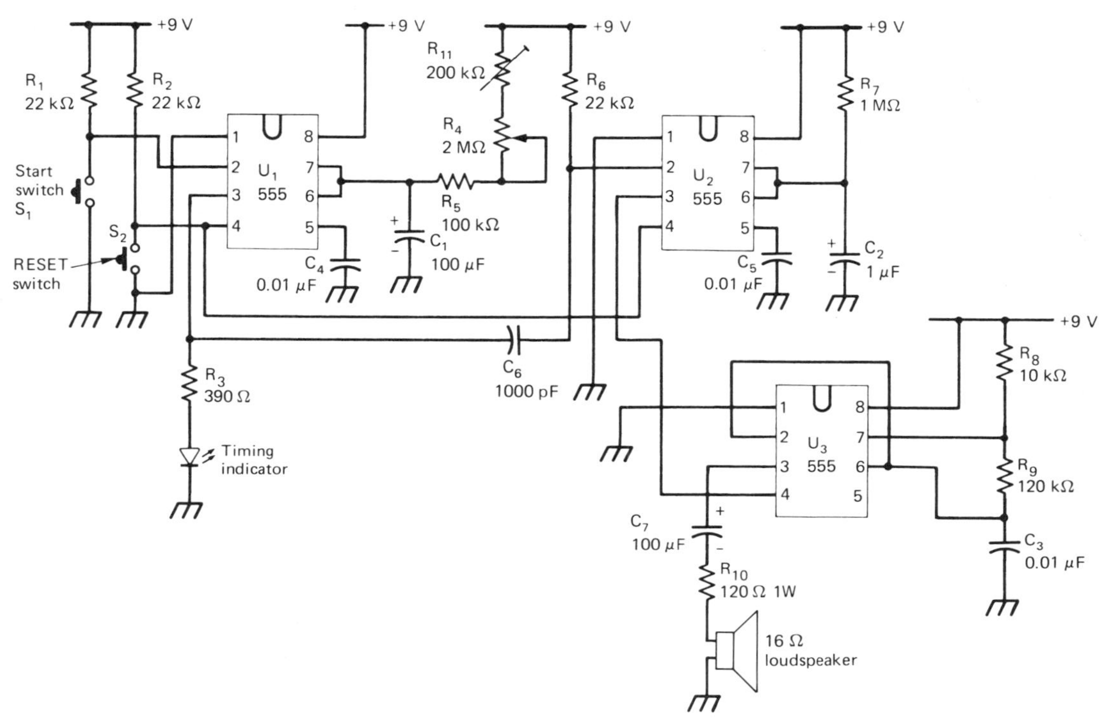

FIGURE 5.16 Timer (20 sec to 4 min)

resistive path R_{11}, R_4 and R_5. At the same time an LED indicates that timing is in progress. The time constant of the monostable can be varied by adjusting R_4; and R_{11}, a preset, allows the 20 sec delay time to be set when R_4 is at minimum. At the end of the time interval the output of U_1 returns to zero and this negative-going edge passes through C_6 to trigger U_2. This is a monostable with a fixed time period set by R_7 and C_2. The output of U_2 goes high for approximately 1 sec. U_3 is wired as an astable multivibrator and, since its reset input (pin 4) is connected to the output of U_2, it can only oscillate when the output of U_2 is high. Thus for one second the U_3 circuit oscillates and produces square waves to drive a loudspeaker via C_7 and R_{10}.

At any time during the timing sequence, the unit can be reset. The reset switch output is connected to pin 4 on both U_1 and U_2 to prevent the gated oscillator from producing the warning signal when the main timer is reset.

QUESTIONS

1. Sketch the time-related waveforms you would expect to measure at the outputs of U_1, U_2 and U_3 following the momentary closure of the start switch.

2. What is the frequency of the gated oscillator U_3?

3. If R_7 has a tolerance of ±5% and C_2 has a tolerance of ±10%, calculate the maximum and minimum time that the warning signal sounds.

4. For the following, state, with a supporting reason, the component (or components) and the type of fault that would cause the symptoms:

   (a) Time delay period is always 240 sec irrespective of the setting of R_4.

   (b) After the end of the time period selected by R_4 the alarm sounds continuously, but can be reset by operating S_2.

   (c) The circuit works normally except for the fact that pressing the reset switch, S_2, during the timing period causes the alarm signal to be given.

5. What would be the symptoms for the following faults?

   (a) An open circuit from pin 3 of U_2 to pin 4 of U_3.

   (b) C_1 short circuit.

   (c) C_6 open circuit.

6. Write a short note on the equipment and measurements required to calibrate and functionally test the unit.

5.9 EXERCISE: WIEN BRIDGE OSCILLATOR (FIGURE 5.17)

A Wien bridge oscillator is a popular method for generating sine waves of low harmonic distortion over the frequency range of 1 Hz to 1 MHz. It is the standard circuit used in the majority of audio and

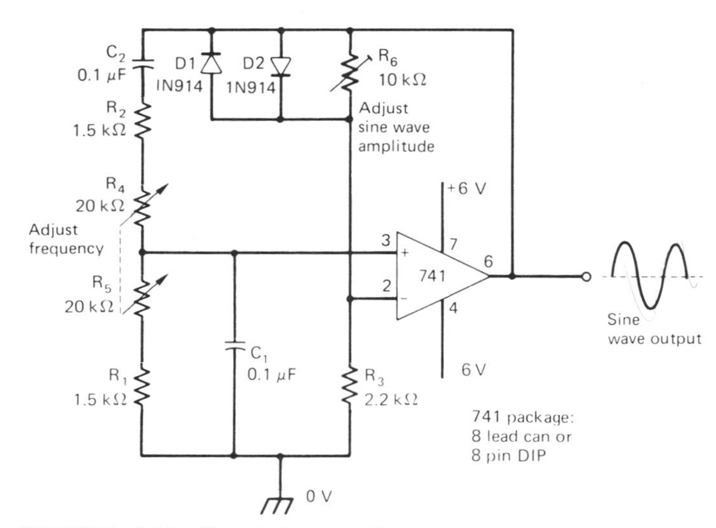

FIGURE 5.17 Wien bridge oscillator.

low-frequency signal generators. In this example, since a 741 op-amp is used, the frequency range is restricted. With the values shown, the frequency can be adjusted from about 75 Hz up to approximately 1 kHz. The bridge arrangement gives two feedback paths, the first via C_2, R_2, R_4, and R_5, R_1, and C_1 gives positive feedback. The other feedback path via R_6 and R_3 is negative and serves to stabilize the overall gain of the circuit. Many oscillator circuits consist of an amplifier with positive feedback, via a frequency-determining network. In the Wien bridge oscillator, the frequency-determining network is the series and parallel *CR* networks. At one particular frequency the phase shift from the output to the non-inverting input will be zero, and the circuit will oscillate at this frequency. Note that if $C = C_1 = C_2$ and $R = R_1 + R_5 = R_2 + R_4$, the frequency is given by

$$f = \frac{1}{2\pi RC}$$

By making R variable and by ganging together R_5 and R_4, the frequency can be continuously changed.

Analysis shows that the overall gain of the amplifier must only be 3 to maintain oscillations. The gain can be set to just above 3 by adjusting R_6 and is held, or stabilized, at this value by the two diodes D_1 and D_2. If the output amplitude increases, the diodes conduct and reduce the overall gain. The distortion of the output waveshape will be found to be slightly more than that obtained by using a thermistor as the stabilizing element. But thermistors suitable for this purpose are expensive and fragile items.

QUESTIONS

In all cases, unless otherwise stated, assume that the oscillator is set at a frequency of 1 kHz.

1. What will be the peak-to-peak amplitude of the output signal?
2. If, with the power off, an ohmmeter was used to measure the resistance from pin 3 to 0 V, what value would be indicated?
3. Sketch a circuit modification to replace the stabilized negative feedback loop with one containing a thermistor of nominal value 5 kΩ (R53 type). Explain how this operates.
4. State the possible faulty component (or components) and the type of fault for the symptoms given in each of the following cases:
   (a) The output becomes almost a square wave with an amplitude of just over 1 V peak to peak.
   (b) The output distorts heavily on its positive half cycle.
   (c) The output becomes almost a square wave with an amplitude of nearly 10 V peak to peak.
5. Describe a suitable method that could be used to measure the percentage distortion of the output signal.
6. Sketch a circuit modification so that another range from 750 Hz to 10 kHz can be provided.

7. The oscillator fails to produce an output signal, and the d.c. output is almost zero. List *all* possible components or faults that could cause this and describe a test procedure to check each one.

5.10 EXERCISE: A $1\frac{1}{2}$ HOUR TIMER WITH AUDIBLE OUTPUT (FIGURE 5.18)

This cicuit uses a ZN1034E timer IC described in section 5.4. The timing elements that determine the period for the internal oscillator of the IC are R_1, R_6 and C_1. Note that pins 12 and 11 are connected together, which means that the internal calibration resistor is used. With R_6 set to its maximum value, the total time delay of the circuit will be approximately 90 minutes. At switch on, when the +12 V is first applied via S_1, the timer is immediately triggered into operation. This is because pin 1 of the IC is connected directly to ground. The output on pin 2 will be low (less than 0.4 V) until the end of the timing period set by R_6. Then the output on pin 2 will rise positive and trigger the thyristor into conduction. This will cause the alarm buzzer to operate and, since the thyristor will continue to conduct, the alarm will sound until S_1 is moved to the Reset position.

QUESTIONS

1. What is the minimum time delay for this circuit?
2. Describe a method for measuring the minimum and maximum time delay.
3. What is the purpose of **(a)** R_5, **(b)** C_2?
4. A fault exists such that the alarm operates at switch on. Voltages

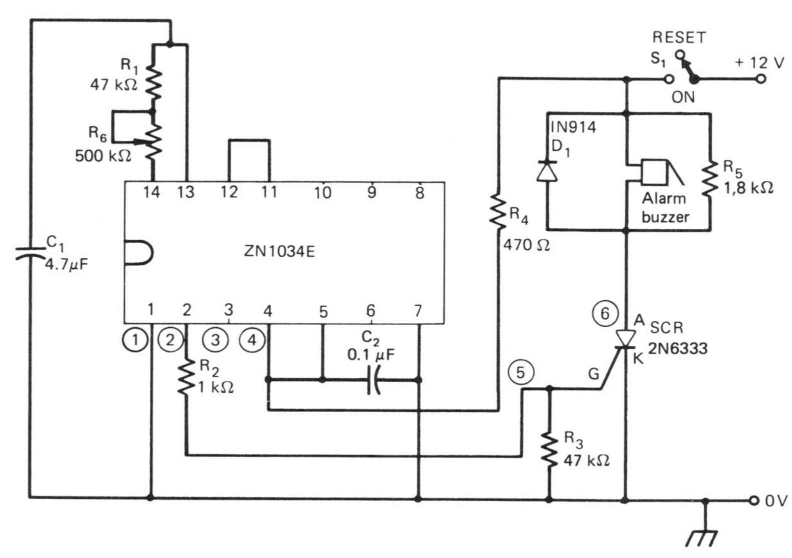

FIGURE 5.18 $1\frac{1}{2}$-hour timer.

measured with a multirange meter are as follows:

Test point (TP)	1	2	3	4	5	6
MR	0 V	0.3 V	3 V	5 V	0.26 V	0 V

State with supporting reasons the faulty component, and the type of fault.

5. In each of the following cases the alarm fails to operate after the time delay. State, with a supporting reason, the component fault.

TP	1	2	3	4	5	6
(a) MR	0 V	3 V	0.2 V	5 V	2.8 V	12 V
(b) MR	0 V	3 V	0.2 V	5 V	0 V	12 V
(c) MR	0 V	0 V	0 V	12 V	0 V	12 V
(d) MR	0 V	0 V	0 V	0 V	0 V	12 V

6. State the symptoms for the following component faults:

   (a) R_5 open circuit.

   (b) R_6 wiper open circuit.

   (c) Pin 1 of the IC to ground open circuit.

5.11 EXERCISE: LOW-FREQUENCY FUNCTION GENERATOR (FIGURE 5.19)

Function generators are oscillators that give simultaneous triangle, square, and sine wave outputs. This circuit, using two op-amps, is a simple example that gives fixed low-frequency outputs at about 1 Hz. U_1 is wired with C_1 as an integrator, and U_2 acts as a comparator circuit. Imagine that the output of U_2 has just gone positive to its saturated positive output level. A portion of this positive level will appear at point x because of the potential divider formed by R_4 and R_5. If R_5 is made about 1.8 kΩ, for example, then the level at point x will be about +700 mV. Since the noninverting input of U_1 is held at ground, the inverting input must also be very near ground. Therefore, C_1 will be charged via R_1 with a current of about 10 μA. The output of U_1 moves negatively as C_1 is charged and, since the charging current via R_1 remains almost constant, the rate of change of U_1's output is linear. When the voltage at point y, the output of U_1, just exceeds a level sufficient to make pin 3 of U_2 go below zero, the output of U_2 will switch negative. Note that U_2 has positive feedback via R_3 and that, while pin 3 is more positive than pin 2, the output will be positive; but when pin 3 is more negative than pin 2, the output will be negative. Since the op-amp has a gain of 100 000 the switch action is rapid. The level at point y that triggers the comparator U_2 is determined by R_3 and R_2. Since the positive saturation voltage of U_2's output is about +4 V then, when point y is about −2 V, pin 3 will just go below zero and the output of U_2 will switch negative.

With U_2's output at −4 V, point x also switches negative to −700 mV. The charging current for C_1 now reverses and point y moves positive. When the level at point y reaches about +2 V, the comparator again switches and the cycle is repeated.

The time period for C_1 to charge from −2 V to +2 V is the time for one half cycle of the oscillator. To obtain an approximate value for this time we can use the formula

$$Q = CV$$

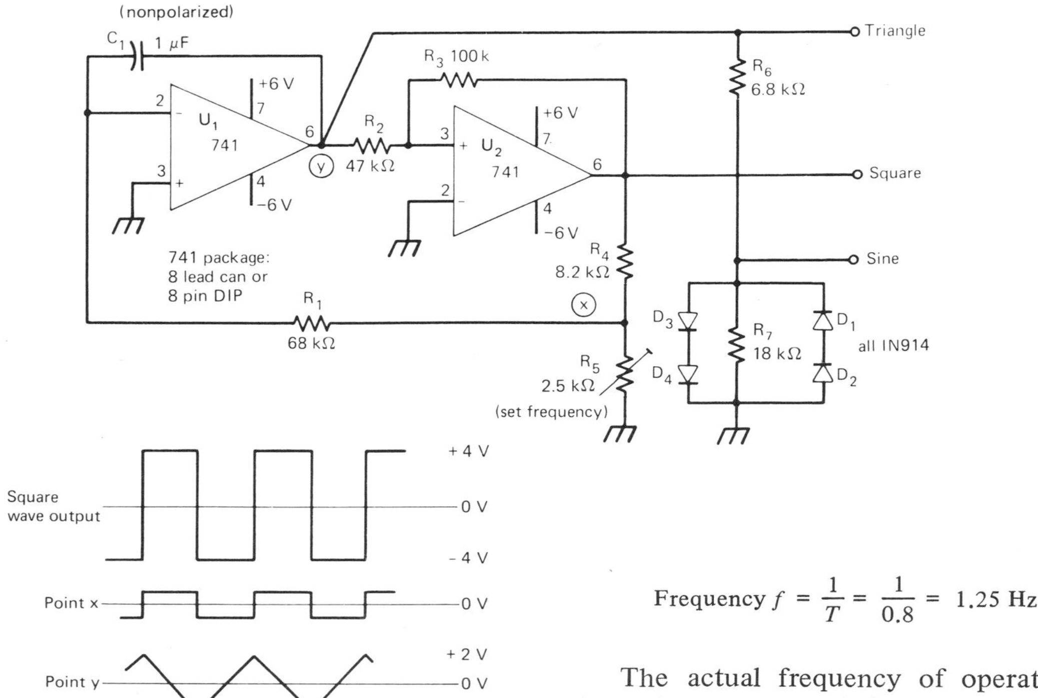

FIGURE 5.19 Low-frequency function generator.

Since the capacitor is being charged by a constant current

$$Idt = CdV$$

$$dt = \frac{CdV}{I}$$

With $C = 1\ \mu\text{F}$, $I = 10\ \mu\text{A}$, and dV = the change in voltage across the capacitor = 4 V

$$dt = \frac{1 \times 10^{-6} \times 4}{10 \times 10^{-6}} = 0.4 \text{ sec}$$

The period $T = 2t = 0.8$ sec.

$$\text{Frequency } f = \frac{1}{T} = \frac{1}{0.8} = 1.25 \text{ Hz}$$

The actual frequency of operation is dependent on several factors such as the actual saturation voltage at the output of U_2, the tolerance of C_1, and the tolerance of R_1, R_2, R_3, and R_4. By making R_5 variable, the frequency can be trimmed to give exactly 1 Hz.

The triangle output is shaped or "bent" into an approximate sine wave by the diode network D_1, D_2, D_3, D_4. R_6 and R_7 act as a potential divider that would cause the output across R_7 to be about 3 V peak to peak. However, the diodes conduct when they are forward biased by about 500 mV and so the resulting sine wave output has an amplitude of 2 V peak to peak. This is a very simple example of a triangle-to-sine convertor, and the resulting distortion present in the sine wave is fairly high. R_6 can be adjusted by test to obtain the optimum result. Use a chart recorder for measurement of the waveforms.

QUESTIONS

Unless otherwise stated, assume the generator frequency is set to 1 Hz.

1. Explain the effects of the following component failures:
   (a) R_5 open circuit.
   (b) R_7 open circuit.
   (c) R_3 high in value to 300 kΩ.
2. The sine wave output becomes distorted as shown in Figure 5.20. State the component failure.
3. During assembly a 100 Ω variable resistor is connected by mistake for R_5. What will be the effect of this on the circuit?
4. Sketch the resulting sine wave output if D_1 became short circuit.
5. The circuit fails to oscillate. The d.c. levels with respect to ground measured with a voltmeter are

U_2 pin 6	+4.2 V
Point x	0.6 V
Point y	-4.25 V

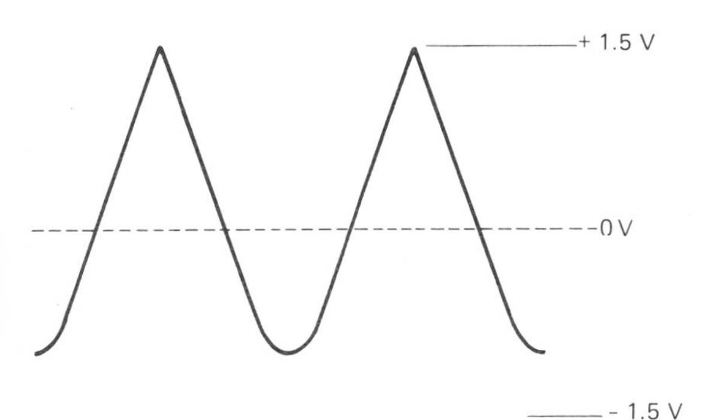

FIGURE 5.20 Fault condition—sine wave distorted.

State with a supporting reason the component (or components) failure that would cause this fault.

5.12 EXERCISE: PULSE GENERATOR USING A 555 (FIGURE 5.21)

The 555 timer is connected as a monostable with the width of the positive going output pulses being controlled by R_{12}. The relaxation oscillator circuit formed by the unijunction transistor produces narrow positive pulses at a fixed frequency set by S_1. These pulses are inverted by Q_2, a transistor switch, and the negative pulses at Q_2 collector are used to trigger the 555.

A unijunction transistor is a very useful device for making simple oscillator circuits that have quite good stability. It is constructed (Figure 5.22) of either a bar or a cube of lightly doped n-type silicon. Ohmic contacts are made at each end, and these are called base 2 and base 1. The resistance between the two base connections is typically 6.5 kΩ for the 2N4891. Somewhere near the center of this bar a p-type region is formed during manufacture, and the connection to this is called the emitter. An equivalent circuit can be drawn as shown to consist of two resistors and a diode. The diode is the pn junction formed between the emitter and the bar. It is important to realize that the cathode connection of the equivalent diode is internal to the UJT and cannot be reached. Imagine the device connected across a d.c. supply with B_2 positive with respect to B_1 and V_{EB1} equal to 0 V. The voltage between B_2 and B_1 is called V_{BB} the *interbase voltage*. r_1 and r_2 form a potential divider, so the voltage

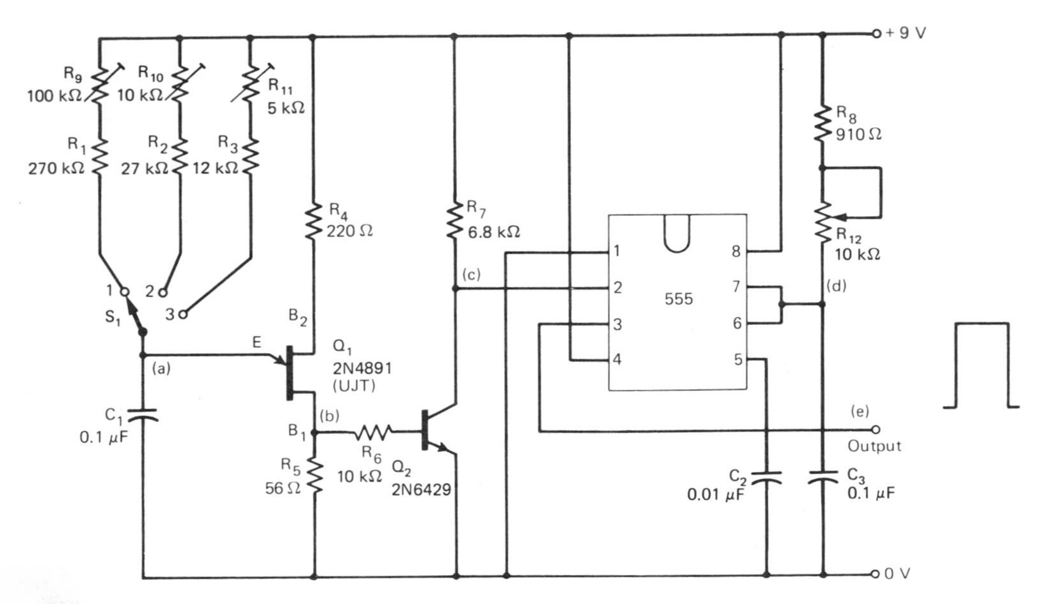

FIGURE 5.21 Pulse generator.

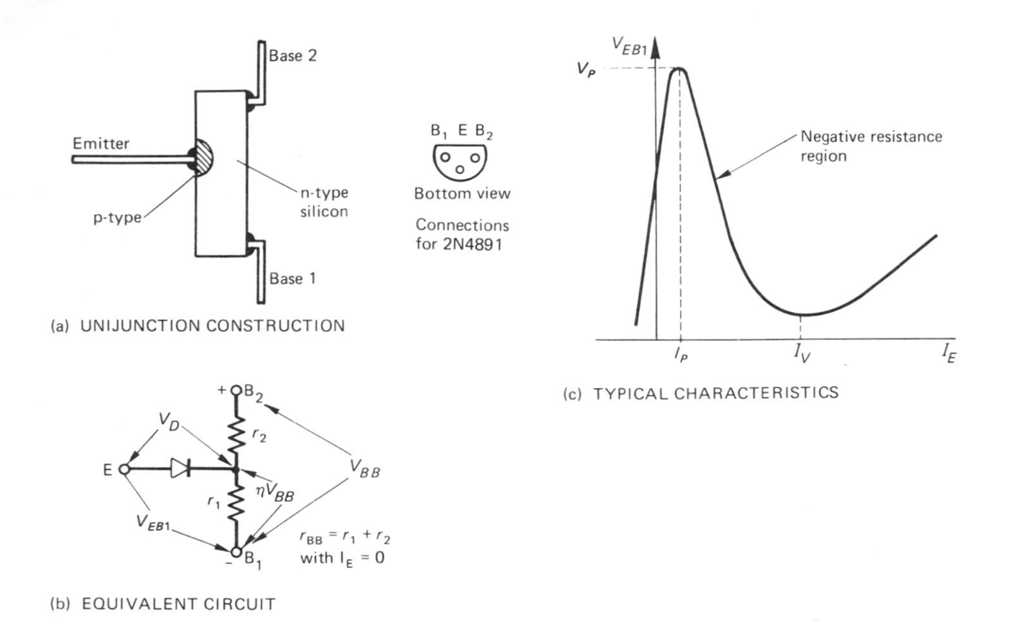

FIGURE 5.22 Unijunction transistor.

across r_1 (internal to the UJT) is a value dependent on the ratio of these resistances. In fact, the voltage across r_1 is called ηV_{BB}, where η, known as the *Intrinsic Stand-off Ratio*, depends on the geometry of the device and is determined by the physical position of the p-region relative to B_2 and B_1. η(eta) has values between 0.55 to 0.82 for 2N4891 devices.

If V_{EB1} is now gradually increased in value, a voltage between emitter and B_1 will be reached when the diode is almost forward biased. This voltage is called V_P, and from the equivalent circuit we can see that

$$V_P = \eta V_{BB} + V_D$$

Once V_{EB1} exceeds V_P the diode conducts, and it injects holes into the B_1 region. The resistance of the B_1 region, that is r_1, falls. However, r_2 remains almost unchanged so the voltage across r_1 also falls in value. This increases the forward bias across the diode causing it to conduct more, and therefore further decreasing the value of r_1. This is a positive feedback effect and so very rapidly the resistance of the r_1 region falls to a very low value. Obviously in any circuit application the current injected at the emitter must be limited or the device will be destroyed. The characteristics for the device are shown in Figure 5.22*c*, where the negative resistance region can clearly be seen. This negative resistance characteristic enables the UJT to be used in simple relaxation oscillators.

Now let us consider the operation of the oscillator section of the circuit in Figure 5.21. At power up C_1 is uncharged so V_E is zero. A small current of just over 1 mA flows through the UJT from B_2 to B_1. C_1 now charges exponentially via R_9 and R_1 toward +9 V. When the voltage across C_1 reaches the trigger voltage $V_P (V_P = \eta V_{BB} + V_D)$, then the unijunction switches to its low resistance state and C_1 is rapidly discharged through R_5. This causes a short duration positive pulse to be developed across R_5. When the capacitor is discharged, the emitter current falls to a value that is insufficient to maintain conduction between the emitter and base 1, and so the unijunction switches back to its high-resistance state. C_1 is again free to charge and the cycle repeats itself. Thus a sawtooth type waveform is generated at the emitter, and short duration positive pulses appear at B_1. The frequency of the oscillations depend on the time constant in the emitter circuit, and V_P, the UJT trigger voltage. With S_1 in position (1) the time constant is $C_1(R_1 + R_9)$. For the 2N4891, since η can have a value of between 0.55 and 0.82, the trigger voltage V_P will also vary from one device to another.

Since $V_P = \eta V_{BB} + V_D$ the values of V_P for this circuit lie between 5.3 V and 7.7 V.

The actual formula for the frequency is

$$f = \frac{1}{CR \log_e \left(\frac{1}{1 - \eta} \right)}$$

where CR is the time constant of the emitter circuit.

Note that V_{BB} does not appear in this formula so that variations in supply voltage should not effect the frequency. However, because of the large variations in η the frequency cannot be accurately set by fixed values of timing components. For each switch position, a variable resistor is included to allow the three frequencies of 30 Hz, 300 Hz, and 600 Hz to be set by test.

R_4 is included as a temperature-compensation component. The interbase resistance of the UJT will fall with increasing temperature, and this will affect the trigger point. With R_4 in the circuit any change in interbase resistance results in a volt drop across R_4, which tends to balance the changes in V_P and I_P.

The rest of the circuit operation is straightforward. Q_2 is normally off and is switched into conduction by the narrow positive pulses at B_1 of the UJT. Pin 2 of the 555 is taken toward 0 V momentarily and this triggers the 555 on. The output rises positive and remains high for a time determined by C_3, R_8, and R_{12}.

QUESTIONS

1. Calculate the minimum and maximum values for the output pulse width.
2. With S_1 in position (3) and R_{12} at maximum, sketch the time-related waveforms you would expect to measure at points (a), (b), (c), (d), and (e). Include typical amplitudes.
3. In each of the following cases no oscillations are observed at point (a). State, with a supporting reason, the component (or components) that is faulty and the type of fault. The voltages were measured with a digital multimeter. S_1 was set to position (1).

TP	(a)	(b)	(c)
(a) MR	0.65 V	0.11 V	9 V
(b) MR	8.72 V	0.07 V	9 V
(c) MR	0 V	0.07 V	9 V

4. Describe the effects of the following faults:
   (a) R_{11} open circuit.
   (b) Q_2 base emitter short.
   (c) C_3 open circuit.
5. Assuming Q_1 oscillator circuit is functioning correctly, how could correct operation of Q_2 be quickly verified?

5.13 EXERCISE: TEMPERATURE-MEASURING CIRCUIT (FIGURE 5.23)

A GM472 bead-type thermistor is used in this circuit as the temperature-sensing element. Thermistors are sensitive devices with relatively large negative temperature coefficients so they are often used over the temperature range −20°C to +150°C. The GM472 has a resistance of about 4700 Ω at 25°C, but at 100°C the resistance will fall to about 500 Ω. By using the thermistor as part of the feedback network in an amplifier, as in this circuit, reasonable linearity over the temperature range +10°C to +70°C is achieved.

A voltage of about −50 mV is developed across R_3 by the potential divider R_2 and R_3 from the stabilized zener voltage. The 741 op-amp is connected as an in-

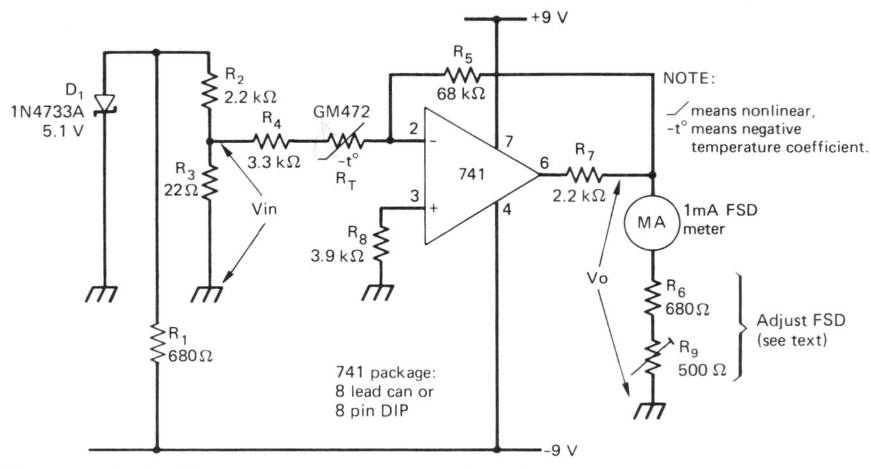

FIGURE 5.23 Linear temperature-measuring circuit.

verting amplifier with negative feedback via R_5.

The voltage gain of the amplifier is

$$A_v = \frac{-R_5}{R_4 + R_T}$$

where R_T is the resistance of the thermistor and the output voltage V_o is

$$V_o = A_v V_{in} = -\frac{R_5 V_{in}}{R_4 + R_T}$$

Thus as R_T varies with temperature the output voltage will also change. When the thermistor is at 100°C, for example, the output voltage will be about +900 mV. R_9 is adjusted so that the 1 mA moving coil meter indicates full scale when the tip of the thermistor is placed in steam or boiling water.

The low end of the scale can be marked by placing the thermistor tip in a mixture of ice and water.

QUESTIONS

Assume that the thermistor is kept at a temperature of 25°C.

1. What is the purpose of R_7?
2. Calculate the current flowing through the 5.1 V zener diode.
3. Why is it necessary to keep the input voltage at a low value?
4. Since the sensing element is likely to be positioned some distance from the rest of the circuit, it and its connections are more liable to damage. Comment on the symptoms that would indicate an open circuit of the thermistor or its connections.
5. List other component failures that would cause the same symptoms as an open circuit thermistor. How could the thermistor be checked?

6. What would be the effect of D_1 going open circuit?
7. A fault exists such that pin 6 of the 741 is stuck at +8 V and the meter pegs the fullscale stop. The IC is known to be within specification. State, with a supporting reason, the faulty component (or components) and the type of fault.

5.14 EXERCISE: BRIDGE CIRCUIT WITH NULL INDICATOR (FIGURE 5.24)

This is a circuit for measuring or comparing resistor values. The Wheatstone bridge circuit consists of R_{13}, R_S, and R_X. R_{13} is a ten-turn helipot fitted with a digital dial. The bridge is supplied with 5.1 V from the zener diode D_2.

At balance

$$\frac{P}{Q} = \frac{R_s}{R_x}$$

where $(P + Q)$ = resistance of R_{13}. Therefore

$$R_x = R_s\frac{Q}{P}$$

R_s is a standard resistance and R_x is the unknown.

With R_s made 1 kΩ, the circuit is capable of measuring resistors in the range of 50 Ω to 50 kΩ with an accuracy dependent on R_{13} and R_S.

Any out-of-balance signal from the bridge is amplified by the differential circuit using the 741, and the output is applied to the indicator circuit. This consists of three LEDs, only one of which is on at any one time. If the ten-turn pot R_{13} is set too high, the output from the op-amp will be positive. A posi-

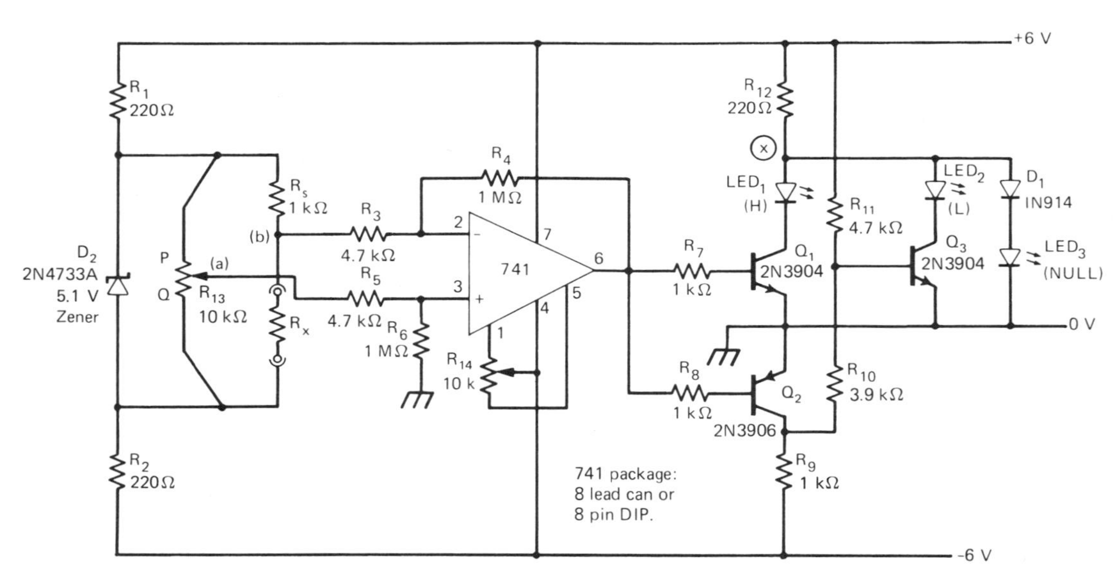

FIGURE 5.24 Bridge circuit with null indicator.

	741 Pin 6	Q_2 Collector	Q_3 Base	Point (x)	State of Indicator Diode LED_1	LED_2	LED_3
(a) MR	+1.4 V	-0.7 V	+0.7 V	+1.65 V	ON	ON	OFF
(b) MR	+0.2 V	-6 V	+0.7 V	+1.65 V	OFF	ON	OFF
(c) MR	+0.2 V	-6 V	+0.1 V	+6 V	OFF	OFF	OFF
(d) MR	-1.2 V	-0.1 V	-0.1 V	+2.5 V	OFF	OFF	ON
(e) MR	+2 V	-6 V	+0.1 V	0 V	OFF	OFF	OFF
(f) MR	+2 V	-6 V	+0.1 V	+2.5 V	OFF	OFF	ON

tive output level greater than about 800 mV will cause Q_1 to conduct and LED_1 will be on, indicating High. Note that under these conditions, since Q_2 and Q_3 are off, LED_2 is off, and also LED_3 is off. If, on the other hand, the ten-turn pot is set too low, the op-amp output goes negative causing Q_2 to conduct. The collector of Q_2 rises to almost zero and this allows the base of Q_3 to go positive. Q_3 conducts and LED_2 will come on, indicating Low. Thus, if the op-amp output is greater than 800 mV in either direction, the High or Low LEDs are on. While one of these conducts, the voltage at point x with respect to ground will be only 1.6 V, and this is insufficient to provide forward bias to D_1 and LED_3. At balance the op-amp output will be almost zero; therefore, Q_1 and Q_2 and Q_3 are nonconducting and LED_3 is forward biased indicating a Null.

R_{14} is an offset zero control for the 741. This is adjusted with points a and b shorted together, to turn on LED_3 (Null).

LED_1, LED_2, and LED_3 may be red, yellow, and green types respectively. In this case, red indicates a high, yellow indicates a low, and green indicates a null.

QUESTIONS

1. If R_X is a 500 Ω resistor, what will be the setting of the digital dial on R_{13} to achieve balance?
2. What would be the effect of R_4 becoming open circuit?
3. In each of the cases at the top of the page a fault exists in the indicator section of the circuit. R_{13} has not necessarily been adjusted to give a Null. State, with a supporting reason, the faulty component (or components) and the type of fault. The voltage readings were taken with a 20 kΩ/volt meter on the 10 V range.
4. R_X is approximately 3 kΩ. No Null can be obtained by variation of R_{13} and the Low indicator remains on. The voltage level at pin 6 of the 741 is -4.8 V. State the component fault.
5. Describe the effects of the following component failure:
   (a) D_2 short circuit.
   (b) R_3 open circuit.

5.15 EXERCISE: ACTIVE FILTER CIRCUITS (FIGURE 5.25)

To conclude this chapter on linear circuits, we will take a brief look at the use of op-

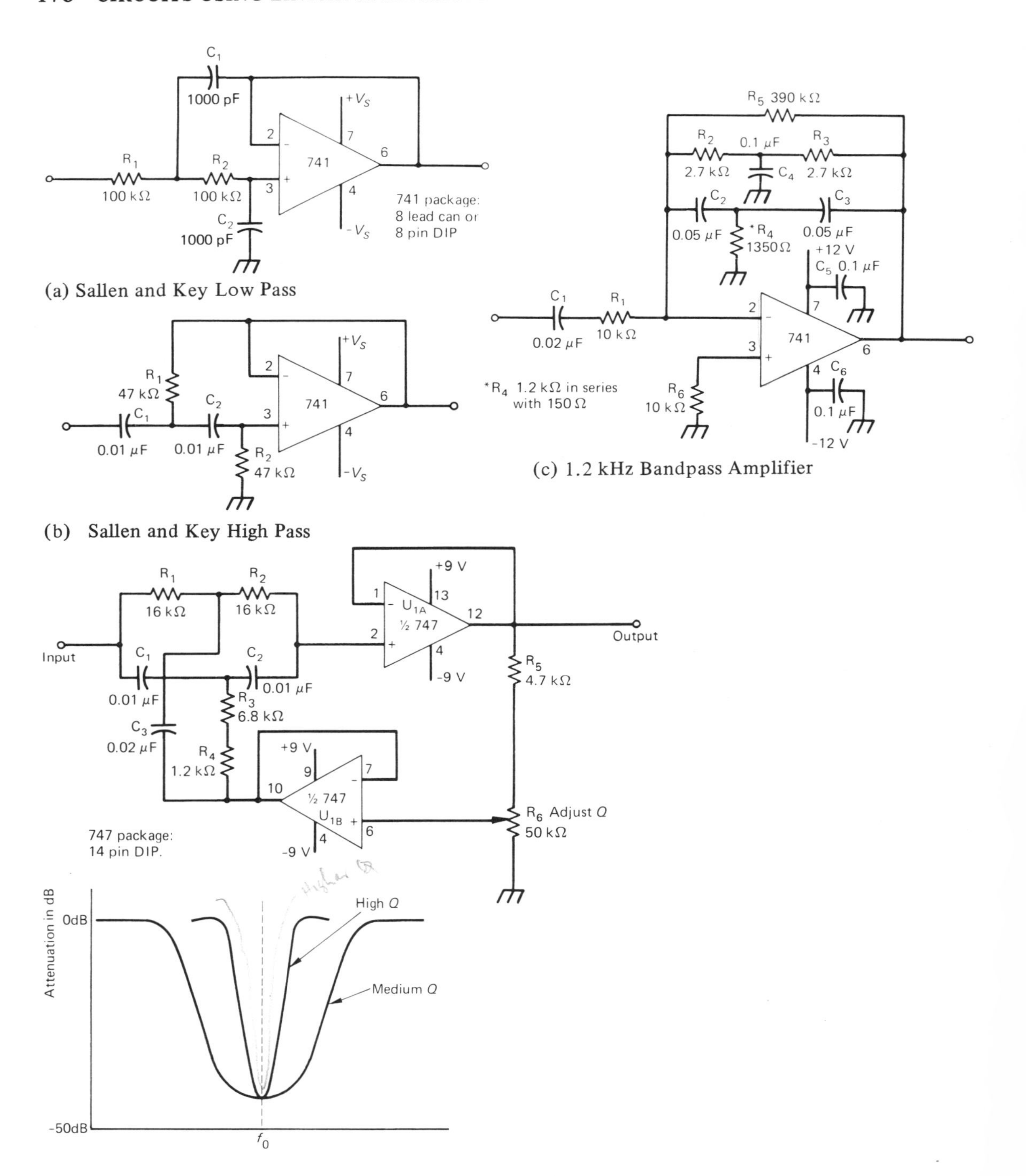

FIGURE 5.25 Active filters

amps in active filter circuits. Simple filters can be created using *RC* networks, since the resistor value remains almost fixed while the reactance of the capacitor will vary with frequency. For a simple *RC* low pass filter the attenuation, or loss, will increase from the cutoff frequency by 6 dB per octave or 20 dB per decade. Cutoff frequency is the frequency at which attenuation is 3 dB. However, an active filter has a transfer function whose equation includes an ω^2 term, and this means a more rapid increase of attenuation above cutoff of 12 dB per octave or 40 dB per decade.

Filters fall into the following categories:

(a) **Low pass** D.C. and low-frequency signals are passed with very little attenuation. At cutoff, the attenuation is 3 dB and for signals of frequency higher than the cutoff the attenuation increases.

(b) **High pass** Signals of frequency higher than cutoff are passed with little or no attenuation. Below cutoff, attenuation increases as signal frequency decreases.

(c) **Band pass** Only a selected band of signal frequencies are passed with little attenuation. Above and below this band, the signals are progressively attenuated. The sharpness of response or Q factor of the circuit is

$$Q = f_0/\Delta f \quad \text{where} \quad \Delta f = f_2 - f_1$$

f_2 is upper cutoff frequency (3 dB)
f_1 is lower cutoff frequency (3 dB)
f_0 is center frequency.

(d) **Band stop or notch** This filter has maximum attenuation at one particular frequency. For signal frequencies above and below a narrow band about the center frequency, the attenuation is very low.

In Figure 5.25 are examples of each of these filters. For the low and high pass active filters the cutoff frequency is given by

$$f_0 = \frac{1}{2\pi\sqrt{C_1 C_2 R_1 R_2}}$$

In each case the op-amp is wired as a voltage follower with 100% negative feedback. The low pass circuit gives unity gain at low frequencies, but as the signal frequency is increased the reactance of the capacitors falls and attenuation increases. Because of the active element, the attenuation after cutoff increases by 40 dB per decade.

The high pass circuit is created simply by interchanging the positions of the capacitors and resistors. In both instances the cutoff frequency can be made continuously variable by making a portion of R_1 and R_2 into a ganged potentiometer.

The band pass circuit in Figure 5.25*c* uses a twin-tee filter network as part of the feedback circuit for the 741 op-amp. The twin-tee is actually a band stop filter having maximum impedance at its center frequency.

$$f_0 = \frac{1}{2\pi CR}$$

where $R = R_2 = R_3 = 2R_4$ and $C = C_2 = C_3 = \frac{1}{2}C_4$.

In this case f_0 is 1.2 kHz. The twin-tee network is used in this circuit as a frequency-selective resistor. At the center frequency its effective resistance is very high, and the gain of the amplifier will be controlled by R_5 and is approximately 35. As the signal frequency is varied above or below the center frequency, the effective resistance of the twin-tee circuit falls and this causes a drop in amplifier gain.

The final circuit (Figure 5.25*d*) shows a notch filter, again using a twin-tee network, but in this case the network is in the input section of the circuit and the Q can be adjusted by the amount of positive feedback applied via the unity gain follower (U_1B). The center frequency is again given by

$$f_0 = \frac{1}{2\pi CR}$$

where $R = R_1 = R_2 = 2(R_3 + R_4)$ and $C = C_1 = C_2 = \frac{1}{2}C_3$.

QUESTIONS

1. Sketch the attenuation characteristics for the low pass and high pass filters in Figure 5.25*a* and *b*. In each case state the cutoff frequency.
2. Explain the effect of the following component failures for the band-pass amplifier in Figure 5.25*c*.
   (a) C_2 open circuit.
   (b) R_5 open circuit.
   (c) C_4 open circuit.
3. Calculate the center frequency for the notch filter **(d)**.
4. In circuit **(d)** a fault exists causing the notch to be exceptionally sharp with amplification of signals at a frequency just below the center frequency. State with a supporting reason the component (or components) fault.
5. Explain the likely effects in circuit **(d)** of
   (a) R_5 open circuit.
   (b) R_1 open circuit.
   (c) R_6 wiper is open circuit.

CHAPTER 6

Power Supply and Power Control Circuits

6.1 PRINCIPLES AND DEFINITIONS

A power unit of some type is essential for the operation of electronic instruments and systems. Because of this, troubleshooting the various types of power supplies in common use are very important areas of study. The power to drive a system or instrument may, of course, be supplied from batteries, but more usually it is derived from the single phase a.c. power line (mains). The purpose of the power unit in this case is to accept the local mains supply (120 V rms at 60 Hz in the United States and 240 V rms at 50 Hz in many other countries) and convert it into a form that is suitable for the internal circuits of the system or instrument. In the majority of cases this means converting the a.c. mains voltage into a fixed stable d.c. voltage. The d.c. output has to remain substantially constant against changes in load current, mains input, and temperature. In addition to this are the requirements of isolation and possibly automatic overload and overvoltage protection. The power unit must effectively isolate the internal circuits from the raw mains, and usually has to provide an automatic current limit or trip if an overload or short occurs. If, in the event of a power supply fault, the d.c. output voltage rises above a maximum safe value for the internal circuits, then the power must be automatically disconnected.

Two methods are used to provide stabilized d.c. voltages. The commonly used type has been the linear series regulator, and this still predominates for modest power requirements. Increasingly, for higher power requirements, switching mode power supplies (SMPS) are being introduced. A switched system is more efficient, wastes less power, and therefore takes up less space than a conventional linear regulator.

Apart from d.c. regulated circuits there are the inverters and converters. These are also examples of switched power systems. An INVERTER is a power unit that produces an a.c. power output from d.c. The frequency of the a.c. output may be 60 Hz, but can be 400 Hz or higher. The d.c. source is typically a battery, and one good example of an inverter is in standby power units that, in the event of an a.c. power line failure, provide a short-term emergency supply of 120 V rms at 60 Hz from say a 24 V battery. The battery is trickle-charged when the a.c. power line is on.

A CONVERTER is basically an inverter followed by rectification or in other words d.c. to d.c. conversion. An example could be the requirement in a portable instrument in obtaining 1 kV at 1 mA d.c. to supply a photomultiplier tube from a 9 V battery.

Since, in practice, the regulated d.c. power supply is the type most test and service technicians come across, it is useful to list the most important parameters and terms used in their description. Some of these terms are, of course, applicable to other types of power supply circuits.

1. **Range** The maximum and minimum limits of the output voltage and output current of a power supply.
2. **Load regulation** The maximum change in output voltage due to a change in load current from no load to full load. The percentage regulation of a power supply is given by the formula:

$$\%\text{ load regulation} = \frac{\text{No-load voltage} - \text{Full-load voltage}}{\text{Full-load voltage}} \times 100\%$$

   This is illustrated in Figure 6.1 where a load regulation graph for a 5 V power unit is drawn.

3. **Line regulation** The maximum change in output voltage as a result of a change in the a.c. input voltage. Often quoted as a percentage ratio, i.e., ±10% mains change to ±0.01% in output voltage.
4. **Output impedance** The change in output voltage divided by a small change in load current at some specified frequency (100 kHz is typical).

$$Z_{out} \approx \frac{\Delta V_o}{\Delta I_L}$$

   At low frequencies, that is for slowly changing load currents, the resistive part of Z_{out} predominates. R_{out} can be read from the load regulation graph (see Figure 6.1) and for a reasonable power unit should be at most a few hundred milliohms.

5. **Ripple and noise** The peak-to-peak or rms value of any alternating or random signal superimposed on the d.c. output voltage with all external operating and environmental parameters held constant. Ripple may be quoted at full load or alternatively at some specified value of load current.
6. **Transient response** The time taken for the d.c. output voltage to recover to within 10 mV of its steady state value following the sudden application of full load.
7. **Temperature coefficient** The percentage change in d.c. output voltage with temperature at fixed

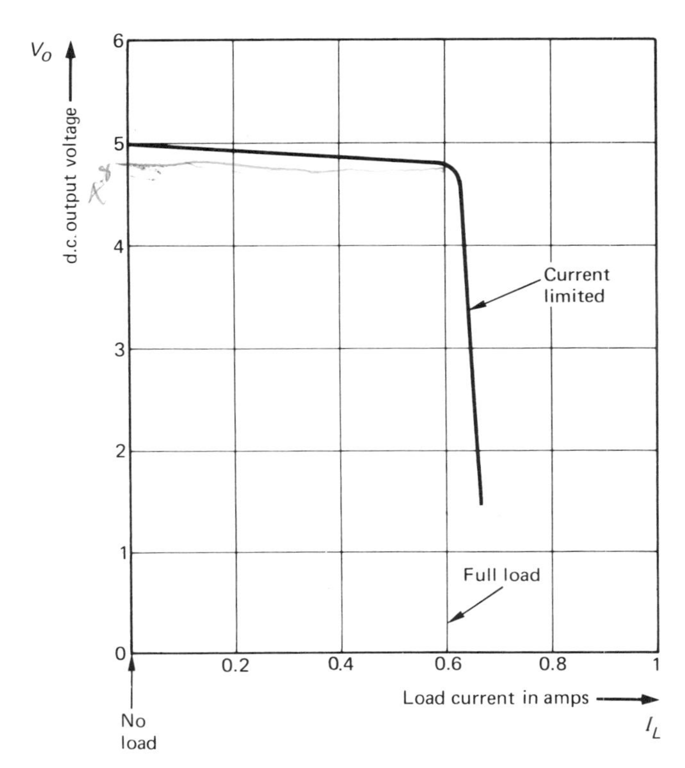

FIGURE 6.1 Example of a load regulation curve for a regulated power supply

$$\text{Load regulation} = \frac{\text{No-load voltage} - \text{Full-load voltage}}{\text{Full-load voltage}} \times 100\%$$

$$= \frac{5 - 4.8}{4.8} \times 100\%$$

$$\text{Regulation} = 4.2\%$$

$$r_{out} \approx \frac{\Delta V_0}{\Delta I_L} = 0.33 \text{ ohms}$$

values of a.c. mains input, and load current.

8. **Stability** The change in output voltage with time, assuming the unit has reached thermal equilibrium and that the a.c. input voltage, the load current, and the ambient temperature are all held constant.

9. **Efficiency** The ratio of output power to input power expressed as a percentage. For example, suppose a power supply of 24 V when loaded to 1.2 A requires an

input current of 400 mA from the 120 V mains. Then

$$\text{Efficiency} = \left[\frac{V_o I_L}{V_{ac} I_{ac}}\right] \times 100\%$$

$$= \frac{24 \times 1.2}{120 \times 0.4} \times 100 = 60\%$$

10. **Current limiting** A method used to protect power supply components and the circuits supplied by the power unit from damage caused by an overload current. The maximum steady state output current is limited to some safe value (see Figure 6.1).

11. **Foldback current limiting** An improvement over simple current limiting. If a preset trip value of load current is exceeded the power supply switches to limit the current to a much lower value (Figure 6.2).

Using these parameters a typical specification for a relatively modest power unit might be, for example:

Input 110 V or 240 V a.c. at 50 Hz or 60 Hz.

Output voltage +24 V.

Output current 1.2 A max.

Temperature range −5° C to 45° C.

Temperature coefficient 0.01%/° C.

Line regulation 10% mains change results in 0.1% output change.

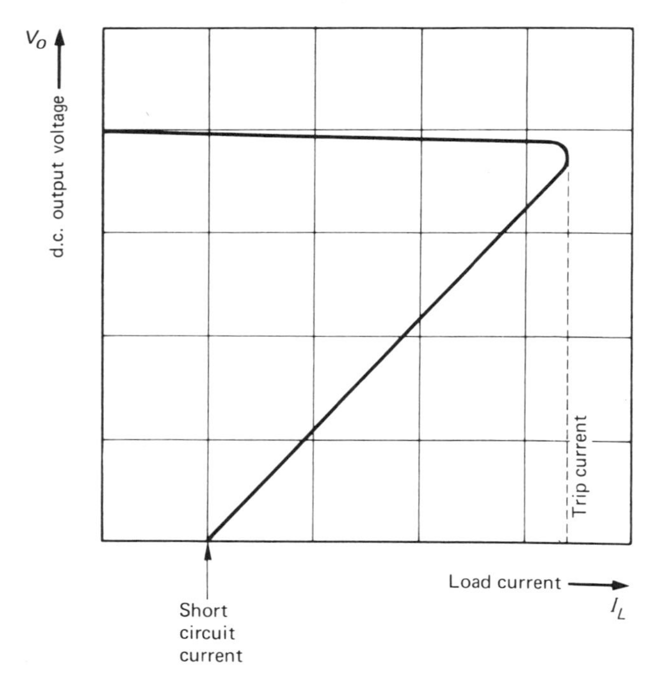

FIGURE 6.2 Foldback current limiting characteristic.

Load regulation 0.2% zero to full load.

From these, the worst-case change in output voltage can be calculated. The percentage change will be the sum of all the changes caused by line regulation, load regulation, and temperature (50°C change). That is,

Worst case change in d.c. output
= 0.1% + 0.2% + 0.5%
= 0.8% i.e., 192 mV

Measuring small changes in power supply output voltages with any reasonable accuracy requires sensitive measuring instruments. A digital voltmeter is the instrument of choice (see Figure 1.7, p. 20, for a typical test setup).

Before studying actual circuits some other points on power supplies are worth consideration. For troubleshooting purposes it is necessary to know how the POWER is distributed throughout a particular system or instrument. In some

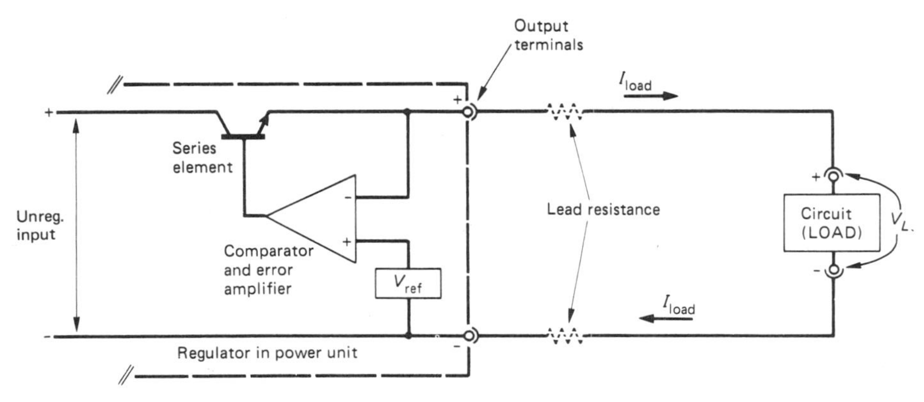

FIGURE 6.3a Load remote from power supply terminals; connecting leads cause V_L to be less than V_o and degrade the regulation.

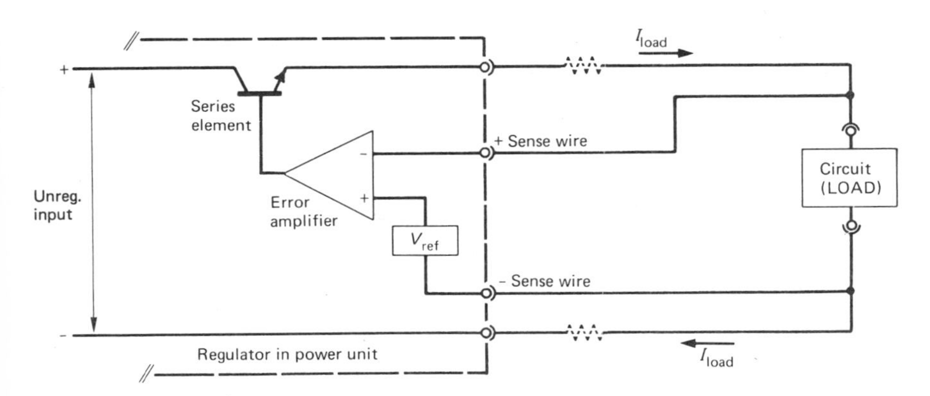

FIGURE 6.3b Remote sensing; includes connecting leads within feedback of regulator and therefore compensates for lead resistance.

situations a power unit may be required to supply its load via fairly long lengths of connecting lead as in Figure 6.3*a*. As the load current flows along the supply and return wires, a voltage drop will be set up causing the voltage across the load to be less than the voltage at the power supply terminals and consequently to have degraded regulation. One technique used to improve this is called REMOTE SENSING in which two extra leads are used to compensate for the effects of supply lead resistance (Figure 6.3*b*). In effect, the technique causes the supply lead resistance to be included within the feedback loop of the regulator. This gives optimum regulation at the load rather than at the power supply output terminals. The current carried by the two sense wires is very small so light gauge wire is used. However, because the two sense wires form the input of the comparator circuit, they have to be shielded to prevent pick up of interference. In practice a shielded pair is used and the shield is connected to chassis ground at the power supply end only. Note that the technique of remote sensing can only be used to give optimum regulation across one load. If the power supply is used to feed a large number of loads in parallel, then some other technique has to be used. Now that IC regulators are readily available, and comparatively cheap, the use of "point of load" regulators or remote regulators is increasing. A simple example is shown in Figure 6.4 where each load is provided with its own regulator circuit. The main power unit supplying the three separate regulators is often unregulated.

In situations where one regulated power unit has to supply several circuits, the arrangement has to be connected so that minimum disturbance is caused to the transmission of the signals from one circuit to another. In Figure 6.5*a*, for example, the parallel connections shown would not be used if circuits C or B are relatively heavy loads, since the currents from these circuits can set up interference signals at circuit A. Single point distribution, as shown in Figure 6.5*b*, if this is possible and economic, is obviously the best solution, since each circuit has its own supply wires. Figure 6.5*c* shows an improved arrangement for Figure 6.5*a*. In this the most sensitive circuit, A is supplied via its own set of connecting leads, which need not be heavy gauge. Circuits B and C are paralleled and positioned near the power supply. As before, one common ground point is provided. The arrangements in a complex system may be quite elaborate, and it is obviously important that the power distribution method used should

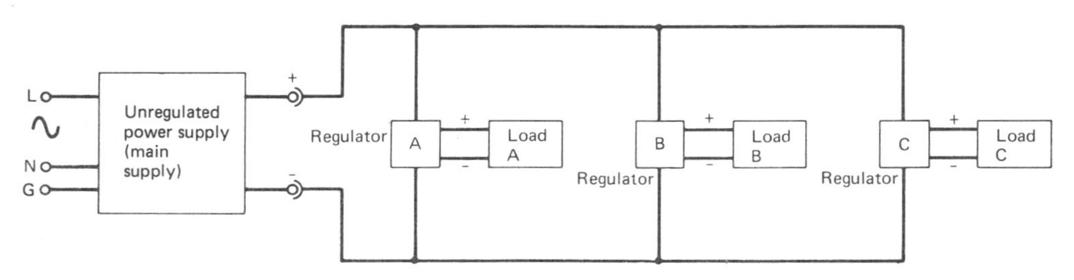

FIGURE 6.4 Use of "point of load" regulators.

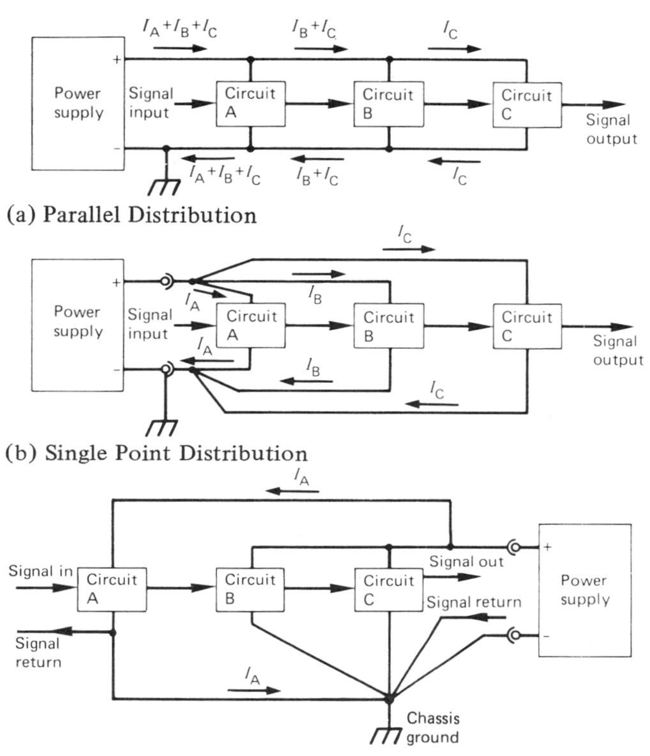

FIGURE 6.5 Power distribution methods using one regulated power supply.

not be altered or interfered with during service or test. The systems' performance will be degraded by repositioning supply leads or changing ground points.

6.2 RECTIFIER CIRCUITS AND UNREGULATED D.C. SUPPLIES

In practically all power supply circuits, some form of rectifier is required to convert the alternating voltage and current into unidirectional voltage and current. To provide an unregulated d.c. output some filter circuit then has to be used to smooth out the pulsating d.c. It is assumed that most readers are familiar with the standard circuits, but they are given in Figure 6.6 for information. In the diagrams the forward voltage drop across the diodes is assumed to be very small in comparison to the peak a.c. voltage of the transformer secondary. In practice about 0.7 V should be subtracted from the d.c. outputs of the half-wave and full-wave circuits, and about 1.4 V for the full-wave bridge circuit. The maximum repetitive peak reverse voltage V_{RRM}, formally called peak inverse voltage, is the voltage across one diode when it is reverse biased. Thus V_{RRM} is the sum of the d.c. output voltage and the peak value of the a.c. secondary voltage. Note that when the secondary voltage of a transformer is quoted, it is the rms value that is given. Naturally the diodes must be able to withstand the peak reverse voltage without breakdown.

The rectifier circuit almost universally used now is the full-wave bridge, since the silicon diodes can be provided relatively cheaply in one encapsulation, and also more importantly the transformer required has only half the number of secondary turns as the full-wave circuit. The half-wave rectifier is rarely used, since it is inefficient and requires relatively large smoothing or filter capacitors. The RIPPLE, that is the a.c. component remaining superimposed on the d.c. output following smoothing, has a frequency of twice the supply frequency for the full-wave circuits and is at the same frequency as the supply for the half-wave rectifier. The amplitude of this ripple depends on the value of the filter components with respect to the load.

Using a "rule of thumb," the minimum value of the required filter capacitor can be calculated from

$$C_{min} \approx \frac{1}{2\sqrt{2} \times f_r k_r R_L}$$

where R_L is the load
k_r is the ripple factor
f_r is the ripple frequency

$$k_r = \frac{\text{rms ripple voltage}}{\text{d.c. output voltage}}$$

As an example assume a 10 V d.c. supply at 500 mA is required from a bridge rectifier and that the rms ripple is to be about 500 mV.

In this instance, $k_r = 50 \times 10^{-3}$, $R_L = 20\ \Omega$, $f_r = 120$ Hz, therefore

$$C_{min} = \frac{1}{2\sqrt{2} \times f_r k_r R_L} =$$

$$\frac{1}{2\sqrt{2} \times 120 \times 50 \times 10^{-3} \times 20} \text{ farads}$$

$C_{min} \approx 2946\ \mu$F (3300 μF is the nearest preferred value).

This, of course, is only a rough approximation, since the ripple waveform is not sinusoidal but a complex shape. The ripple voltage will cause a ripple current to flow through the capacitor. A simple calculation will show that the reactance of the 3300 μF at 120 Hz is about 0.4 Ω. Thus the a.c. current flowing through the capacitor is about 1.25 A. A physically large electrolytic with a ripple current rating well in excess of this must be used to prevent the possibility of the capacitor overheating. The ripple current causes I^2R losses in the capacitor. The resistance in this case is the equivalent series resistance (ESR) of the capacitor.

A typical unregulated d.c. power supply is shown in Figure 6.7, and this will be used to discuss some possible fault conditions. Both the line and neutral supply leads contain fuses. Antisurge (slow blow) fuses are used because, with a capacitive input filter, a large current may be required at the instant of switch on. This is because the capacitor is uncharged and acts almost as a short circuit. The a.c. power on/off switch S_{1A} and S_{1B} is a double-pole single-throw type (DPST). The line and neutral wires are both switched for safety reasons,

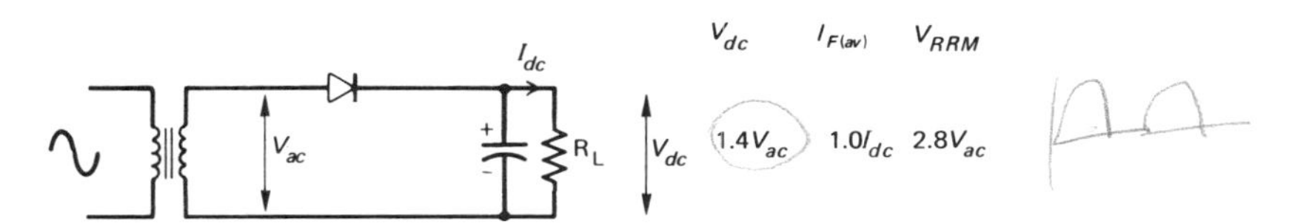

(1) Half-Wave–Capacitive Input Filter

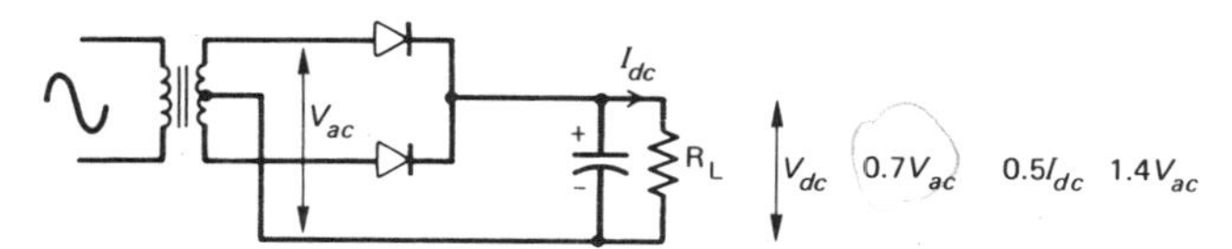

(2) Full-Wave–Capacitive Input Filter

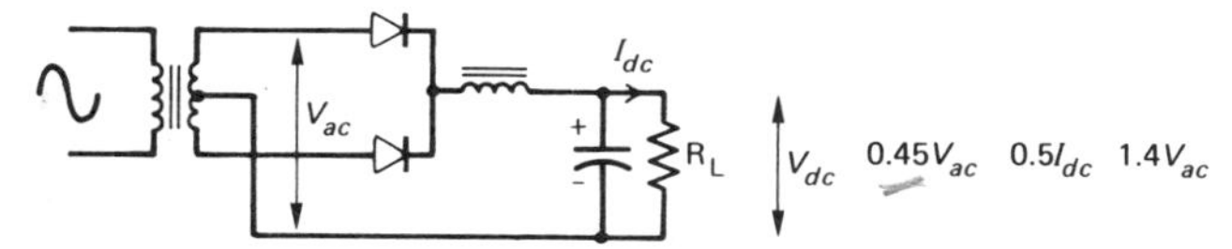

(3) Full-Wave–Choke Input Filter

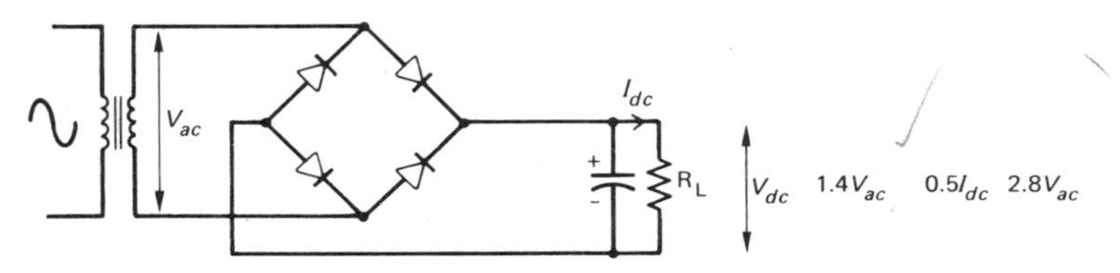

(4) Full-Wave Bridge–Capacitive Input Filter

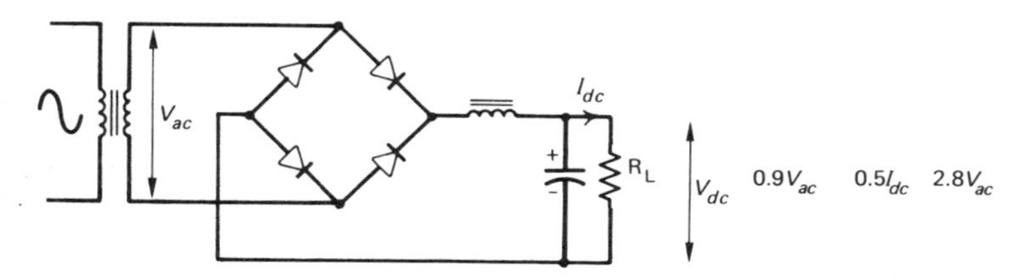

(5) Full-Wave Bridge–Choke Input Filter

FIGURE 6.6 Basic rectifier circuits

$I_{F(av)}$ = d.c. current through diode
V_{ac} = rms secondary voltage
V_{RRM} = maximum repetitive peak reverse voltage

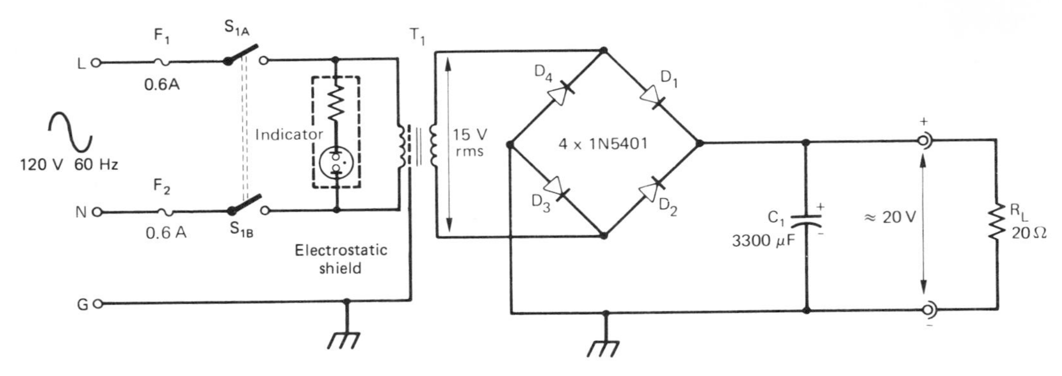

FIGURE 6.7 Unregulated power supply to give 20V at 2A.

so that, if an error is made in connecting the supply leads causing the line and neutral wires to be crossed, the on/off switch will still isolate the live a.c. power from the primary.

The transformer has an electrostatic screen wound between the primary and secondary coils, which consists of a layer of copper foil extending over the primary winding. The ends of this screen are insulated from each other to prevent it acting as a single shorted secondary. The screen is connected to the a.c. power line ground and therefore reduces a.c. power line interference as well as providing additional isolation and protection.

A variety of faults could occur. With either the *primary or secondary open circuit* there would obviously be no d.c. output although the neon indicator would be on. Suppose, however, the unit failed with zero d.c. output and the neon indicator off. The next step would be to check the fuses for continuity with an ohmmeter. If F_1 is blown and F_2 is still intact, this indicates that in all probability the *primary winding is shorting to the screen*. In this case F_2 would not blow because the neutral wire of the supply is usually connected back to the a.c. power line ground at the substation. A fault such as C_1 *short* or *any diode in the bridge short* would cause excessive primary current to flow which should blow both fuses.

Shorted turns on either the primary or secondary winding of the transformer is another possible fault. A symptom for both these faults is that the transformer will be overheating. Shorted turns on the primary will cause the secondary a.c. voltage to rise, thereby increasing the d.c. output voltage and current. Shorted turns on the secondary will reduce the output voltage.

Should any one *diode become open circuit*, then the rectifier acts as a half-wave type giving a lower d.c. output voltage and increased ripple voltage at 60 Hz instead of 120 Hz.

6.3 LINEAR REGULATORS

To provide improved performance the unregulated power supply must be followed by some form of regulator. The linear series regulator is a circuit generally used for medium power requirements, and even quite simple circuits are capable of excellent performance. Basically it is a high gain control circuit that continuously monitors the d.c. output voltage and automatically corrects the output to hold it constant irrespective of changes in load current and unregulated input voltage. As shown in Figure 6.8*a* the output is compared with a stable reference voltage and any difference, or error, between the output and the reference is amplified and fed to the base of the series control element. The series element is a power transistor connected as an emitter follower providing a low output impedance to drive the load. The performance of the circuit depends on the stability of the voltage reference source and the gain of the error amplifier.

A typical example of a series regulator using discrete components is shown in Figure 6.8*b*. This unit should provide 10 V at 1 A from a 15 V unregulated supply. Here the control element is formed by the Darlington connection of Q_2 and Q_3. The full load current of 1 A flows through Q_3

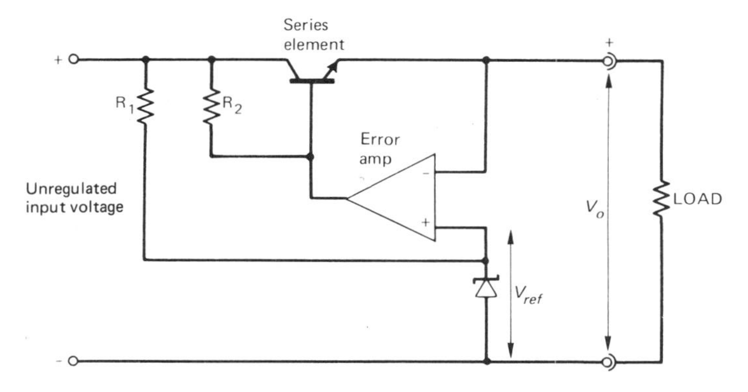

(a) Block Form

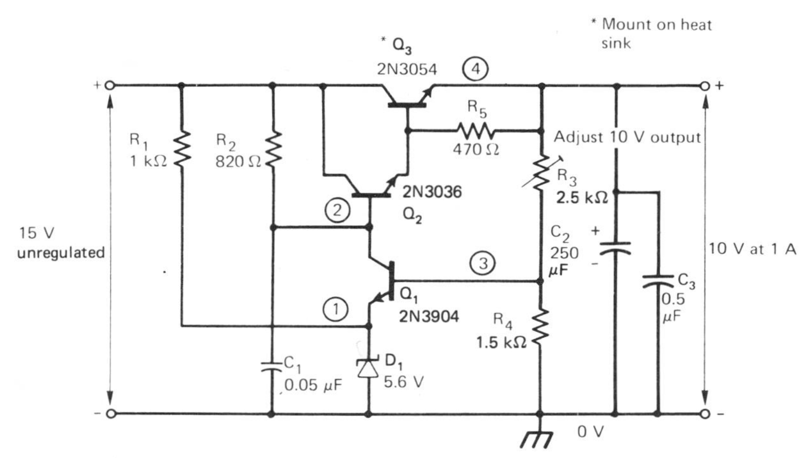

(b) Typical Circuit Using Discrete Components

FIGURE 6.8 Basic diagram of series regulator.

and, since its value of current gain h_{FE} may be relatively low, the base current required by Q_3 may be as high as 40 mA. This current is supplied by Q_2, which only requires a base current of between 1 and 2 mA. The error amplifier is Q_1, the inverting input being the base and the noninverting input the emitter. The latter is held constant by the 5.6 V zener. Under normal conditions the base voltage of Q_1 will be about 0.6 V higher than its emitter at 6.2 V. Therefore, if the voltage across R_4 is 6.2 V then, if R_3 is adjusted to a value of 1 kΩ, the total volt drop across R_3 and R_4 should be 10 V. If the output voltage drops in value, a portion of this fall appears on Q_1 base. Since Q_1 emitter is held constant by the zener reference voltage, the base/emitter voltage of Q_1 will decrease in value. Q_1 collector voltage rises increasing the forward bias to Q_2 and Q_3, which thus tends to correct (i.e., increase) the output voltage. This process is, of course, automatic.

The basic series regulator as shown suffers from a few drawbacks. The series power transistor is unprotected from an overload current. If an accidental short circuit occurred across the power supplies' output terminals, a very large current would flow through Q_3 and in a few milliseconds Q_3 would burn out. It is unlikely that a normal fuse in the primary or secondary circuits of the unregulated supply would protect Q_3. A current-limiting circuit is an essential addition. Another point is that the output is not protected from an overvoltage. For example, if Q_3 failed collector/emitter short, the output would then be 15 V. If the absolute maximum voltage allowed across the load is say 12 V, then the overvoltage caused by the Q_3 short could damage the circuit being supplied. In many circuits an overvoltage trip circuit is included. Both these protection circuits are explained later.

The efficiency of the series regulator is not high. The load current flowing through Q_3 causes it to heat up, and usually the series transistor has to be mounted on a fairly large heat sink so that a safe value of junction temperature for Q_3 is not exceeded. In the example the total power dissipated by Q_3 is

$$P_T = P_{CE} + P_{BE}$$
$$= V_{CE}I_C + V_{BE}I_B$$

Assuming $h_{FE} = 25$, then

$$P_T = 5 \times 1 + 0.7 \times 40 \times 10^{-3} = 5.028 \text{ W}$$

The voltages at the various test points measured with a 20 kΩ/V multimeter will be as follows. The unit is fully loaded to 1 A.

TP	1	2	3	4
MR	5.6 V	11.3 V	6.2 V	10 V

Since the circuit is completely d.c. coupled, a component fault will probably affect every test point voltage. Most faults produce a unique set of voltage readings as the symptoms. For example, suppose D_1 becomes short circuit, the voltage readings would then be

TP	1	2	3	4	
MR	0 V	2.5 V	0.7 V	1.1 V	[D_1 short circuit]

No other component fault would give this set of voltages.

Let's assume that R_3 becomes open circuit. This means that there is now no base current supply to Q_1. Test point 2 must rise, increasing the forward bias to Q_2 and Q_3, causing the output voltage to increase. An additional symptom would be that the load regulation would be poor and output ripple would be higher. The test voltages are

TP	1	2	3	4	
MR	5.6 V	14.4 V	0 V	13.1 V	[R_3 open circuit]

Note that the actual output voltage will depend on the current gains of Q_2 and Q_3.

A set of test readings as follows:

TP	1	2	3	4
MR	5.6 V	0 V	0 V	0 V

would indicate that no forward bias is available for Q_2 and Q_3. The fault is R_2 open circuit *or* possibly C_1 short.

The difference in test readings for say Q_2 base/emitter open circuit from those caused by the previous fault are that nearly 15 V would be indicated at TP 2:

TP	1	2	3	4	
MR	5.6 V	15 V	0 V	0 V	[Q_2 base/emitter open circuit]

Work out the symptoms for various other faults such as Q_3 collector/base short, Q_1 base/emitter short, or R_4 open circuit.

Protection of Series Regulators

The basic circuits for providing protection to the series regulator are shown in Figure 6.9. Simple circuits using few components are preferred, since they do not then degrade the overall reliability of the supply. In Figure 6.9*a*, R_m is a load current monitoring resistor. If, because of an overload, the voltage across R_m increases to 600 mV, Q_2 conducts and diverts base current away from Q_1, the series pass transistor. The characteristic will be as drawn in Figure 6.1. If, for example, R_m is 1 Ω, when the load current is about 600 mA the voltage across R_m is sufficient to turn on Q_2. Making R_m a 2 Ω resistor would limit the maximum output current to 300 mA and so on. The resistor is inside the feedback loop and, therefore, does not degrade the load regulation.

FOLDBACK CURRENT LIMITING is a useful feature, since the power supply will switch to give almost zero output voltage if the preset value of load current is exceeded. The current monitoring resistor R_m is placed in the return line, and the voltage developed across it is used to switch on thyristor Q_2, i.e., a silicon controlled rectifier (SCR). As soon as the overload trip current is exceeded, the thyristor is triggered on and the voltage across it falls to approximately 0.9 V. This will be insufficient to forward bias the diode D and Q_1 so the output voltage will be zero. Once triggered on, an SCR remains conducting, so the fault has to be removed and the power supply switched off from the a.c. line (or unregulated d.c.) before the SCR will turn off. An LED is sometimes used to indicate that an overcurrent fault has occurred. Foldback current limiting is very effective in preventing damage to the series pass transistor. In both cases R_m can be made adjustable or an amplifier can be added to make the trip point accurate and sensitive.

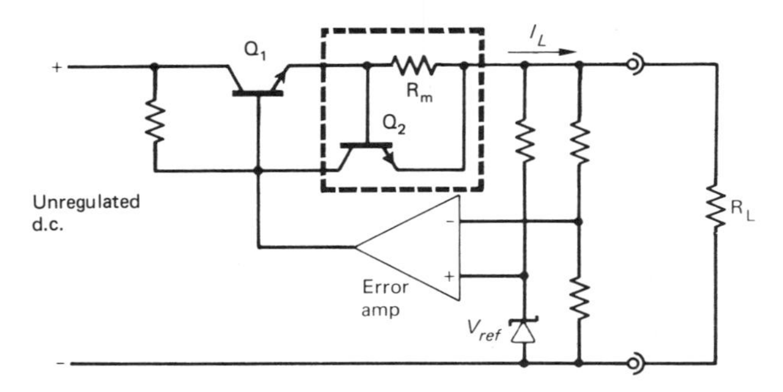

(a) Simple Current Limit

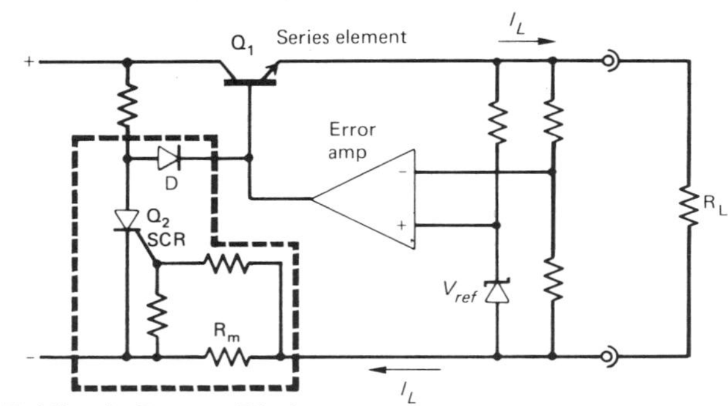

(b) Foldback Current Limit

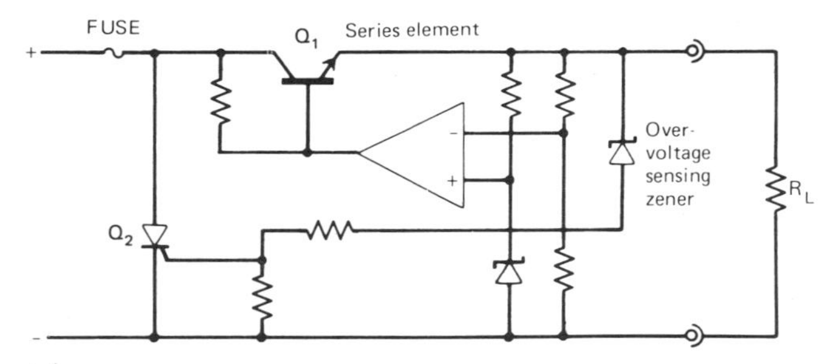

(c) Overvoltage Protection

FIGURE 6.9 Protection circuits for series regulator.

OVERVOLTAGE protection is vital when a series regulator is supplying a load made up of sensitive digital ICs, such as TTL. With TTL, if the power line exceeds 7 V, damage may occur to the ICs. In Figure 6.9*c* a zener diode is used to sense the voltage across the power supply's output terminals. If the d.c. output voltage rises so that this zener conducts, then thyristor Q_2 is turned on and the voltage at Q_1 collector falls rapidly to zero causing the fuse to blow. Alternatively the anode of Q_2 may be connected to Q_1's base as in the previous circuit. A circuit such as this is called a CROWBAR.

Many modern power supplies now use MONOLITHIC IC REGULATORS. These have simplified the design, reduced component count, and made troubleshooting slightly easier. An understanding of the circuit operation, even though it is mostly inside the IC, is still an advantage when troubleshooting. There are naturally several different ICs available, but it would not be useful to detail them all here. Instead we shall look at one of the most popular, relatively inexpensive and versatile IC regulators, the μA723. This IC is available in a 14-pin DIP encapsulation or in a metal can version with ten leads. The pin configuration for the 14 pin DIP together with the equivalent circuit are shown in Figure 6.10. The internal circuitry contains a reference supply, error amplifier, series pass transistor, and a current limiting transistor. The connections to the various sections are brought out to the IC pins allowing the user flexibility in designing a regulator to suit his or her requirements. The stable, temperature compensated, voltage reference source gives a voltage at pin 6 of 7.15 V ± 0.2 V, and this can be used either directly connected to the noninverting input or via a potential divider. Two basic circuits are shown in Figure 6.11. The first gives output voltages from 2 to 7 V and the second gives output voltages from 7.2 V to 37 V. In the first circuit a current limiting resistor R_{SC}, of 10 Ω, is shown. This will limit the maximum output current to 65 mA.

The maximum current that can be taken via the series pass transistor is 150 mA, but the safe maximum current in a particular application depends on the value

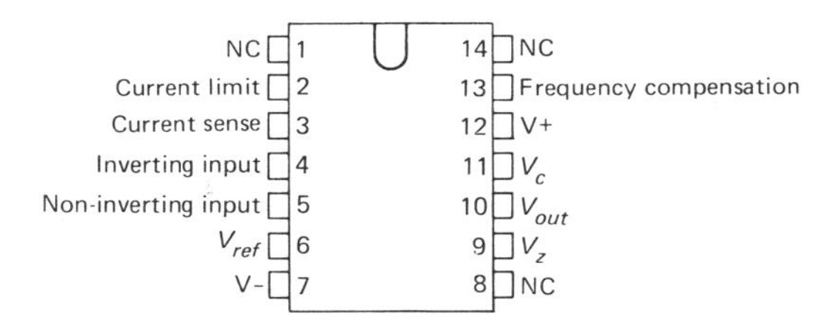

(a) Pin Configuration

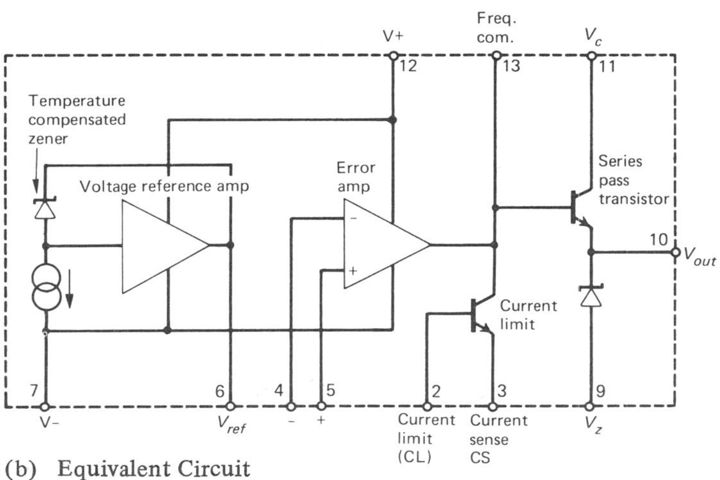

(b) Equivalent Circuit

FIGURE 6.10 μA723 IC regulator.

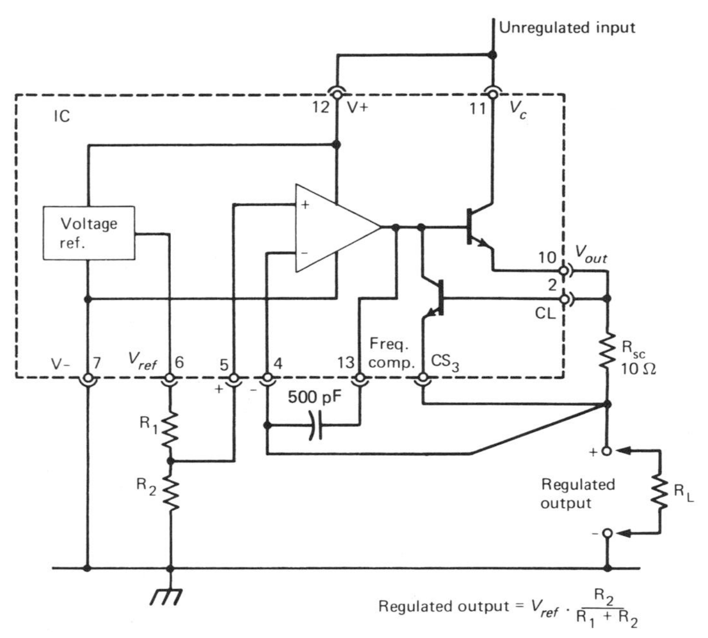

(a) 2 V to 7 V regulator

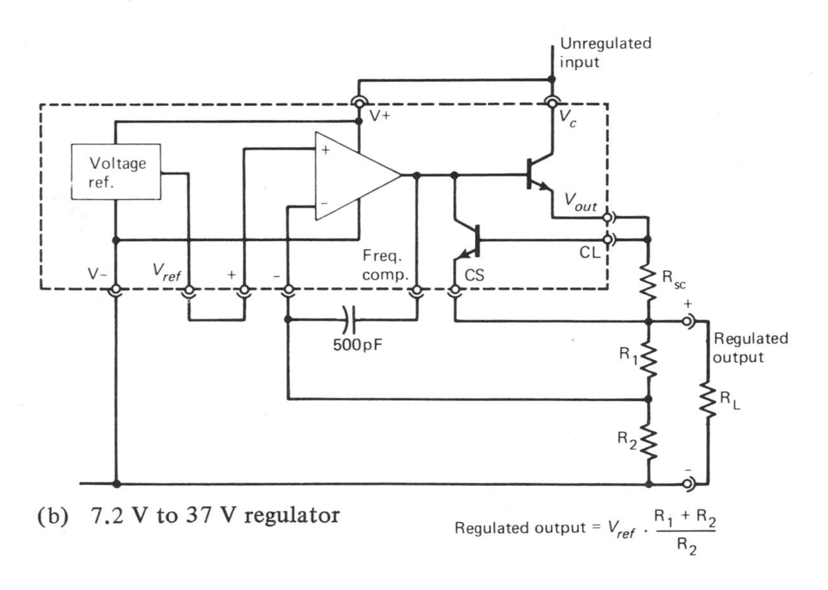

(b) 7.2 V to 37 V regulator

Regulated output = $V_{ref} \cdot \frac{R_1 + R_2}{R_2}$

FIGURE 6.11 Basic uses of μA723.

of the unregulated input. At an ambient temperature of 25°C the maximum power dissipation of the IC is 660 mW. Therefore, the safe current limit when the output is short circuited is given by

$$\text{Maximum current limit} = \frac{P_{max}}{V_s}$$

Thus if V_S, the unregulated input, is 20 V, the maximum safe current under short circuit conditions will be 33 mA. R_{SC} should be made a 22 Ω, and then the current limit will be about 30 mA.

The output current of the regulator can be increased by using an external power transistor. The internal pass transistor then supplies the base current to the external transistor. An example later shows this.

Briefly the specification for the μA 723 is as follows:

Parameter	Test Conditions	Typical Value
Line regulation	V_{in} = 12 V to 15 V	0.01%
	V_{in} = 12 V to 40 V	0.02%
Load regulation	I_L = 1 mA to 50 mA	0.03%
Ripple rejection		86 dB
Reference voltage		7.15 ± 0.2 V
Long-term stability		0.1% per 1000 hrs
Input voltage range		9.5 V to 40 V
Output voltage range		2 V to 37 V
Average temperature coefficient	V_{in} = 12 V to 15 V I_L = 1 mA to 50 mA	0.002% per °C

Two other points concerning the μA723 are:

1. The input voltage must always be at least 3 V greater than the output voltage. This, of course, means that some power is dissipated as heat in the series pass transistor.
2. A low value capacitor has to be connected from the frequency compensation pin to the inverting input. This ensures that the circuit does not oscillate at high frequencies.

6.4 SWITCHING POWER SUPPLIES

Switching power systems and switching mode regulators are used for their high efficiency. Intensive development has taken place over the last few years to produce power supplies of maximum efficiency and small size and weight. Many of these circuits are developments from the basic inverter shown in Figure 6.12. An inverter is a device that converts d.c. to a.c. In the circuit this is achieved by the switches S_1 and S_2, which alternately reverse the d.c. connection to the transformer primary. The transformer has to be center-tapped. On one half cycle current flows through the top half of the primary winding and on the other half cycle, when the switches change, the current flows in the opposite direction through the lower half of the primary. The result is that a.c. will be produced at the secondary. The switches

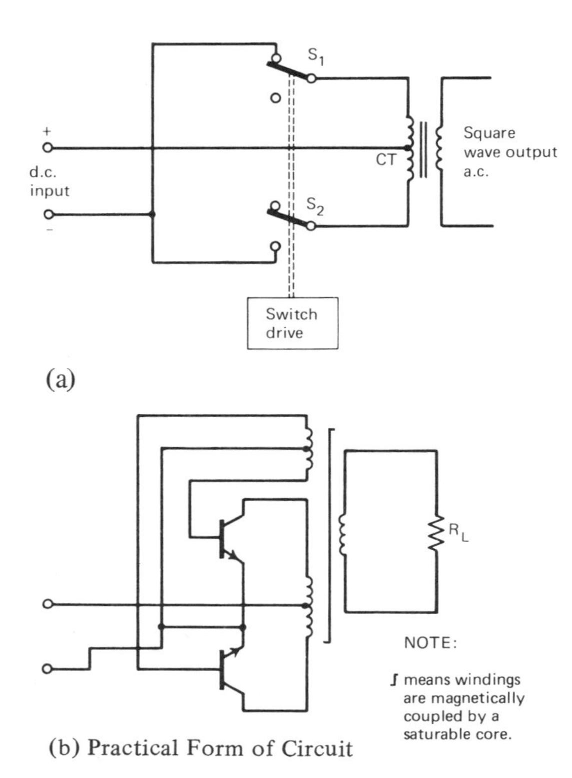

FIGURE 6.12 Basic inverter circuit.

are usually special-purpose transistors or thyristors driven by some form of square wave or pulse oscillator. Another method is to have feedback windings on the primary so that the inverter transistors form a self-oscillating circuit. The frequency of the switching signal, especially if the inverter is used as part of a regulator, is typically in the range 5 kHz to 25 kHz. A high frequency is used because the transformer and any subsequent filter components will be relatively small. Manufacturers and users also prefer a frequency that is just above the audio range (15 kHz) for obvious reasons. The upper limit of operating frequency is set by the core losses in the transformer and the switching times of the transistors. At too high a frequency the efficiency starts to fall off.

By following an inverter with a rectifier circuit and filter, a converter, that is d.c. to d.c., is created. If a feedback loop is then added that senses the d.c. output, compares it with a reference level, and feeds a signal that can modify the switching time of the transistors, a type of switching regulator results. This is shown in Figure 6.13. This circuit uses a principle called PRIMARY SWITCHING. The a.c. power line is rectified and smoothed giving a d.c. level of about 170 V. This d.c. voltage is switched, at a frequency above audio, by high voltage transistors to provide an alternating waveform to the transformer primary. The secondary a.c. is rectified and smoothed to give a d.c. output voltage across the load. This d.c. output is regulated by comparing it with a zener reference supply. The difference or error signal is used to alter the duty cycle of the switching transistors. If the d.c. output should fall when the load current increases, then the error signal causes the pulse width modulator to switch the transistors on for a longer time than they are off during each cycle of the 20 kHz oscillator. More power is provided via the transformer to the load, and the output voltage rises to very nearly its previous value. The opposite occurs if the load current is reduced. Primary switching is the method used in most SMPSs of high power, since the transformer, operating at 20 kHz is much smaller than a 60 Hz type. However, it is possible to replace a conventional linear regulator with a switching type, using secondary switching (Figure 6.14). When the series transistor is switched on, current is allowed to flow to the *LC* filter. When the transistor switches off, the

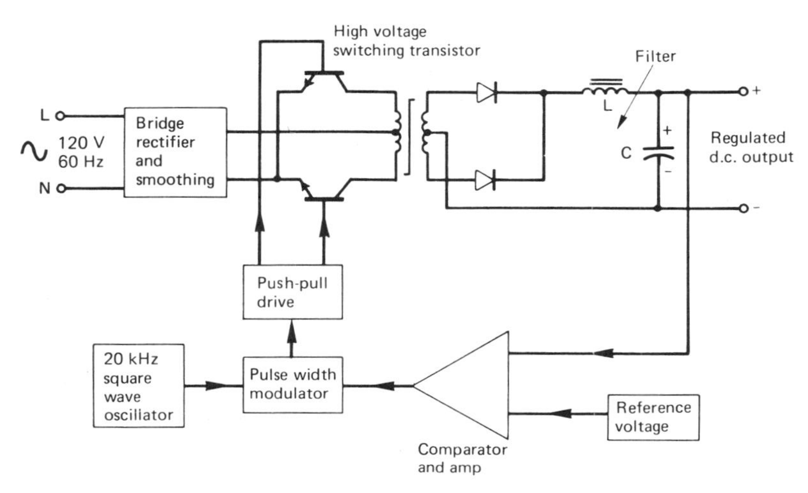

FIGURE 6.13 Switching mode regulator using primary switching.

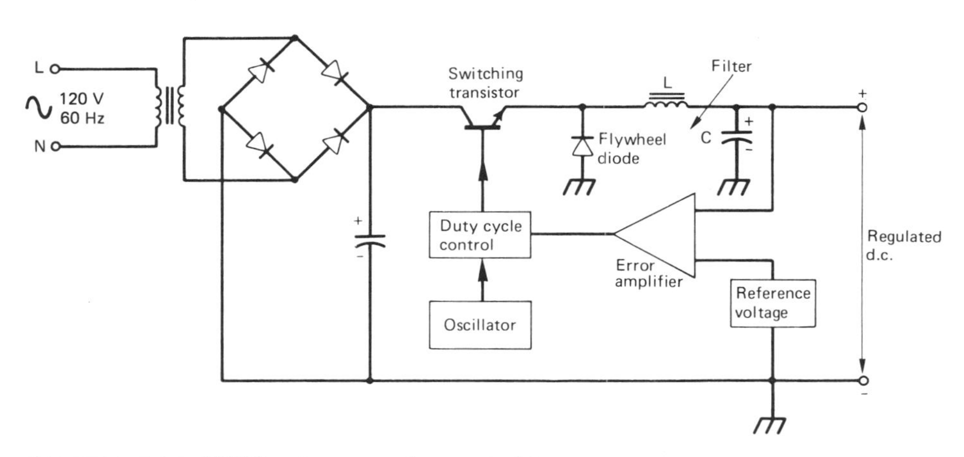

FIGURE 6.14 SMPS using secondary switching.

inductor keeps the current flowing with the "flywheel diode" acting as a return path. Various methods can be used to regulate the d.c. output. The duty cycle of the switching waveform, or the frequency of the oscillator can be varied, or a mixture of both methods can be used. Since the transistor is being operated as a switch, it is either off or on, and in both these cases the power dissipated by the transistor will be low.

Although SMPSs are more efficient and take up less space than linear regulators, the SMPS cannot match the regulation performance of the linear circuit. SMPSs find their main use in applications that require large currents at low and medium voltages.

6.5 HIGH VOLTAGE D.C. POWER SUPPLIES

There are many applications that require a well regulated high voltage (HV) or extra high voltage (EHV) supply. Among the most common are:

(a) *Cathode ray tubes* used in oscilloscopes, radar displays and video display units (VDUs).
Typical voltage 2 kV up to 20 kV
Typical current a few milliamps
Typical regulation 1%

(b) *Photomultiplier tubes* used in scintillation counters, flying spot scanners, and low level photometry.
Typical voltage 1 kV to 3 kV
Typical current 0.5 mA to 5 mA
Typical regulation 0.1% or better

The voltage supply to the tube must be held constant since the electron gain of a photomultiplier tube is dependent on the applied voltage. A change of only 1% in voltage resulting in approximately 10% change in tube gain.

(c) *Lamp supplies* for photocopiers
Typical voltage 5 kV to 10 kV
Typical current 5 mA
Typical regulation a few percent

In addition to those listed above, HV supplies are used for image intensifier tubes, camera tubes, X-ray machines, insulation testers, electron microscopes, and other specialized tubes and instruments. A service technician therefore needs to be familiar with some of the more commonly used HV circuits.

To produce a d.c. output above 1 kV it is possible to use an a.c. power line transformer with a secondary wound with the necessary number of turns, and this is often the solution used in some oscilloscopes and X-ray systems where a simple fixed voltage unit at low cost is required. The drawbacks to this method are that the unit is rather bulky, the transformer and smoothing (filter) capacitors required are very large, and the HV generated at the transformer secondary is highly dangerous, since it has a relatively low impedance. Smoothing capacitors in such units, because of their relatively high value, may also store a lethal charge for some time after the a.c. power line is switched off. It is always wise, after disconnection from the a.c. power line, to discharge any capacitors before carrying out any investigations. Discharging should be done by connecting a 1 MΩ high-voltage resistor across the capacitor for a few seconds. A special probe can be made for this purpose. This precaution applies to all HV units and will prevent the possibility of receiving an electric shock when the power supply is off.

A preferred method to produce high voltage d.c. is to use a converter or switched system. A low-voltage (often unregulated) d.c. is switched across the primary winding of a ferrite-core transformer. Fast-switching transistors driven from an oscillator circuit are used with frequencies typically in the range 10 kHz to 30 kHz. The transformer secondary is wound with many turns so that a high-voltage a.c. appears at the secondary. Voltage multiplying rectifiers then convert this secondary a.c. to an even higher value of d.c. output. The d.c. is then filtered and regulated. Regulation can be effected by taking a portion of the d.c. output via a potential divider and comparing this portion with a low-voltage reference level. Any error is amplified and is used to either adjust the duty cycle of the primary switching waveform or the d.c. voltage applied to the transformer primary. Since a high frequency is used, the transformer and smoothing components are relatively small.

VOLTAGE MULTIPLIER circuits, of which there are several types, are very useful for producing high values of d.c. voltage at low current. Some typical examples are shown in Figure 6.15. The simple doubler circuit (Figure 6.15*a*) is basically two half-wave rectifiers with the outputs summed by the two capacitors. D_1 conducts on the positive half cycle to charge C_1 to V_P (the peak value of the a.c. secondary), and then on the negative half cycle D_2 conducts to charge C_2 also to V_P. As with other rectifiers of this type, the regulation is poor so that only a relatively light load of a few milliamps can be supplied. Another doubler commonly used is also shown. In this circuit C_1 must be a larger capacitance value than C_2. On negative half cycles, D_1 conducts and C_1 charges to V_P. On the next positive half cycle, the left-hand plate of C_1 goes positive and D_2 conducts carrying C_2 to almost 2 V_P. Typically C_1 is twice the value of C_2 to allow correct charge-sharing between the two capacitors. A voltage tripler can be constructed by adding an extra half-wave rectifier as shown (Figure 6.15*b*). Alternatively a tripler and quadrupler can be made by using a few sections of the well known Crockroft-Walton "ladder." Because the output impedance of multipliers increases rapidly with the number of multiplier stages used, the rectifier in HV circuits is usually limited to a doubler circuit. The output impedance of a voltage multiplier increases approximately by the cube of the number of stages. A quadrupler has an output impedance about eight times greater than a doubler. High values of output impedance limit the available output current and degrade the load regulation.

HV power units are specified in the same way as other d.c. power supplies. Parameters such as load and line regulation, output impedance, ripple, temperature coefficient and stability being the most important. The measurement of high values of d.c. voltage does, however, require careful attention. The meter and measuring leads used must have high values of insulation resistance and breakdown voltage ratings well in excess of the voltage being measured. Polytetrafluoroethylene or PTFE covered wires are often used for measuring leads. It is also important that the loading effect of the meter should be slight, for in some cases the HV supply may be designed to provide only a few hundred microamps to the load. One useful meter that has a very high impedance and therefore takes negligible current from the HV being measured is the ELECTROSTATIC METER. These simple, sturdy meters are constructed of metal vanes. One set is fixed and the other set is pivoted and free to move against a restraining spring. When a high voltage is applied across the vanes, the electrostatic field forces the moving set to rotate. These moving vanes are attached to the pointer, and the scale can be calibrated in kV. Unfortunately, the scale is non-linear, being cramped at the lower voltages, and the meter accuracy is not high (typically ±5%). However, electrostatic meters can be very useful for checking high volt-

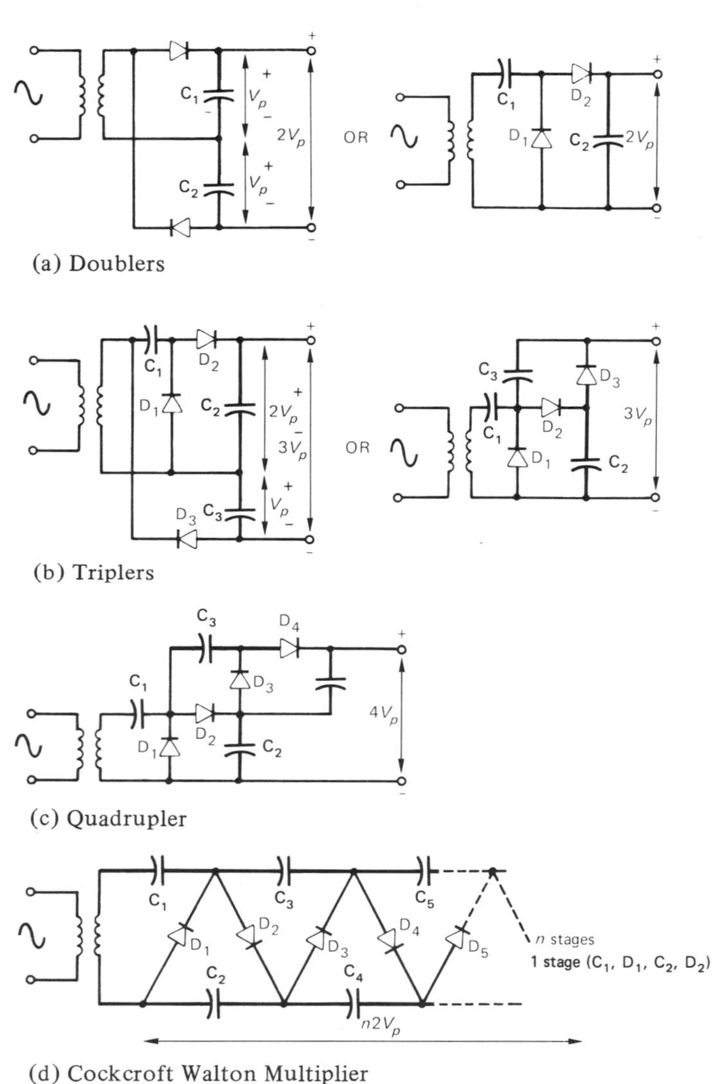

FIGURE 6.15 Voltage multipliers.

ages up to 50 kV. Some moving coil multimeters have ranges that go up to 3 kV, but for full-scale deflection the current required may be 50 μA. This loading may be too high for some HV supplies. Alternatively the meter may be used on its lowest current range (50 μA) with a multiplier resistor chain of several megohms.

Measurement of regulation requires that small changes in the HV should be detected. A potential divider using precision resistors can be used as shown in Figure 6.16 to bring the HV within the range of a digital voltmeter. R_1 must be

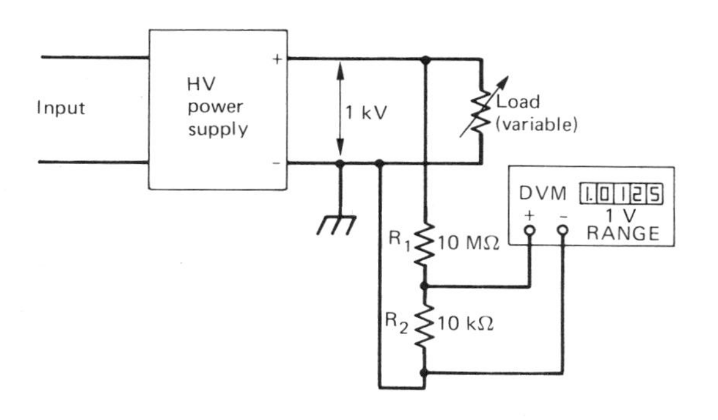

FIGURE 6.16 Measurement of HV regulation.

either made up of a series of medium value resistors (ten 1 MΩ, for example), or a special type of high-voltage resistor can be used. In both cases care must be taken to ensure that the body of R_1 is well insulated from the chassis or any low potential point.

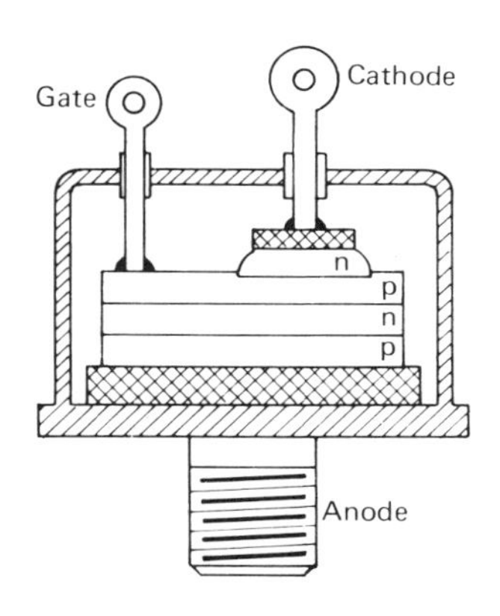

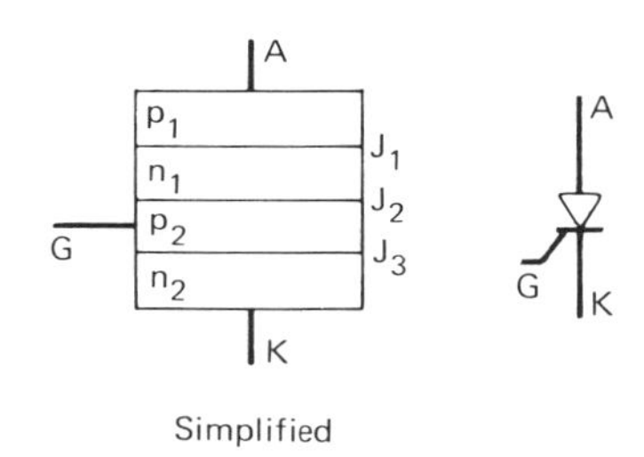

FIGURE 6.17 Typical structure of medium power thyristor.

6.6 THYRISTORS AND TRIACS

Thyristors (silicon controlled rectifiers) and triacs are semiconductor devices that are now being used extensively in power control circuits. They are particularly suited for a.c. power control applications, such as lamp dimmers, motor speed control, temperature control, and inverters; and also are commonly used as overvoltage protection elements in d.c. power supplies.

Thyristors

The construction of a typical medium power thyristor is shown in Figure 6.17. It is basically a four-layer pnpn structure. The gate connection to the p_2 region enables it to be switched from a nonconducting (forward blocking) state into a low-resistance forward conductivity state. Once triggered into forward conduction, the thyristor remains on unless the current flowing through it is reduced below the holding current value or it is reverse-biased. The characteristics (Figure 6.18) show that in the reverse direction the thyristor is off, and only a small leakage current flows unless the reverse breakdown voltage is exceeded. This is because junctions J_1 and J_3 have increased depletion regions. When the anode is made positive with respect to the cathode, and with no gate signal applied, junctions J_1 and J_3 will be forward-biased but J_2 will be reversed, and its depletion region will be increased. Therefore, only a small forward leakage current flows, and the thyristor is said to be forward blocking or off. It will remain in this

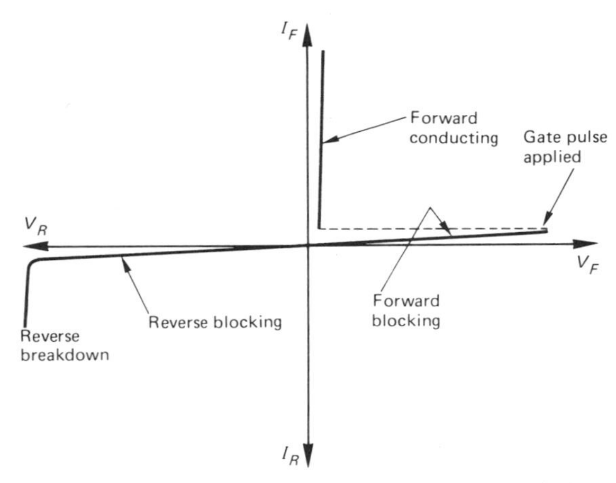

FIGURE 6.18 Typical thyristor characteristics.

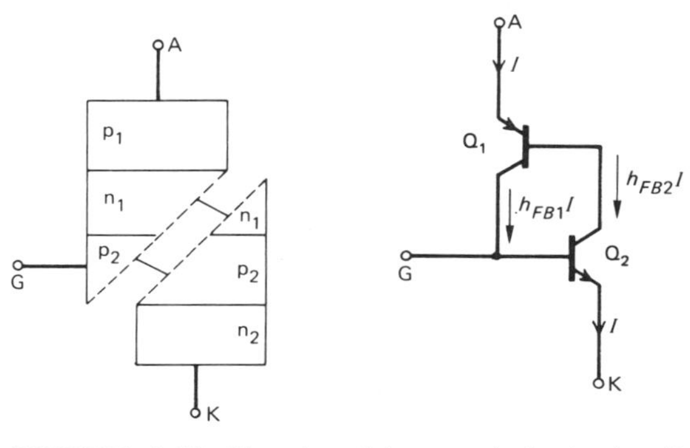

FIGURE 6.19 Two-transistor equivalent circuit for thyristor.

high-resistance off state unless the voltage between anode and cathode is made to exceed the forward breakover voltage, or until a positive pulse is applied at the gate. Triggering by exceeding the forward voltage rating is undesirable, and applying a gate signal is the normal preferred method to switch the thyristor on.

The reason why the thyristor switches on and remains on even when the gate signal is removed can be more easily understood by using the two-transistor equivalent circuit of Figure 6.19. Here the top $p_1n_1p_2$ region is considered as a pnp transistor and the bottom $n_1p_2n_2$ region as an npn transistor. Imagine that no gate signal is applied and that a voltage (less than breakdown) is applied, making the anode positive with respect to the cathode. The small current that flows in at the anode connection must flow out from the cathode. For Q_2 the emitter current I is the sum of its base currents and its collector current plus any leakage.

$$I = h_{FB1}I + h_{FB2}I + I_{CBO}$$

where $h_{FB} = \dfrac{I_C}{I_E}$ and I_{CBO} is leakage current.

$$I = I(h_{FB1} + h_{FB2}) + I_{CBO}$$

Therefore

$$I = \frac{I_{CBO}}{1 - (h_{FB1} + h_{FB2})}$$

Now the common base current gain of a transistor, h_{FB}, is highly dependent on the value of collector current. Although the current flowing through the transistors remains low, the sum of the two current gains $(h_{FB1} + h_{FB2})$ remains less than unity. For example, if $(h_{FB1} + h_{FB2}) = 0.9$, then $I = 10\ I_{CBO}$. This will be an insignificant value of current, and a high resistance will exist between anode and cathode. If, however, Q_2 is forward-biased by a gate signal, then the increase in current through Q_2 raises the current gain and rapidly $(h_{FB1} + h_{FB2})$ will approach unity. Then the two transistors switch into a conducting state. They are connected in a positive feedback configuration, the collector currents of each transistor supplying the base current to the other, and therefore switch-on is very rapid. The gate signal can be removed and the two transistors will remain conducting, since the current flowing through them is high enough to ensure that the sum of h_{FB1} and h_{FB2} exceeds unity.

Note that, when switched on, the thyristor will pass large values of current limited only by the external load, with only a small voltage dropped across the anode to the cathode.

Thyristors can be used in d.c. circuits such as in lamp flashers and high-speed trip circuits. A simple example is shown in Figure 6.20a. When S_1 is momentarily pressed Q_1 is triggered into conduction, its anode voltage falling to about 0.8 V, and nearly the full d.c. supply voltage will appear across R_L. The commutating capacitor will charge, with its right hand (r.h.) plate being at $+V_s$, via R_2 and Q_1. When a trip signal is received at Q_2's gate, Q_2 is triggered on, its anode falling rapidly from $+V_s$ to 0.8 V. This negative step is transmitted directly through C, and the anode of Q_1 is taken negative. This switches off Q_1. The voltage at Q_1's anode rapidly returns to $+V_s$ as C is charged in the opposite direction via R_L and Q_2. The commutating capacitor must therefore be a nonpolarized type.

In a.c. circuits the thyristor turns off automatically every time the a.c. supply reverses, and hence relatively simple a.c. power control circuits can be constructed as in Figure 6.20*b*. Here, phase control is used to give smooth control of a.c. power in the load. The average power in the load can be varied by adjusting the time position of the gate trigger pulse relative to the supply waveform, and conduction angles from 10° to 170° are possible. As the supply voltage goes positive, C_1 is charged via the variable resistor R_2 and R_1. The resistor-capacitor network provides a variable phase shift network and a variable potential divider. Increasing the value of R_2 increases the phase shift between the voltage across the capacitor and the supply, and also reduces the amplitude of capacitor voltage. The diac (abbreviation for a.c. trigger diode) is a bilateral device that has approximately the same reverse breakdown characteristics in either direction. This three layer, two-junction device is similar in construction to a bipolar transistor without a base lead attached. When the potential across C_1 exceeds about 35 V, the trigger diode conducts and supplies a gate pulse to the thyristor. The thyristor switches on and applies power to the load.

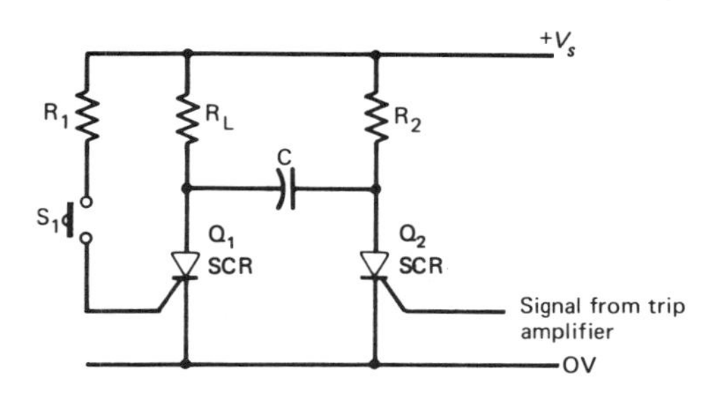

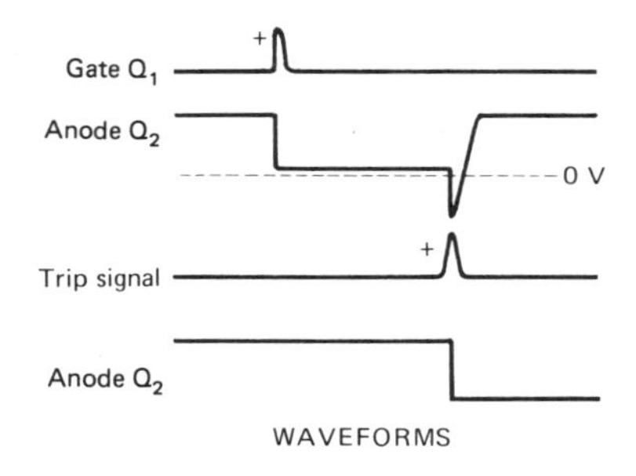

FIGURE 6.20*a* D.C. control circuit using a thyristor and its waveforms.

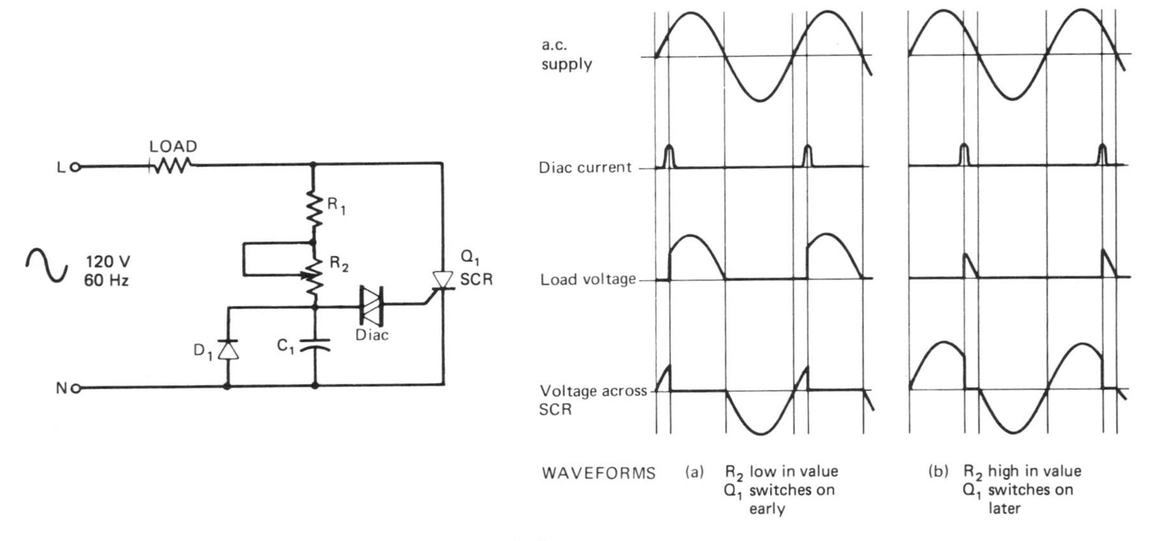

FIGURE 6.20*b* Half-wave a.c. power control circuit using a thyristor and its waveforms.

By adjusting R_2, the time position of the trigger pulse can be controlled, and this in turn controls the power dissipated by the load. The diode D_1 prevents reverse bias being applied across the thyristor gate cathode junction, and at the same time it ensures stable triggering by discharging C_1 on each negative half cycle.

Because of its relatively low efficiency this simple half-wave circuit is not suitable for most applications. A later troubleshooting example shows a more typical circuit.

Triacs

The triac is similar in operation to two thyristors connected in reverse parallel, but with a common gate connection. This means that the device can pass or block current in both directions. Also it can be triggered into conduction in either direction by applying either positive or negative gate signals. Triacs are mostly used in full-wave a.c. control circuits in preference to two thyristors or to a bridge rectifier and thyristor, because simpler heat sinks and more economical trigger circuits can be used.

Failures of thyristors and triacs, caused mostly by thermal effects such as excessive temperature or a high rate of temperature cycling, can be typically open circuits or shorts between connections. These are, of course, complete failures. Partial failures such as impaired gate sensitivity giving erratic triggering or reduced forward breakover voltage can also occur.

When switched on, a thyristor should have a voltage between anode and cathode of approximately 1 V and a voltage between gate and cathode of about 0.7 V. Fault conditions would give the following symptoms:

(a) Anode to cathode open circuit: no current flow from anode to cathode. Measured anode to cathode voltage always high.

(b) Anode to cathode short circuit: the thyristor will conduct in both directions with no gate signal applied. Measured voltage from anode to cathode will be zero volts.

With the power switched off it is possible to measure for short circuit anode to cathode or for open or short circuit gate to cathode with an ohmmeter. The gate cathode of a thyristor has similar characteristics to a diode. A low resistance (typically a few hundred ohms) should be indicated with the gate positive with respect to the cathode, and a high resistance (greater than 100 kΩ) with the gate negative with respect to the cathode. A high resistance should be indicated in either direction for the anode to cathode connections. Naturally any other components in parallel with the gate or anode will affect the readings. If there is any doubt it is always wise to unsolder one connection before making the measurement.

6.7 EXERCISE: SIMPLE D.C. POWER SUPPLY (FIGURE 6.21)

This supply is designed to provide the following outputs:

(a) −15 V nominal at 800 mA—unregulated

(b) −7.5 V ± 10% at 50 mA—zener regulated.

The transformer with a secondary rating of 12 VA gives an a.c. secondary voltage of 12 V rms when the d.c. load is 800 mA. A bridge rectifier is used to give the nominal −15 V, and an input (or smoothing) capacitor of 3300 μF is used to keep the ripple at a reasonable low value.

A 1N5922B zener diode, which has a power rating of 1.5 W at 25°C, provides the regulated −7.5 V line. This simple type of shunt regulator provides a good performance and has the advantage of being

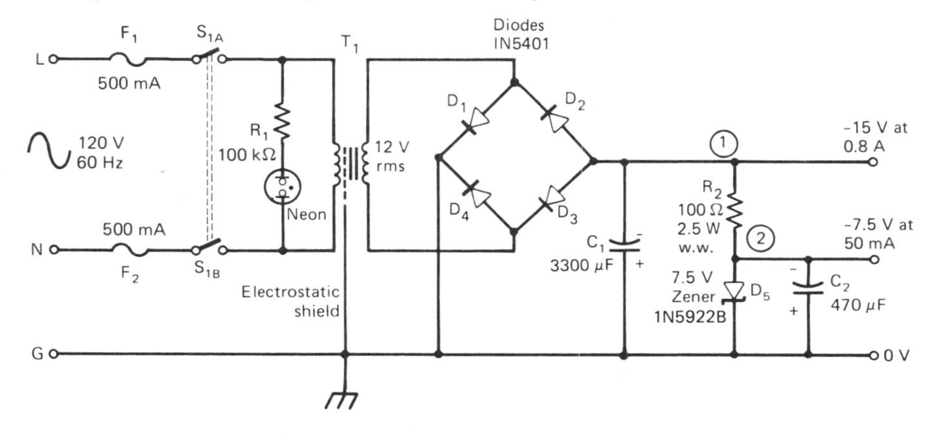

FIGURE 6.21 Simple power supply.

short-circuit proof. If the −7.5 V line is shorted to 0 V, the zener current becomes zero and the load current is limited by R_2 to a maximum of 150 mA. The power rating of R_2 must be high enough to be able to withstand a possible short, and this is why a 2.5 W wire wound (w.w.) resistor is used.

QUESTIONS

In all cases, unless otherwise stated, assume that the power supply is fully loaded.

1. Calculate the approximate peak-to-peak value of the ripple on the −15 V line.
2. What is the value of the zener current when the load on the −7.5 V line is 50 mA?
3. What is the power dissipated by R_2 and D_5 when the −7.5 V line is
   (a) Supplying the load with 50 mA.
   (b) Unloaded?
4. In the following questions the power supply fails with the symptoms given. In each case state with a supporting reason the component or components that are faulty and the type of fault.
   (a)

TP	1	2
MR	−15.1 V	−9.1 V

   (b) Both fuses blown. The following resistance measurements were made using the ohms range of a standard multimeter
   Transformer primary resistance: 43 Ω.
   Transformer secondary resistance: 2 Ω
   Resistance TP1 to 0 V; 250 Ω (−15 V load removed).
   (c) Zero output voltage on TP1 and TP2 neon indicator on.
5. State the symptoms for the following points
   (a) C_2 short circuit.
   (b) D_2 open circuit.
   (c) R_2 open circuit.

6.8 EXERCISE: DISCRETE LINEAR REGULATOR (FIGURE 6.22)

This regulator circuit will provide a 9 V output at 100 mA with a load regulation of about 0.3%. A reference voltage of 5.6 V is set up by diode zener D_1, and the difference between this reference and a portion of the output voltage, from the potential divider R_5 and R_4, is amplified by Q_2. The amplified error is used to control the conduction of the series transistor Q_1. The maximum available current is limited by R_2 and Q_3 to just over 120 mA. Even if a short exists across the output the 2N3036 will remain undamaged as long as it is provided with a small heat sink.

QUESTIONS

Assume the unregulated input is 15 V d.c. and the output load current is 75 mA.

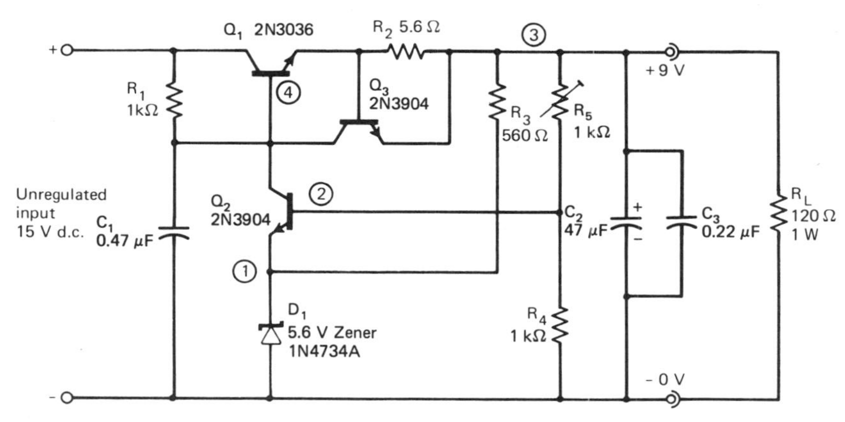

FIGURE 6.22 Linear regulator.

1. In each of the following cases the unit fails and the voltages at the test points with respect to 0 V are measured with a standard 20 kΩ/volt multimeter. State, with a supporting reason, the component (or components) that are faulty, and the type of fault.

	TP	1	2	3	4
A	MR	5.7	0	11	12.4
B	MR	0	0.62	0.95	1.65
C	MR	11.2	7.9	11.2	12.5
D	MR	0	0	0	1.3
E	MR	1.4	1	1.4	2.5
F	MR	0	0	0	14.9

2. Calculate the power dissipated by Q_1 under conditions of:
   (a) Normal load (100 mA).
   (b) Output short circuit.
3. What is the maximum and minimum value to which the output voltage can be set by adjusting R_5?
4. Describe the symptoms that result from the following faults
   (a) Q_2 base/emitter short.
   (b) R_4 open circuit.
   (c) Q_1 collector/base short.
   (d) C_1 short circuit.

6.9 EXERCISE: OP-AMP POWER SUPPLY (FIGURE 6.23)

This unit is suitable for supplying power to op-amp circuits which often require a dual supply with a common ground. In this example the outputs are fixed at approximately +9 V and −9 V and will supply load currents up to 100 mA with reasonable regulation. The a.c. voltage from the center tapped secondary of 12V–0–12V is rectified and then is smoothed by C_1 and C_2 to give +16 V d.c. at TP1 and −16 V d.c. at TP2. Shunt regulators are used to provide the fixed outputs of ±9 V. An 8.2 V zener diode provides the input to an emitter follower (Q_1 or Q_2), and in this way changes in zener diode current, which would cause small changes in zener voltage

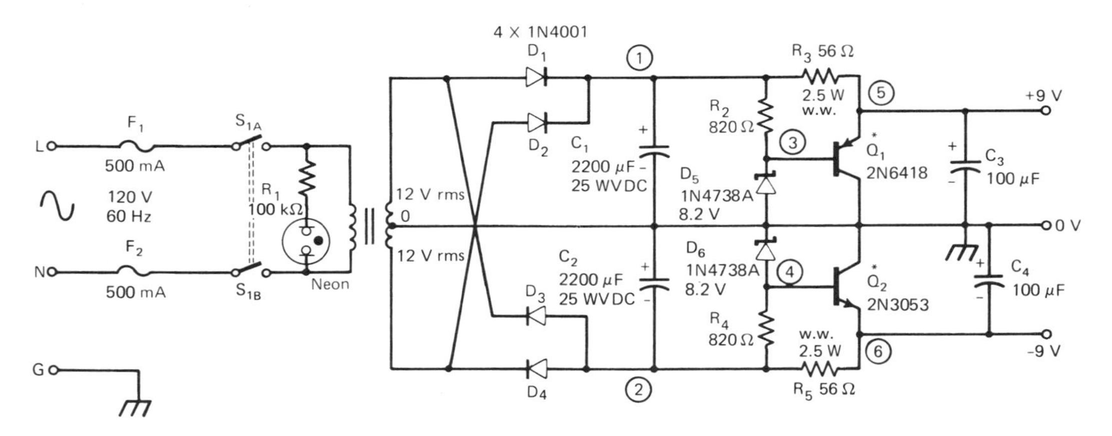

FIGURE 6.23 Op-amp power supply.

and degrade the regulation, are minimized. With the base of Q_1 held at +8.2 V, its emitter will be forced to about +8.8 V and a current will flow through R_3. If the +9 V line is unloaded, this current must all flow through Q_1, and therefore Q_1 must be provided with a small heat sink. When load current is taken from the +9 V line, the current through Q_1 falls by an almost equal amount. In this way the current via R_3 is divided between the load and Q_1.

Shunt regulators can be used in low power applications, since they have the advantages of simplicity and of being inherently short circuit proof. Suppose the +9 V line is shorted to 0 V, the current through Q_1 will fall to zero and the transistor will be completely protected. The maximum output current under these short circuit conditions is limited by the series resistor R_3 and, provided that this resistor can physically dissipate the power caused by the short, no damage will occur. R_3 and R_5 must therefore be wire-wound types with power ratings of 2.5 W. The disadvantage of shunt regulators is that power is dissipated in the regulator when the unit is not loaded.

QUESTIONS

1. Describe a test to determine the line and load regulation for this circuit.
2. Assuming that the unit is not loaded and that the following voltages are present

TP	1	3	5
MR	+ 16 V	+8.3 V	+9 V

Calculate

(a) The current flowing in Q_1.

(b) The base current of Q_1 if its h_{FE} is 40.

(c) The current through D_5.

3. The unit fails with the symptoms given. State, with a supporting reason, the faulty component (or components) and the nature of the fault. Loads of 100 mA are connected.

	TP	1	2	3	4	5	6
A	MR	+16.1	-16	+8.2	0	+8.8	-0.75
B	MR	+16.1	-16	+16	-8.3	+10.8	-8.9
C	MR	+8	-16	+7.9	-8.3	+5.2	+8.9
D	MR	+16.0	-15.3	+8.2	-8.3	+8.8	0
E	MR	+16.1	-16	+10.3 V	-8.3	+10.8	-8.9
F	MR	+16.1	-16	+8.2	-8.3	+8.8	-10.7

6.10 EXERCISE: AN INVERTER (FIGURE 6.24)

This is an example of a low-power inverter using readily available components. A 6 V d.c. input is switching, at a frequency determined by the astable multivibrator (Q_1 and Q_2), across the center tapped winding of the transformer. An iron-core transformer is used, since this can be more easily obtained than the ferrite-core type. In this case the low-voltage center-tapped winding is used as the primary and the "a.c." winding is used as the secondary. The transformer has a center tapped primary marked as 12–0–12 and a secondary of 120 V.

The square wave signals from Q_1 and Q_2 collectors are used to drive Q_4 and Q_3

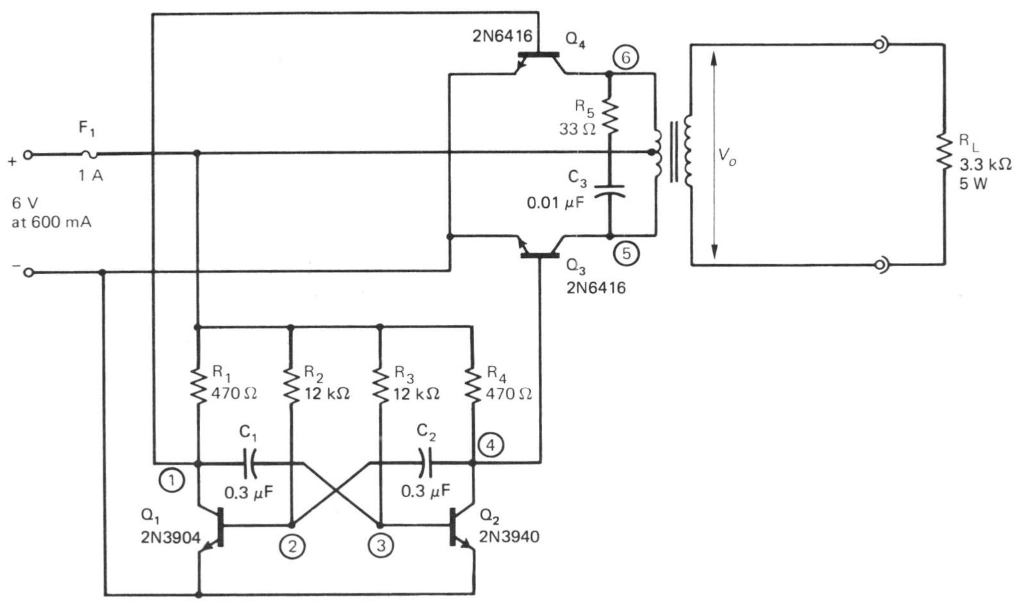

FIGURE 6.24 An inverter.

alternatively into conduction. For example, when Q_1 turns off, its collector voltage rises and base current flows into Q_4 via R_1. Q_4 switches on and current flows through the top half of the primary winding. On the next half cycle of the astable, Q_1 turns on and Q_4, therefore, switches off. At the same time Q_2 goes off causing Q_3 to conduct. Current now flows in the opposite direction through the lower half of the transformer primary. In this way an alternating current is set up in the primary, and this induces an a.c. voltage in the secondary. An output of about 50 V rms is obtained when the load current is 30 mA.

Because the bases of the two switching transistors (Q_3 and Q_4) are directly connected to the collectors of the transistors of the astable (Q_2 and Q_1), the waveforms and the operating frequency of the astable are modified. The voltages at TP1 and TP4 are only allowed to rise to about +0.8 V, just sufficient to drive the 2N6416s into conduction. The switching frequency is therefore about 800 Hz.

R_5 and C_3 are included as filter components to reduce the amplitude of any spikes generated when the transistors switch.

QUESTIONS

Assume the inverter has a secondary load of 3.3 kΩ.

1. Sketch the waveforms you would expect to measure with an oscilloscope at TP1, TP2, and TP6.
2. In each of the following cases a fault exists so that there is no a.c. secondary output. State, with a supporting reason, the faulty component (or components) and the type of fault.

	TP	1	2	3	4	5	6
A	MR	0.15	0.7	0.7	0.15	6	6
B	MR	0	0.7	0.7	0.15	6	6
C	MR	0.15	0.7	0.7	0.05	6	6
D	MR	0.75	0	0.7	0.15	6	4.8

3. What would be the symptoms for the following faults?
   (a) Q_3 base/emitter open circuit.
   (b) Secondary winding open circuit.
   (c) Q_2 base/emitter open circuit.
4. Design a simple rectifier and smoothing filter circuit to convert the a.c. output into d.c. What value of d.c. voltage would result from your circuit arrangement?

6.11 EXERCISE: A GENERAL-PURPOSE LABORATORY POWER SUPPLY USING A μA723 REGULATOR (FIGURE 6.25)

The μA723 voltage regulator IC U_1 described in Section 6.3, is used as shown to provide a variable output voltage from 3 V to 25 V at a maximum current of 1.5 A. The regulator is supplied from a nominal 36 V d.c. source (similar to Figure 6.7 except that the transformer must have a rms secondary of 25 V). The current output from the regulator IC is boosted by the

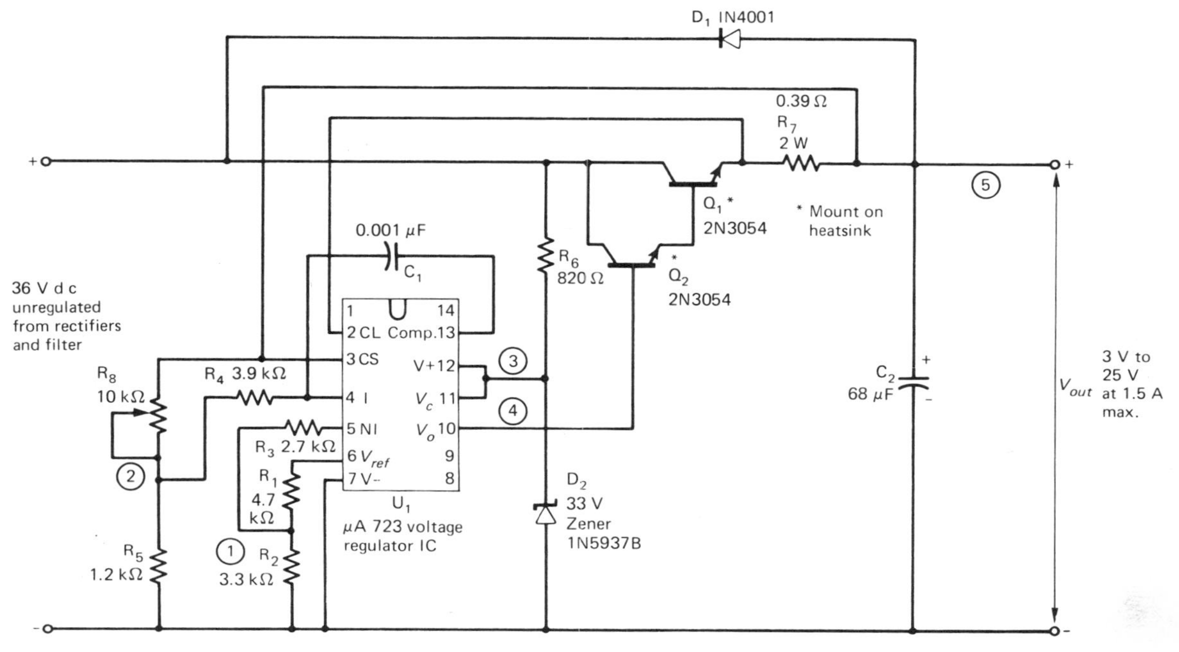

FIGURE 6.25 Power supply using a μA723.

Darlington pair (Q_1 and Q_2). Note that both these transistors must be mounted on heat sinks. The supply to the μA723 IC is limited to 33 V by the zener D_2, this giving protection to the IC should the unregulated d.c. input rise above a safe value. The reference voltage from the μA723 is reduced by the potential divider R_1 and R_2 to 3 V, and this voltage is applied to the noninverting input (pin 5) via R_3. If R_8 is set to zero, then TP2, and the inverting input to the error amplifier internal to the IC, must also be at 3 V. Since the output is connected back to TP2 via R_8, then if R_8 is set at zero ohms the output voltage will also be +3 V. By adjusting the value of R_8, the output voltage can therefore be set anywhere between 3 to 25 V. For example, if R_8 is set to a value of 1 kΩ, that is the same value as R_5, the output voltage will be 6 V.

The current limit is set by R_7. The voltage developed across this resistor by the load current flowing through it is applied to the current limiting transistor internal to the μA723 IC. If the load current exceeds about 1.7 A, then the voltage across R_7 will be about 660 mV, and this forward-biases the internal transistor, which then causes the base current drive to Q_1 and Q_2 to limit. If the output is short circuited, the power dissipated by Q_1 will be high, and therefore this transistor and Q_2 must both be mounted on adequate heat sinks.

D_1 is included as a protection device for the IC so that C_2 is rapidly discharged when the unit is switched off.

QUESTIONS

Assume the power supply is not loaded unless otherwise stated.

1. In the following cases the unit fails with the symptoms stated. Explain which component has failed and give details of the type of fault.

	TP	1	2	3	4	5	Other Symptoms
A	MR	3	3	32	28.9	27.5	(a)
B	MR	7.05	7.05	32	8.6	7.2	(b)
C	MR	0	0	0	0	0	(c)
D	MR	3	3.25	32	1.4	36	
E	MR	3	3	32	29.3	2.7	

   (a) No control with R_8.

   (b) 7.2 V is the minimum voltage that can be obtained.

   (c) R_6 overheating.

2. Calculate the power dissipation of the collector of Q_1 under the following conditions:

   (a) Output set to 25 V, load current 1.2 A.

   (b) Output set to 3 V, load current 1.2 A.

   (c) Output short circuit.

3. What would be the symptoms for the following faults:

   (a) A short between pins 5 and 6 of the IC.

   (b) R_8 open circuit track.

   (c) R_7 open circuit.

6.12 EXERCISE: AN HV POWER SUPPLY (FIGURE 6.26)

The brief specification for this unit is as follows:

D.C. output voltage: adjustable from about 500 V to 2 kV.

Typical load current: 2 mA.

Load regulation: better than 0.25%.

Output ripple: less than 250 mV d.c. at an output of 1 kV and with a 1 mA load.

This type of power supply can be used to provide the high d.c. voltage required for a photomultiplier tube. The current taken by such a tube is not large, but the stability and regulation of the HV needs to be reasonably good. This is because the gain of a photomultiplier tube depends directly on the value of the voltage supplied.

The circuit consists of an oscillator section formed by Q_2 and Q_3 that is supplied with power via the series transistor Q_1. Suppose the voltage at Q_1 emitter is −6 V, then this voltage is switched across the primary of the transformer alternately by Q_2 and Q_3. Feedback windings provide the necessary positive feedback to keep the oscillator going. This circuit is basically an

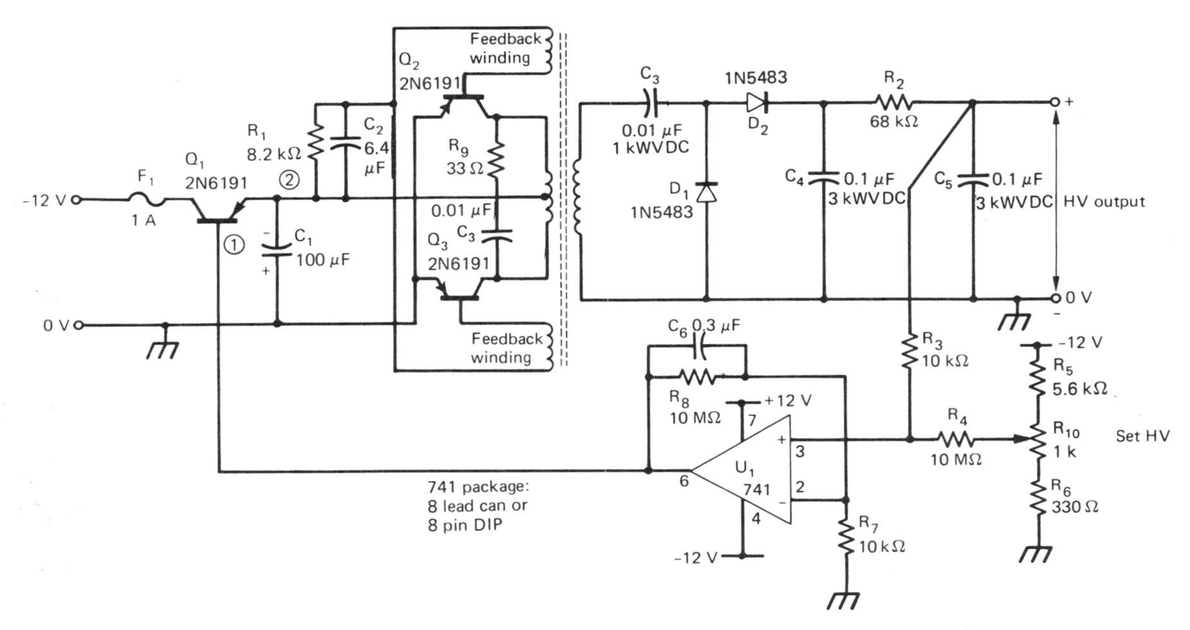

FIGURE 6.26 (i) Use a suitable heat sink for Q_1, Q_2, and Q_3, (ii) R_3 must be either a high voltage type or made up of several resistors in series.

inverter. The step-up transformer (about 100:1) produces an a.c. secondary voltage which is rectified and smoothed by the voltage doubler circuit of C_3, D_1, D_2, and C_4. Further filtering is provided by R_2 and C_5. Since the natural frequency of the oscillator, determined by the transformer winding, is about 3 kHz, the values of the smoothing capacitors can be quite low. The 0.1 μF is perfectly adequate. However, these capacitors must have a high d.c. working voltage (WVDC) rating. Ceramic types are available for this purpose.

The output voltage is regulated and held stable by U_1 the 741 error amplifier. A portion of the HV output is applied to the noninverting input via R_3 and R_4. The other end of R_4 is set to a low d.c. voltage by the potentiometer R_{10}. The 741 is wired as a noninverting amplifier, its output directly driving the base of Q_1. Since the junction of R_3 and R_4 is virtually at ground (the gain of the 741 is very large), then when the wiper of R_{10} is set, to say, -1 V the other end of R_3, which is the HV output, must be very nearly +1 kV. Note that R_8 and R_7 are in the same ratio as R_3 and R_4, giving an overall gain of about 1000 for the noninverting amplifier. The output of the 741 is driven negative to a value that allows Q_1 to conduct to adjust the HV to +1 k V. By varying R_{10} the HV output can be adjusted over the range of 500 V to about 2 kV, this being sufficient for most photomultiplier tubes.

Actually the circuit can be seen to be a

version of a switched type regulator (see Figure 6.13), except that the inverter section is self-oscillating and that control over the output voltage is achieved by varying the d.c. voltage applied to the inverter via Q_1 instead of varying the duty cycle of the inverter switching waveform. Imagine the output voltage dropping in value as a load is applied. This fall in voltage is applied to the input of the noninverting amplifier causing the 741 output to go more negative. Q_1, the series element, is forced to conduct more, and the d.c. voltage applied to the center top of the primary winding increases. This naturally causes the amplitude of both the primary a.c. waveform and the secondary a.c. to increase, and therefore the d.c. output is automatically corrected to very nearly its original value.

Construction of this type of circuit requires care in the choice of components and in the layout. The most critical component is the transformer. The core can be made up from two ferroxcube U-cores (Mullard type FX2380). These are clamped together, when the windings have been fitted, to form a square-shaped core. A gap of about 0.005 mm should be allowed between the two faces by fixing a piece of paper of that thickness between them. The windings can be constructed as follows (Figure 6.27):

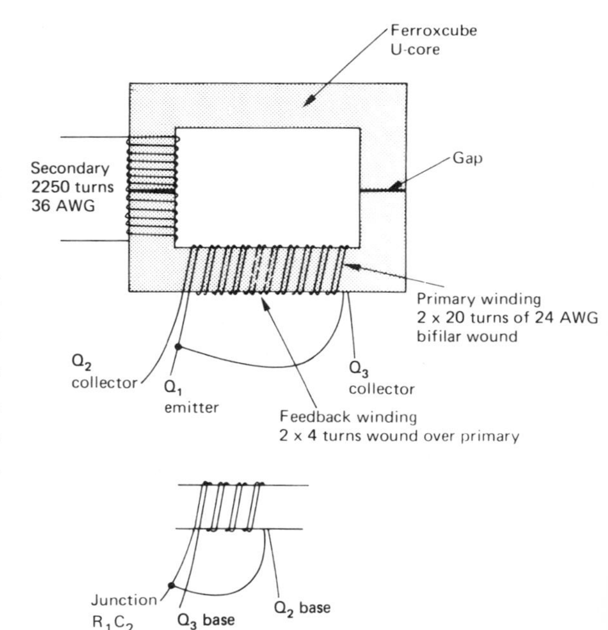

FIGURE 6.27 Winding details for HV transformer.

1. **Main primary winding**: 2 X 20 turns bifilar wound in one single layer of 24 American wire guage (AWG) enameled copper wire. The winding must then be covered with polyvinal chloride (PVC) tape.
2. **Feedback windings**: 2 X 4 turns of bifilar wound 24 AWG enameled copper wire. This winding must be wound directly over the primary winding.
3. **Secondary winding**: approximately 2250 turns of 36 AWG enameled copper wire insulated from the core by a single sheet of polythrene and interleaved with waxed paper between layers.

The components used for the voltage doubling rectifier and the filter are best mounted either on good quality glass fiberboard or by using PTFE insulated standoff connectors.

QUESTIONS

1. Draw a block diagram to indicate the main features of the supply.
2. Explain briefly the purpose of
   (a) R_1 and C_2.
   (b) R_2 and C_5.
3. The unit fails with the following symptoms. In each case state the probable cause and the type of component fault:
   (a) The d.c. output is at maximum (2 kV) and no control can be affected by varying the setting of R_{10}.

TP	1	2
MR	- 10.2 V	- 9.5 V

   (b) The HV output is zero and the oscillator is off.

TP	1	2
MR	- 10.2	0 V

   (c) The HV output is zero but the oscillator is on.

TP	1	2	
MR	- 10.2 V	- 9.5 V	The d.c. voltage across C_4 is 2.15 kV.

4. State the symptoms for the following faults:
   (a) Q_3 feedback winding open circuit.
   (b) D_2 open circuit.
   (c) Q_1 collector/emitter short.

6.13 EXERCISE: REMOTELY CONTROLLED 5 V LOGIC SUPPLY WITH OVERVOLTAGE PROTECTION (FIGURE 6.28)

This 5 V d.c. supply, used to power TTL logic, is arranged to be switched from some remote control point. When +12 V is applied at the control input, the unregulated 16 V across C_1 is switched by Q_5 to provide approximately 7.5 V to the input of the IC regulator U_1 (μA7805). U_1 regulates the logic supply and holds it constant at +5 V. The load current is 500 mA.

Without +12 V applied, the astable circuit formed by Q_1 and Q_2 is off and no switching signal is applied to the base of Q_5. As soon as the control voltage is applied, the astable oscillates (approximately 15 kHz), and the square wave present at the collector of Q_2 is used to switch the darlington pair Q_3 and Q_4 on and off. When these two transistors are on, base current drive is supplied to Q_5, and this series transistor conducts. On the next half cycle of the astable oscillator, Q_3 and Q_4 turn off consequently turning off Q_5. The switched voltage at the collector of Q_5 is filtered by L_1 and C_6 to provide a d.c. voltage of about +7.5 V at the input to the voltage regulator. D_3 acts as a commutating diode to provide a current path for L_1 when Q_5 switches off on every other half cycle.

The μA7805 is a monolithic fixed voltage IC regulator giving an output of +5 V ± 200 mV with a regulation of 0.2%. It also has internal foldback current limiting and is therefore protected against short circuit at the output. Since the input voltage to

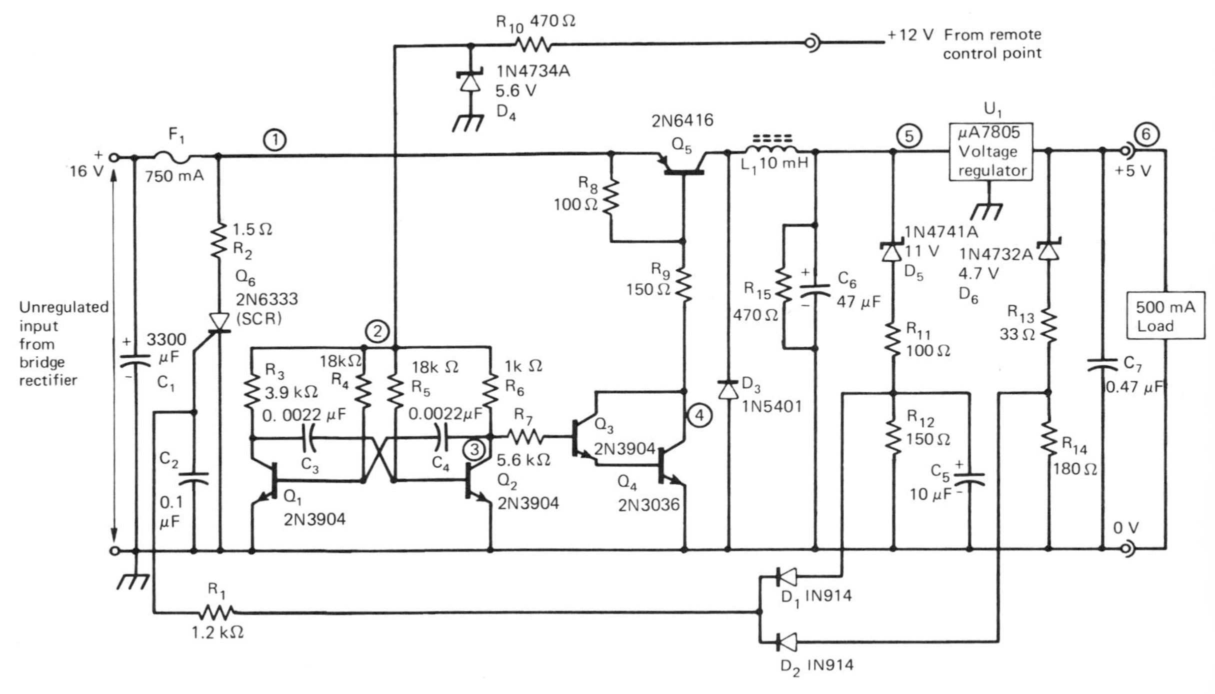

FIGURE 6.28 Remotely controlled logic supply.

the IC is quite low (7.5 V), its power dissipation is also low and it only requires a small heat sink.

Overvoltage protection for both the TTL logic load and the μA7805 is provided by the crowbar circuit of Q_6. The output voltage level is sensed by D_6, a 4.7 V zener, and the potential divider R_{13} and R_{14}. If a fault occurs in U_1 causing the output voltage to rise to just greater than 6.5 V, then the voltage across R_{14} will be 1.5 V. This will be sufficient to forward-bias D_2 and to trigger Q_6 into conduction. The anode voltage of Q_6 will fall to just less than 1 V and F_1 will blow. This will disconnect the unregulated input voltage from the rest of the circuit thereby protecting the TTL logic load from an overvoltage. The thyristor Q_6 is a relatively sensitive type requiring a minimum gate trigger voltage V_{GT} of 0.8 V at a gate trigger current I_{GT} of 0.2 mA. The sensing circuit of D_5 and R_{11}, R_{12} provides protection for IC_1 and the TTL load in the event of a collector/emitter short occurring in Q_5. In the same way as described above the SCR will be triggered on and the fuse will blow.

Troubleshooting a circuit such as this is made easier by visualizing the circuit in block diagram form; for example:

Q_1, Q_2	15 kHz gated astable
Q_3, Q_4	switch drive
Q_5	series switch
D_3, L_1, C_6	filter
U_1	5 V regulator

D_5 sense for Q_5 short failure
D_6 sense for U_1 short failure
Q_6, F_1 crowbar circuit.

Imagine a fault such as R_7 open circuit. The astable will oscillate correctly when the +12 V control level is provided but no signal will be available to switch Q_3 and Q_4. These would remain off and therefore hold off Q_5. The voltages at the various test points would be:

TP	1	2	3	4	5	6
MR	16.5	5.6	2.4	16.5	0	0

Since TP3 is oscillating the d.c. meter will indicate the mean value. Note that similar symptoms will occur for a base/emitter open circuit in either Q_3 or Q_4.

An oscilloscope used to measure the waveform at TP3 and TP4 would also rapidly point to the fault, since the square wave at TP3 will have a full amplitude of 5.5 V instead of rising to about +4.5 V. Furthermore no square waves will be present at TP4.

Investigating faults that cause the crowbar to operate requires resistance tests to be made, since the fault should be cleared before a new fuse is inserted. Apart from short circuits on Q_5 or from the input to output of U_1, other component faults will cause the crowbar to operate. For example, a short on Q_4 collector emitter will hold Q_5 on causing TP5 to rise above the trip point of the D_5 circuit. Either D_5 or D_6 short circuit will also cause the crowbar to operate.

QUESTIONS

1. Explain briefly the possible advantage in using an oscillator switching circuit to provide the d.c. input to the voltage regulator U_1.
2. Why is an operating frequency of 15 kHz more suitable than say 5 kHz?
3. Design a simple circuit using an LED and one resistor to give an indication that F_1 has blown.
4. Give two reasons why the +12 V control signal is reduced to 5.6 V by R_{10} and D_4, and explain the likely effect of D_4 going open circuit.
5. Calculate the voltage level at TP5 that will cause the crowbar to operate (neglect the effect of I_{GT}).
6. The power supply fails with the following symptoms. For each state the component or component faults and add a supporting reason. Assume that the +12 V line is present.

	TP	1	2	3	4	5	6
A	MR	16.5	5.6	0.7	16.5	0	0
B	MR	16.5	5.6	2.2	8.3	12.5	0
C	MR	16.5	5.6	2.2	8.3	0	0
D	MR	16	5.6	2.3	0.2	0	0

7. Describe a simple method whereby the operation of the crowbar could be tested without blowing fuses or damaging other components.
8. List the possible component faults that could cause the crowbar to fail to operate on an overvoltage present at

   (a) Either TP5 or TP6.

   (b) TP5 only.

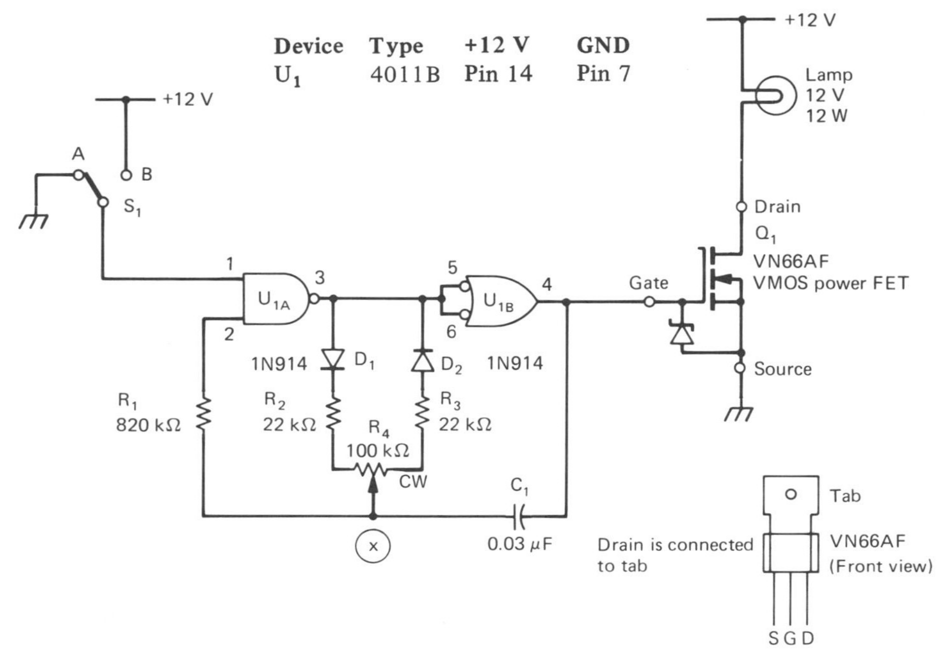

FIGURE 6.29 Lamp-dimmer using VMOS power FET.

9. A fault exists such that the crowbar operates as soon as the +12 V control signal is applied. U_1 has not failed short circuit and, without U_1 in circuit, the crowbar does not operate. State with a supporting reason the cause of the fault.

6.14 EXERCISE: LAMP BRILLIANCE CONTROL USING A VMOS POWER FET (FIGURE 6.29)

The VMOS power FET is a device that can be used as a direct interface between CMOS logic and a load. This is because the VMOS, a voltage controlled device, has a very high input impedance and requires negligible current at its gate input. In this example a bulb of 12 V and 12 W is connected in the drain circuit of the VMOS, and its brightness is controlled by varying the mark-to-space ratio of the switching waveform.

Two NAND gates from a 4011B CMOS IC are connected as a square wave oscillator and the output is applied directly to the VMOS gate. With S_1 in position A, the oscillator is inhibited, since one of the inputs of U_{1A} will be held at logic 0. When S_1 is moved to position B, both inputs to U_{1A} become logic 1 for a short time, and the oscillator will run at a frequency of about 166 Hz. The duty cycle of the output waveform can be varied by R_4. For example, suppose R_4 is fully clockwise. When the output of U_{1B} goes positive causing pin 2 of U_{1A} to go positive and thereby driving U_{1A}'s output to zero, C_1

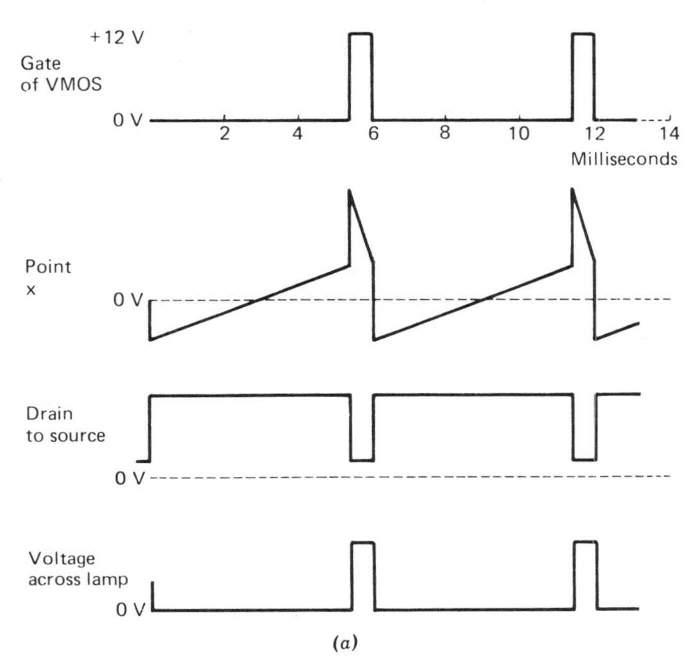

FIGURE 6.30*a* Waveforms with R_4 set fully clockwise (voltage across lamp has a 10% duty cycle).

discharges via R_3, D_2 and the output circuit of U_{1A} toward 0 V. When the level at point (x) falls to the threshold of U_{1A}, U_{1A}'s output switches to logic 1 forcing U_{1B}'s output to logic 0. C_1 now charges via R_4, R_2, D_1 and the output of U_{1A} towards +12 V. This is illustrated in Figure 6.30*a* showing the waveforms at various points for R_4 fully clockwise. Under these conditions the VMOS FET will be switched on for a much shorter time than it is held off, and the resulting brightness of the lamp will be low. If, on the other hand, R_4 is fully anticlockwise then the waveform will be as shown in Figure 6.30*b*, and the lamp brilliance will be at maximum. Since the VMOS is acting as a switch its dissipation will be relatively low and only a small heat sink is required.

QUESTIONS

1. If $V_{DS(on)}$, the drain source saturation voltage, is typically 2 V and the drain current is 1.25 A, calculate the mean power dissipated by the VMOS FET under full-brightness conditions.
2. What is the purpose of R_1?
3. Which instrument would be most suitable for troubleshooting this circuit?
4. A fault occurs causing the lamps brilliance to be at maximum when S_1 is in position B. With S_1 in position A the lamp is off. State the possible component fault.

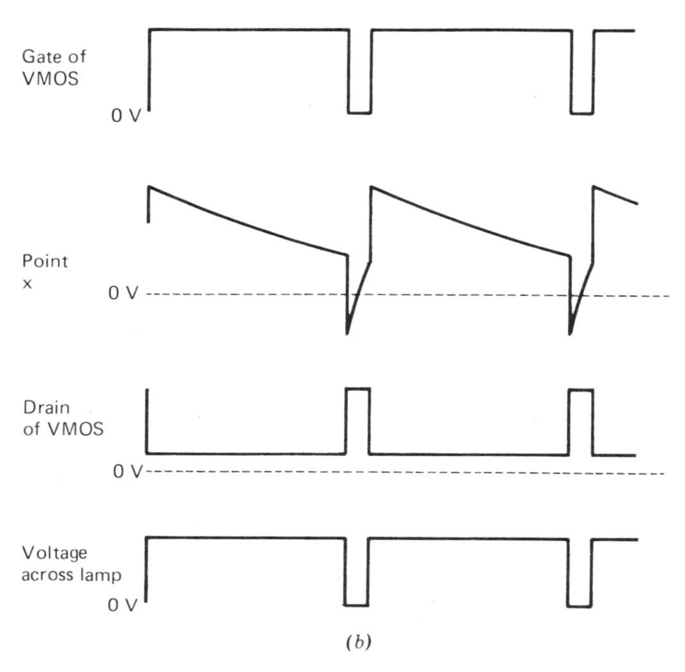

FIGURE 6.30*b* Waveforms with R_4 set fully anticlockwise (voltage across lamp has a 90% duty cycle).

5. State the symptoms for the following faults:
   (a) D_2 short circuit.
   (b) Drain to source short on Q_1.
   (c) Gate to source short on Q_1.
   (d) C_1 open circuit.
   (e) Open circuit from pin 7 of U_1 to 0 V.

6.15 EXERCISE: FULL-WAVE A.C. POWER CONTROLLER (FIGURE 6.31)

This shows a typical circuit arrangement for controlling the a.c. power dissipation in the load, as for example in a stage lighting system. One of the advantages of this circuit is that the actual control element is isolated from the mains supply. This is achieved by the use of the two transformers T_1 and T_2. T_1 provides a 40 V rms a.c. voltage to the bridge rectifier, and the 55 V peak amplitude full-wave rectified waveform at the bridge output is limited, on each half cycle, to +10 V by R_1 and D_5. At the beginning of each half cycle, C_1 is charged via the resistive network R_2, R_7, and R_8 so that the emitter voltage of the unijunction transistor rises positive. The rate of change of this voltage is controlled by the setting of R_7. When the voltage across C_1 reaches the peak point of the UJT Q_1, Q_1 is triggered into conduction

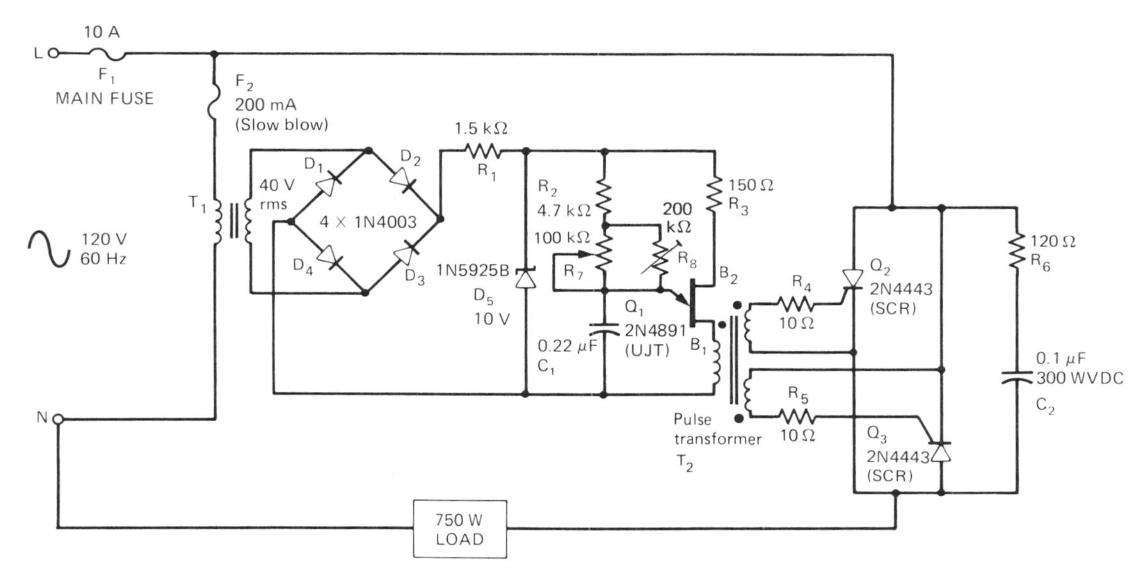

FIGURE 6.31 Full-wave a.c. power controller.

and C_1 is rapidly discharged through the primary winding of the pulse transformer T_2. Positive pulses are induced in the secondary windings and the thyristor (Q_2 or Q_3), which has its anode positive with respect to its cathode, will be triggered on. A portion of half cycle power will be applied to the load. On the next half cycle, as C_1 is again charged and discharged, the pulse generated will turn on the other thyristor. As the mains goes through zero, the thyristor that had previously been conducting is reverse-biased and therefore turns off.

The amount of power dissipated by the load is controlled by the time position of the trigger pulses relative to the start of each half cycle of the mains waveform. Thus with R_7 at a low value, C_1 is charged very rapidly, the trigger pulses occur very early in each half cycle, and maximum power is applied to the load. On the other hand, with R_7 at maximum, C_1 is charged relatively slowly, the trigger pulses occur very late in each half cycle, and only a small portion of power is applied to the load. R_8 is a preset pot that can be adjusted to give a set value of minimum power.

QUESTIONS

1. What is the purpose of R_6 and C_2?
2. The unit develops a fault that results in power control from minimum to only about 50% being possible. List the component failures that could cause this fault.
3. Describe the effects of the following component faults:

   (a) D_1 short circuit.

   (b) R_1 open circuit.

(c) C_1 open circuit.

(d) R_8 open circuit.

(e) D_5 short circuit.

4. What service instrument would be most useful for troubleshooting on this circuit?

5. A fault develops so that only minimum power is dissipated in the load irrespective of the setting of R_7. State the faulty component.

CHAPTER 7

System Maintenance and Troubleshooting

7.1 MAINTENANCE PRINCIPLES

A system can be defined as "anything formed of component parts connected together to make a regular and complete whole." Therefore, any electronic instrument or piece of equipment can be considered as a system. Take, for example, the block diagram of the audio frequency (a.f.) signal generator shown in Figure 7.1. The overall system of the generator is made up of the subsystems of the various circuit blocks: the oscillator, amplifier, squarer, attenuator, and power supply. Each of the subsystems must be working correctly to allow the overall system to function. Visualizing a complete system in block diagram form is an essential aid for troubleshooting systems.

Any further reference, in this chapter, to a system can be considered to cover individual instruments or separate pieces of equipment, as well as instances where several instruments are themselves interconnected to perform some overall task.

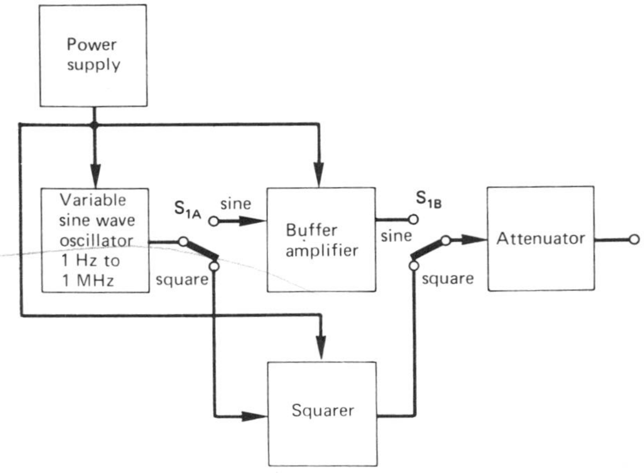

FIGURE 7.1 Block diagram of an a.f. signal generator.

The purpose of maintenance is to achieve a satisfactory level of system availability at a reasonable cost and maximum efficiency. Availability is discussed in Chapter 2, Section 2.7, and is defined as

$$\text{Availability} = \frac{\text{MTBF}}{\text{MTBF} + \text{MTTR}}$$

where MTBF is the mean time between failures and MTTR is the mean time to repair.

To achieve high levels of availability, i.e., those approaching unity, the MTTR value must be low, and this implies that the system can be maintained relatively easily. Maintainability is defined as the probability that a system that has failed will be restored to a full working condition within a given time period. The mean time to repair and the repair rate (μ) are measures of maintainability:

$$\mu = \frac{1}{\text{MTTR}}$$

and

$$\text{Maintainability } M(t) = 1 - e^{-\mu t} = 1 - e^{-t/\text{MTTR}}$$

where t is the time allowed for the maintenance action.

EXAMPLE

In a system the average time to repair any fault is 2 hours. Calculate the value of maintainability for a time of 4 hours.

$$M(t) = 1 - e^{-t/\text{MTTR}} = 1 - e^{-4/2} = 1 - 0.135 = 0.865$$

Therefore, the probability M of the system being returned to a working state within 4 hours is 0.865 (86.5%).

In the same way that values for system reliability can be calculated for a given operational time, the value of maintainability can also be predicted. In both cases it is the probability of success that is calculated for the given time. The time t for reliability is the system operational period, while t for maintainability is the allowed maintenance time. Prediction of maintainability involves establishing a value for the system MTTR.

MTTR is the average of the times taken to repair any fault in the system, and accurate assessment of this is understandably difficult. The designer can aim for a low value of MTTR by paying close attention to the accessibility of components, building in fault display panels, and by providing internal test facilities.

The maintenance policy adopted for a particular system will depend on several factors, such as the type of system, its location and operating and environmental conditions, the required levels of reliability and availability, the overall competence of a skilled maintenance staff, and the provision of spares. For certain types of systems the maintenance policy may include in its program details of recalibration and preventive actions. Recalibration, often carried out at 90-day intervals on measuring instruments such as the oscilloscope and DVM, is really a type of preventive maintenance, since the recalibration task is to first check the amount of drift of some parameter or characteristic from the specified figure, and then to correct for any partial failure that may have caused the measuring instrument's performance to be outside its tolerance limits. But, in practice, no components are replaced. True PREVENTIVE MAINTENANCE is a policy of replacing components or parts of a system that are nearing the end of their life, and are therefore wearing out. The replacement is carried out before the component actually fails. Failures of components entering the wear-out period or subjected to continuous wear are not random and can be predicted. The reliability of a system can therefore be improved by replacing items that are wearing out before they fail. Examples of this are: components with moving parts that are continuously in use such as servo potentiometers, motors and motor brushes; or contacts on relays and switches, especially those subjected to arcing when switching inductive or capacitive loads. Filament lamps are yet another example.

Suppose a system uses a large number of indicator lamps with a failure graph as shown in Figure 7.2. In this case the failures follow a Gaussian distribution with the peak failures occurring at 1000 hours. The probability of any one indicator lamp failing before 1000 hours is 0.5. Thus if all lamps were replaced after 1000 hours, the probability of any one lamp failing during that time is 0.5. Suppose, however, that all the lamps are replaced at a time equal to one standard deviation before the mean life. This would be at approximately 800 hours, and the probability of each lamp failure before replacement is 0.159. This is a considerable improvement and will result in better reliability for the system.

In other cases it is either difficult to accurately predict the point at which components enter the wear-out period or it becomes uneconomic to carry out preventive maintenance. A further disadvantage is that the disturbance created during any preventive maintenance action may itself cause failures.

CORRECTIVE MAINTENANCE or "replace as failed" is the service action that is normally required for the majority of electronic systems since, during the useful

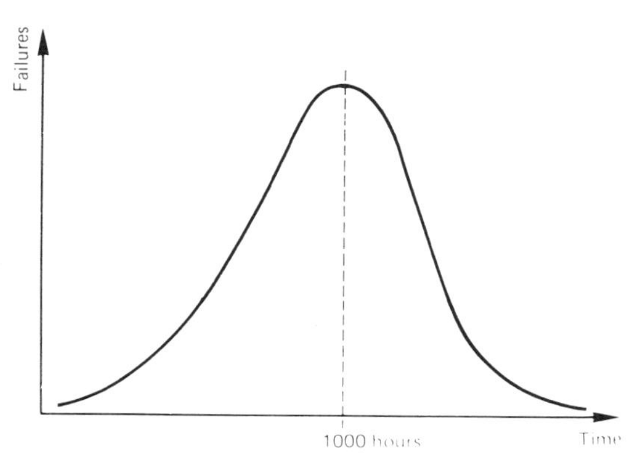

FIGURE 7.2 Possible failure rate of filament lamps.

life of the system, failures of the component parts will be entirely random. These failures cannot be predicted and cannot therefore be prevented by service checks. In fact, carrying out such checks may be the cause of faults rather than the prevention. Making routine checks on equipment where faults are entirely random can only result in degraded reliability. Corrective maintenance is concerned with the detection, location, and repair of faults as they occur. The exercises later in this chapter are concerned with developing skill in this area, which requires a good understanding of system troubleshooting methods in addition to an understanding of overall system and circuit operation.

As shown above there are three distinct phases in the corrective maintenance task:

1. **Fault detection** The presence of a fault must be established and all symptoms must be accurately noted. This means that a functional test, checking the system's actual performance against its specification, must be made. Only by doing this will a full set of fault symptoms be obtained. In some cases a

system may be reported as faulty but, in fact, the failure is caused by incorrect operation, or in other cases system failure may be reported with either very little or misleading information. The functional test will enable the fault to be detected and as much information as possible on the fault to be presented.

2. **Fault location** *or troubleshooting* The task now is to narrow down the search for the cause of the fault, first to one block (or subsystem) within the system, and finally to one component within that block. This task is simplified by the use of one or a mixture of fault location methods outlined in the next section.

3. **Fault rectification** The faulty component or part is repaired or replaced. A functional check must then be carried out on the whole system.

7.2 TROUBLESHOOTING SYSTEMS

As shown in the previous chapters, when a component fails in an individual circuit a certain set of troubleshooting symptoms result. These symptoms, often unique to the fault, will be changes in circuit operation, in d.c. bias levels, and changes in output signals. By interpreting the symptoms it is possible to pinpoint the faulty component. With a complete system, however, the task of finding a single faulty component among several thousand is made more difficult because of the size and complexity of the system. The problem can be tackled by considering the system in block diagram form.

The system is divided up into several functioning circuit blocks and, by measurement, the portion or block that has failed can be located, and then detailed measurements can be made on that block to find the actual faulty component. The block diagram is an essential aid to system fault location and in addition assists in helping the understanding of the operation of complex systems. In a service manual the block

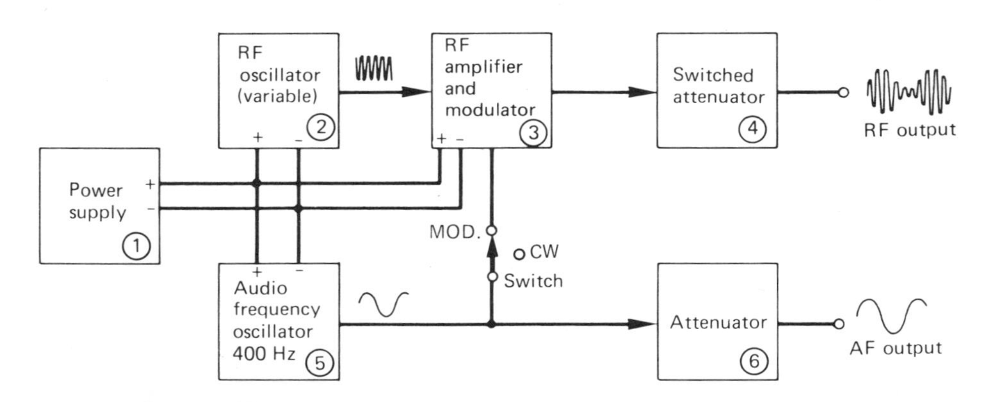

FIGURE 7.3 Block diagram for an r.f. signal generator.

diagram can often, at first, prove more useful than the full circuit diagram. Before looking at the various methods for fault location, consider the block diagram for a radio frequency (r.f.) signal generator shown in Figure 7.3. In its basic form there are six blocks to consider. The variable r.f. sine wave oscillator feeds a high-frequency signal to the r.f. amplifier and modulator. The r.f. output via the attenuator can be either amplitude modulated at 400 Hz or unmodulated (continuous wave) according to the setting of the switch.

In this instrument there are two output signals and two possible output states for the r.f. output. The outputs that are present can be used as symptoms to locate a faulty block if the generator fails. Suppose, for example, that the r.f. output is correct in both the modulated and continuous wave switch positions, but that no 400 Hz a.f. output is present. Then the fault must be in the attenuator or its connections. On the other hand, if the generator failed giving no outputs at all, the fault would almost certainly be in the power supply. This is because it is unlikely, although possible, that both oscillators would fail simultaneously.

These two examples demonstrate the use of the block diagram and the sort of logical approach that is required for troubleshooting. However, because two outputs are present and a switch can be used to modify the state of one output, the location of a faulty block is relatively simple. For more complex systems some general method must be used.

One powerful method which is being increasingly used is NONSEQUENTIAL fault location. This uses automatic testing, based, for example, on the theoretical analysis of the transfer characteristics (response of output to input) of the system. This method is better suited to computer-aided fault analysis rather than the individual service technician. The faulty system, linked to computer-controlled test gear, would have its whole system checked, and the results would be matched with those resulting from typical fault conditions held in the computer's memory.

The individual service technician, when faced with a faulty system, usually has to select one or a mixture of SEQUENTIAL troubleshooting methods. The possible sequential methods are shown in Figure 7.4. Note that here actual electrical measurements and tests are being considered, but a visual inspection for broken wires, cold solder joints, solder bridges, damaged p.c.b. traces, or burnt and damaged components can also be very worthwhile. Visual checks over the mechanical structure are best conducted systematically, moving in sequence from one area to the next and so on.

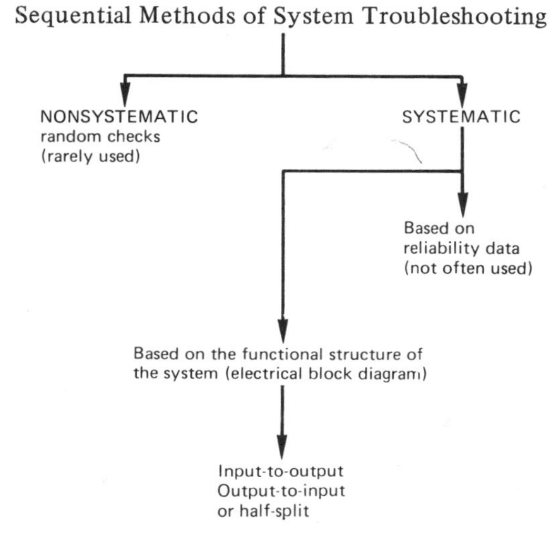

FIGURE 7.4

In system fault location it is, of course, possible to use a completely random series of tests to find which block is faulty, testing the circuits in any order. Although such a method may sometimes produce quick results, it is not generally recommended. One of the systematic logical approaches should be preferred, and by "systematic" is meant a method governed by a set of rules. The rules could be determined by the reliabilities of the various circuit blocks. For example, if it is known that circuit (x) has a failure rate 10 times higher than any other circuit, then checking (x) first could be considered a reasonable action. One could then check the next least reliable circuit and so on. This method is not often used, since a large amount of service data must be available for the assumption of circuit reliabilities to be reasonable.

The most popular systematic troubleshooting methods are:

- **(a)** Input to output.
- **(b)** Output to input.
- **(c)** Half-split.

The first two methods are fairly straightforward. A suitable input signal, if required, is injected into the first block and then measurements are made sequentially at the output of each block in turn, either from input to the output or vice versa, until the faulty block is found. The HALF-SPLIT method, on the other hand, is extremely useful when the system is made up of a large number of blocks in series. As a good example, consider the frequency divider chain of a digital frequency meter shown in Figure 7.5. Here the frequency of a stable crystal-controlled oscillator is divided down by decade counters to give the various timing pulses. With the eight blocks shown, it is possible to divide the unit into two equal halves (half-split), test to decide which half is working correctly, then split the nonworking section into half again to locate the fault. Assume that block (7) has failed; the sequence of tests would be as follows:

1. Split the whole into half by measuring the output from block (4). The output from (4) will be found to be correct at 1 kHz, showing that the fault lies somewhere in blocks (5) to (8).

2. Split blocks (5) to (8) in half by measuring the output from block (6). Again, this output will be found to be correct at 10 Hz.

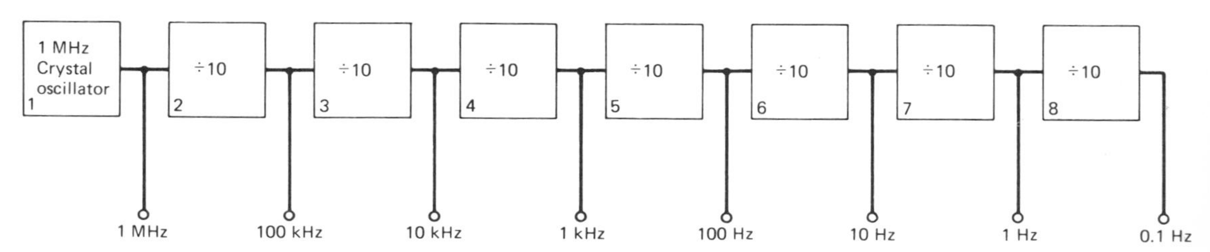

FIGURE 7.5 Frequency divider chain; example for half-split method of troubleshooting.

3. Split blocks (7) and (8) by measuring output of block (7). There will be no output proving that the fault is in block (7).

In practice the number of measurements necessary to locate any faulty block in the frequency divider chain using the half-split method is always three. On average, more measurements would be required using input to output checking. For the input to output or output to input method, the number of checks on a series system is given by the formula

$$C = \frac{1}{2n}(n - 1)(n + 2)$$

and for half-split by

$$C = 3.32 \log_{10} n$$

where n is the number of blocks or units and C is the mean number of measurements required. Note that these formulas apply to series-connected circuits only.

As can be appreciated the half-split method can be really powerful in troubleshooting a long chain of series circuits. For example, if $n = 100$ then the number of checks C is only 7. There are, however, several assumptions made for the half split:

(a) That all circuit blocks are equally reliable.

(b) That only one fault exists.

(c) That all measurements are similar and take the same amount of time.

In practice these assumptions do not restrict the use of the half-split method. The first assumption, that all blocks are equally reliable, can be seen to be reasonable in most instances. Unless other information is available, it is reasonable to assume that components with higher failure rates are spread evenly through the system. If this is not the case then troubleshooting based on reliability data might be a suitable alternative.

The second assumption is that only one fault exists. Again, this is a reasonable assumption to make, but even where multiple faults do exist the half-split will still be the most powerful method to use.

Finally, the third assumption can be seen to be perfectly valid in our example, since all the series blocks are identical (decade dividers), but this will not be true in the majority of electronic systems. Some checks and measurements on parts of complex systems will be more difficult, may require special test equipment, and be more time-consuming than others. Therefore, in practice, the type of check and the time required to make it must be considered.

Most systems do not consist of only series-connected blocks but have parallel branches and possibly feedback loops. The connections that complicate troubleshooting methods are:

(a) Divergence: an output from one block feeding two or more units.

(b) Convergence: two or more input lines feeding a circuit block.

(c) Feedback: which may be used to modify the characteristics of the system or, in fact, be a sustaining network.

DIVERGENCE is a commonly encountered situation. Two examples exist in Figure 7.3. The power supply has to supply d.c. power to blocks 2, 3, and 5, and the output from the audio frequency oscillator has to feed its 400 Hz sine wave signal to both blocks 3 and 6. The rule for any divergent arrangement is to check each output in turn and then to continue the search for a faulty block in the area that is common to the incorrect outputs. A possible divergent arrangement is shown in Figure 7.6. Suppose signals for w, x, and z are correct but y is incorrect, then the fault must be situated in block C.

The usual arrangement for CONVERGENCE is that two or more inputs are required at a particular circuit block for the output of that block to be correct. This is similar to the AND function in digital logic circuits and is referred to as *summative*. All such inputs must be checked one by one at the point of convergence. If all are correct the fault lies beyond the convergent point, but if any one is incorrect then the fault must lie in that input circuit. This is illustrated in Figure 7.7 where block D requires the three inputs x, y, and z for correct operation. If, for example, all three inputs are correct then the fault can only be in block D. But if input y is incorrect then the fault lies in the circuits producing the y signal.

Systems with FEEDBACK LOOPS, which connect the output of some block with the input of an earlier block via some network, present one of the more difficult problems in troubleshooting. The output signal, or some portion of the output, is fed back in some way to the input of an earlier block, thus causing a closed loop to exist around the system. This makes it more difficult to locate any faulty block within the loop, since the outputs of all blocks will probably appear at fault. This is similar to a completely d.c. coupled system where an incorrect voltage at one point causes all other voltages to be incorrect.

First of all the type of feedback being used in the system and its purpose must be understood. The feedback may only be used to *modify* the characteristics of the

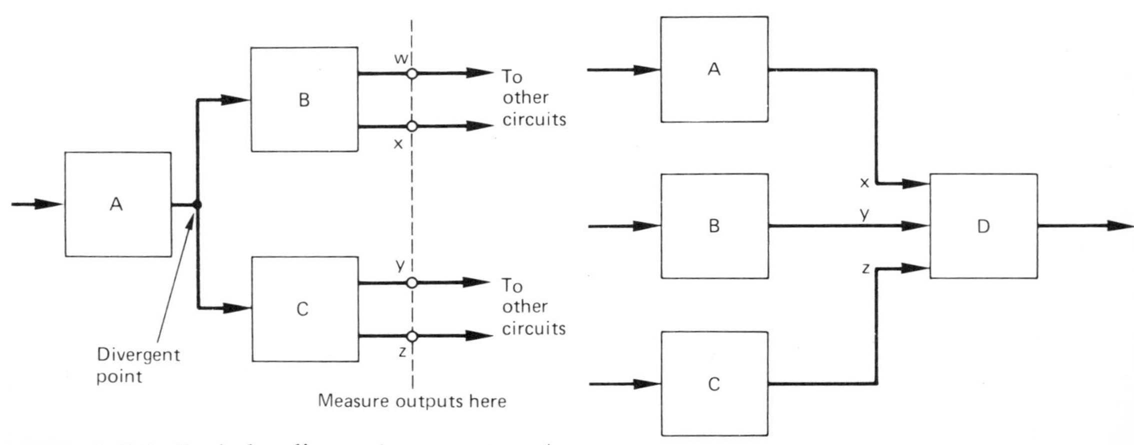

FIGURE 7.6 Typical divergent arrangement within a system.

FIGURE 7.7 An example of convergence.

system as is the case in automatic gain control circuits used in superheterodyne radio receivers; or the feedback may be totally essential for an output to exist. This latter type of feedback is called *sustaining*, since a feedback signal has to be present to maintain an output of either some oscillation or fixed level. Sustaining feedback is used in many position-control systems, where the feedback signal, proportional to the position of some output device, is used to cancel the effect of an input reference level. As the output motor is driven, the feedback signal moves toward the same value as the reference input and the error signal reduces to zero. In this way the output is fixed and held at the desired position. Any fault causing a break in the feedback loop would result in the output being driven to one or other of the extreme limits, i.e., to reach its end stop.

Having decided the type of feedback, the way in which it is connected, and its purpose, then the best course of action can be taken to locate any fault. With modifying feedback it may be possible to break the feedback loop, thus enabling each block to be tested separately without a fault signal being fed around the loop. The feedback is best disconnected at the input block end, but naturally care has to be taken with any changes like this, since the feedback may be providing both d.c. bias and an a.c. modifying signal. In this case the a.c. portion of the feedback can be eliminated simply by decoupling to ground via a suitable capacitor.

If sustaining type feedback is disconnected from the input, it may be possible to inject a suitable signal in its place and then check circuit block outputs, including of course the feedback element. Since such

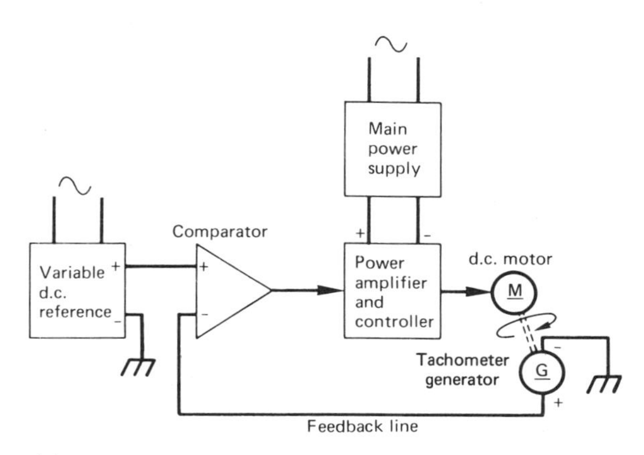

(a) Speed Control System

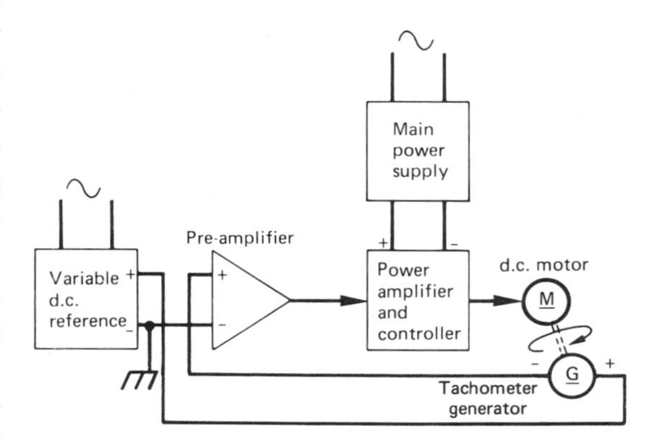

(b) Alternative Method For Connecting Feedback

FIGURE 7.8

a wide variety of feedback circuits exists, no standard rule for troubleshooting can be used. Knowledge of the system, understanding of the operation, and a logical approach are essential. For example, consider the block diagram of a motor speed control system shown in Figure 7.8. The speed of the d.c. motor is set by the level from the reference supply and held constant by the feedback applied via the tachometer generator. A tachometer generator is a device that produces a d.c. output voltage proportional to its speed of rotation. When the motor has reached the desired speed, the d.c. feedback signal from the tachometer generator balances the input reference voltage. The difference signal, after amplification, is just sufficient to keep the motor running constantly at the desired speed. Imagine a break in the feedback lead. This would cause the feedback signal to the comparator to be zero, and the motor would tend to run at maximum speed irrespective of the setting of the reference voltage. Faults causing the motor to run at maximum speed could be in either the comparator, controller, the tachometer generator (open circuit) or an open circuit feedback line. To locate the fault the sequence of tests would be:

(a) Measure tachometer generator output. A relatively large d.c. output should be present, since the motor is running at high speed. If this is correct then:

(b) Measure feedback signal at the inverting input of the comparator. This should be the same as the d.c. level measured in (a). If this is correct then:

(c) Check the comparator output, which should only be a low-value d.c. level when the variable reference is set to near minimum. If this last test is correct, then the fault can only lie in the power amplifier and controller.

To illustrate the changes in troubleshooting conditions if the feedback is connected differently, study the block diagram in Figure 7.8*b*. In this case the d.c. output from the tachometer generator is connected in series with the input reference supply. When a reference d.c. is applied the motor runs, and as its speed increases the output signal from the tachometer generator also increases. This signal is subtracted from the reference d.c. input so that the pre-amplifier receives an input that is just sufficient to keep the motor running at the required speed. The operating characteristics for this method of connection will be almost identical to that of the previous circuit, but a fault causing a break in the feedback line will result in zero input to the preamplifier and the motor will not run at all. If such a fault did occur, a simple test to check system operation would be to inject a small d.c. voltage at the preamplifier input. If the motor then runs, the fault must lie in the tachometer generator, the feedback lines, or in the input reference supply.

Further examples of systems with feedback loops are discussed later in this chapter.

7.3 SYSTEM TROUBLESHOOTING AIDS

When any electronic system fails, it is the service technician's task to verify, locate, and then repair any fault. This, of course, is usually required to be completed in the shortest possible time. To achieve rapid troubleshooting and repair so that system down-time is low, the technician needs to be provided with aids to back up his or her troubleshooting skill. Among the most important of these aids are:

(a) Maintenance manuals and troubleshooting guides.

(b) Test instruments.

(c) Specialized tools.

Note that aspects of the overall design such as accessibility of components, provision of fault display, and built-in test circuits cannot be considered as troubleshooting aids in this context. These aspects are desirable and should be provided by the designer to achieve high values of maintainability. What is considered as a direct aid is any provision of information and of test gear and tools that assists in the troubleshooting task.

Before considering specific aids in more detail, it should be stated that aids such as maintenance manuals or troubleshooting guides are not always available. The technician can then rely only on his or her own knowledge, skill, and experience to deal with system faults. Experience of other similar systems can allow the technician to repair faults without manuals, and in some circumstances it may be possible to compare the faulty system with an identical working model. However, where a system is unknown and no other model exists, it is always wise to seek information before starting any tests to locate a fault. Proceeding with tests without knowing exactly how the system works could lead to incorrect conclusions, cause confusion between an operational error and an actual fault, or in the worst case cause extra faults.

An important aid is the MAINTENANCE MANUAL. The preparation requires a high degree of skill, for it is essential that only necessary information, i.e., information strictly related to maintenance, is provided. Otherwise it will be a hindrance rather than a help. What are the important aspects of a good maintenance manual? The essential features listed, in the usual order of appearance, are:

1. Description of the system with an explanation of its use.
2. Performance specification.
3. Theory of operation.
   - (a) System (refer to block diagram).
   - (b) Individual circuits (refer to circuit diagrams).
4. Maintenance
   - (a) Preventive (if required): i.e., replacement of parts subject to wear, recalibration, and lubrication.
   - (b) Corrective:
     - (i) Methods for dismantling, including any safety procedures.
     - (ii) List of test instruments and special tools required.
     - (iii) Test instructions.
     - (iv) Troubleshooting guides and suggested troubleshooting procedures.
5. Circuit diagrams.
6. Spare parts list.
7. Mechanical layout: either photographs, line drawings, or possibly exploded views of the mechanical structure.

To appreciate the importance of the maintenance manual, it is a good idea to obtain or borrow a copy of a manual for a test instrument such as a CRO. If possible, manuals from different manufacturers could also be compared. In a manual, irrelevant material should not be included and any written instructions must be clear and concise so that ambiguous meanings are not implied. Diagrams should follow ANSI Y32.2-1975 and ANSI Y32.14-1973. In general, a diagram should be arranged so that a reader is able to quickly and easily understand its meaning. This is achieved by

(a) Using correct symbols.

(b) Good relative arrangement of symbols.

(c) Careful routing of interconnections.

Where there is a clear sequence from cause to effect, such as the signal flow from input to output in a tv receiver, this should be drawn from left to right. If a diagram contains a number of circuits arranged between common supply lines, the circuits should be arranged in functional groups, preferably in the order in which they operate. Additional information such as voltage levels and typical waveforms is often provided to further assist troubleshooting. These will be given next to the lines representing conductors or signal paths.

It is in the CORRECTIVE MAINTENANCE SECTION of the manual that the most useful information to assist in troubleshooting will be found. Instructions on dismantling and safety will be given, followed by test and troubleshooting guidance. The safety instructions should always be noted as they will lower the risk of accidents to the service staff and point out precautions that must be taken to protect any sensitive components. Since most electronic systems have an a.c.-derived power source and possibly high internal voltages, proper safeguards must always be achieved. The section will also contain a list of the necessary test gear together with details on any setup and adjustment procedures. This is obviously useful when a fault has been repaired and the system has to be checked for satisfactory operation, but in the case of a fault the most useful aid is some form of troubleshooting guidance. This may take several forms. For example, tables may be provided showing typical symptoms for various fault conditions, together with the most probable causes and/or suggested courses of action to narrow down the area of the fault. Probably the most useful aid is the "troubleshooting guide," which is actually a series of programmed steps shown in block diagram form, starting from a particular set of troubleshooting symptoms.

To fully appreciate this type of guide, an example is given in Figure 7.10, which is a programmed troubleshooting guidance chart for the digital weighing system shown in block form in Figure 7.9. By obtaining an answer of yes or no to a question at each stage, mostly by measurement or observation, the technician performing the troubleshooting is guided to the faulty component or part. In the example a total inoperative condition for the system is considered and this naturally suggests a power failure of some kind. An experienced technician would not require such guidance, since he or she would automatically work through the tests suggested. This is because of the skill acquired through

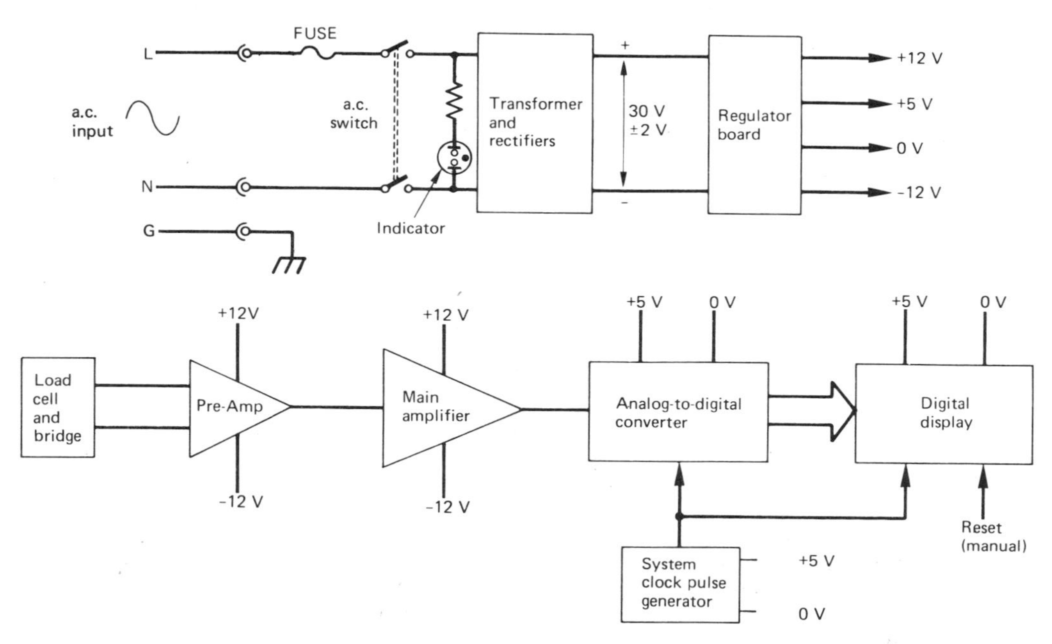

FIGURE 7.9 System block diagram for example of troubleshooting guide: electronic weighing system.

experience. However, such techniques can be extended to cover complex systems and faults in systems that are difficult to find. Guides can then be very helpful in speeding up the location and repair of faults. Preparing troubleshooting guides is itself a useful exercise. Try writing a troubleshooting guide for a failure of the main amplifier in the weighing system. The symptoms would be that the display would be "stuck," probably in the reset position (reading all zeros), and that no change in display would take place when the load cell is operated.

TEST INSTRUMENTS of some kind are essential aids for troubleshooting and maintenance. Apart from any specialized instrument that may be required for, say, communication systems or complex digital instruments, a large majority of system faults can be located using only three standard test instruments:

(a) Multirange meter (either analog or digital).

(b) Oscilloscope.

(c) Signal generator.

Intelligent use of one, or a combination, of these three instruments can speed the troubleshooting process, provided that the performance and limitations of the type of instrument being used are fully appreciated. This means understanding the accuracy, resolution, loading effect, and bandwidth of the test instrument. The performance specifications for some of these instruments are discussed in Chapter

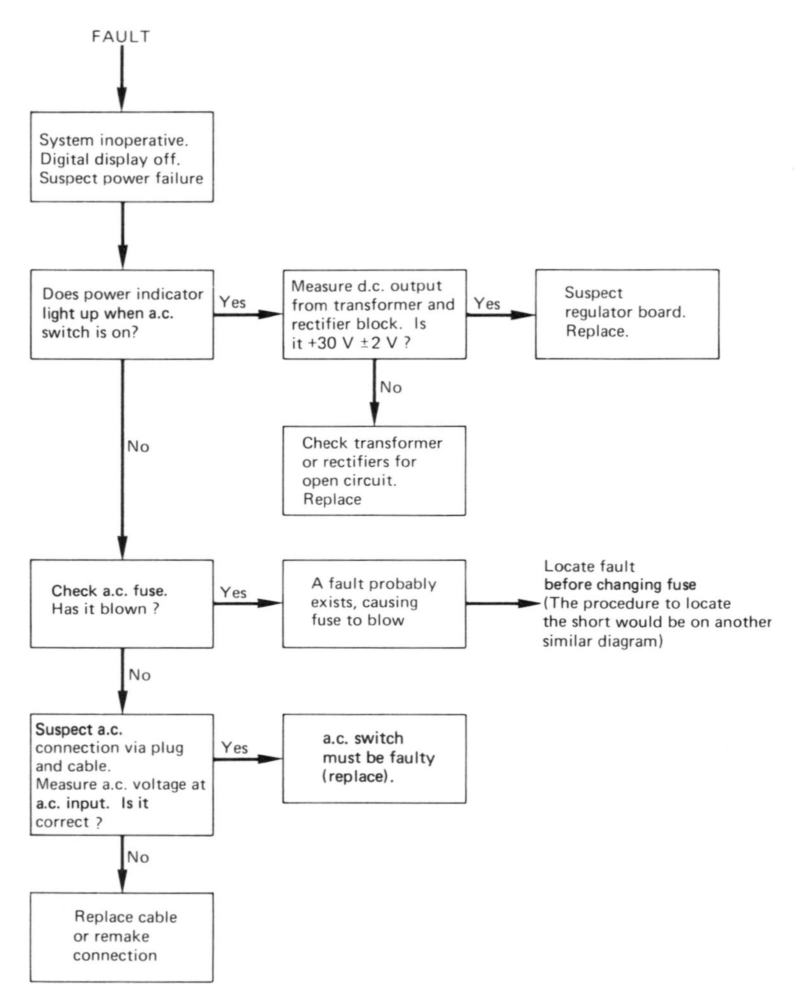

FIGURE 7.10 Troubleshooting guide.

1, and typical specification figures are given in Table 1.4. For many service tests, high accuracy is not required, since the measurement is most likely to be of a d.c. bias level that may have wide tolerance limits or to check that a signal is present. However, it is worth bearing in mind any error likely to be introduced by the meter and its test leads.

For example, suppose the voltage at some test point is written in the maintenance manual as +30 V ± 2 V when it is measured with a 20 kΩ/V MULTIMETER. This statement, although taking into account the loading effect of the meter on the circuit, may not allow for any instrument error. The accuracy of an analog meter is typically ±3% of full-scale deflection. This means that on the 100 V d.c. range a reading of 30 V would have an accuracy of ±3 V. Thus even if the voltage measured at the test point was +33 V, this could be within specification, and a fault condition is not necessarily indicated. It is more the presence of a value near the one stated in the manual that is required rather than a measured value that is exactly as predicted. For this reason an analog multimeter can often be used in preference to a digital type, since an indication of approximate voltage level can be quickly observed on the analog instrument. A digital meter, of the type that has a low sampling rate, can in some instances give an indication that is misleading. The digital multimeter, however, is invaluable when high accuracy is required, or when very small changes in a level have to be detected. In addition, the digital meter has the advantage of a high input impedance, typically 10 MΩ, so that its loading effect is slight.

The CATHODE RAY OSCILLOSCOPE (CRO) is a versatile and extremely useful instrument. With it, it is possible to measure both d.c. values and a.c. waveforms. The sensitivity is usually high, typically 10 mV/div, and the loading effect is slight, since the input impedance is usually greater than 1 MΩ. The frequency, shape, and time period of a single waveform can be determined, or waveforms can be displayed in time or phase relationship to one another. This is easily achieved either with a single-beam CRO or dual-beam type, since the reference signal can be used to directly trigger the CRO's time base. The accuracy of both Y (amplitude) and X (time) channels is at best ±3%. At low frequencies the voltage signal to be measured can be taken directly to the Y-input via suitable wires or a coaxial cable. For high frequencies, to avoid the possibility of signal degradation, a fully shielded probe should be used. This is because a simple coaxial lead will behave like a badly matched transmission line between the test point and the CRO Y-input, causing attenuation and phase distortion. The cable capacitance of coaxial lead is typically 50 pF per meter, and this will be placed in parallel with the CRO's input capacitance across the test point further degrading the signal. The use

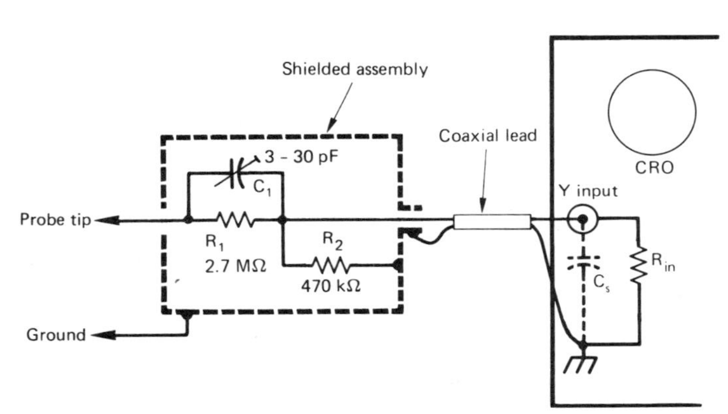

FIGURE 7.11 Passive probe unit for an oscilloscope.

Probe gives 10:1 attenuation.

Time constant R_1C_1 is made equal to the input time constant of the CRO $R_{in}//R_2 \cdot C_s$. Where C_s is coaxial lead capacitance plus input capacitance of CRO. When properly adjusted, the probe presents low capacitance to the circuit being measured ($C_1 \cong 10$ pF) and it operates as a simple resistive divider.

TABLE 7.1 Properties of General-Purpose Measuring Instruments

Instruments	Typical Use	Accuracy	Comments
Analog multirange meter (moving coil)	Measurement of d.c. and a.c. voltage and current Ohms range 1 Ω to 20 MΩ Bandwidth: 15 Hz to 15 kHz Impedance d.c. ranges 20 kΩ per volt a.c. ranges 10 kΩ per volt	±3% FSD	Sturdy and well-proven instrument. Good ranges (3 V to 3 kV). Loading effect must be taken into account in medium and high impedance circuits.
General purpose digital multirange meter ($3\frac{1}{2}$ digits)	Measurement of d.c. and a.c. voltage and current. Ohms range 1 Ω to 20 MΩ. Impedance: 10 MΩ in parallel with 100 pF. Bandwidth: 45 Hz to 10 kHz	±0.3% of reading ±1 digit.	Good sensitivity and resolution. Easy to use and read. Loading effect can be neglected in most cases.
General purpose cathode ray oscilloscope	Measurement of d.c. levels, a.c. voltage, frequency, waveshape, rise and fall times. Comparison of time or phase relationship between signals. Bandwidth: d.c. to 10 MHz. Impedance: 1 MΩ in parallel with 20 pF.	±3% amplitude and time.	Versatile. Gives direct visual information on waveform of signal. A probe unit must be used for best results at higher frequencies.

of a properly compensated probe unit will reduce these effects considerably. A simple probe is basically a resistive attenuator with capacitive compensation as shown in Figure 7.11.

Comparisons of the measuring instruments described are given in Table 7.1.

SIGNAL or FUNCTION GENERATORS are used in maintenance when it is required to inject some suitable test signal into the system. The complexity and performance characteristics of the instrument is usually dictated by the system under test, but a very useful aid is a small handheld signal injector. For analog systems this is usually a simple fixed-frequency battery-powered oscillator running at 1 kHz, with its output available from a metal prod and a lead with an alligator clip provided for ground connection. A simple device like this can be easily constructed and is always available for use, since it is carried around in the pocket. In the same way logic pulsers and logic state sensors can be built and used for checking digital systems (see Chapter 4, Section 4.5). The design for the 1 kHz signal injector can be extended to create a *continuity tester*. This is a battery-powered 1 kHz oscillator with an audible output provided by a small loudspeaker. When the two output leads are shorted together, or connected via the low resistance of a cable wire, the oscillator's output is fed to the loudspeaker. A small tester like this will prove very useful in checking the continuity of cables, connecting wires, and p.c.b. traces.

Naturally, in some instances, it will be necessary to use other more sophisticated types of test instruments for maintenance. The use of frequency counters, spectrum analyzers, storage oscilloscopes, chart recorders, XY plotters, and microprocessor-controlled troubleshooting instruments will increase. However, it is the understanding of the use of basic aids that is essential as in some cases they are the only instruments available. In the exercises that follow only a multimeter and CRO are required.

7.4 EXERCISE: LIGHT-CONTROLLED FLASHING UNIT (FIGURE 7.12)

This simple system, with a full circuit shown in Figure 7.12, will serve as an introduction to system troubleshooting methods. The lamp automatically flashes on and off, at a low frequency, if the ambient light falls below a reference level. By picturing the circuit as separate blocks as shown, the operation can be more easily followed and troubleshooting will be speeded up.

A light-dependent resistor (l.d.r.), or photoconductor, R_{10}, is used as the transducer that senses the ambient light level and the changes in resistance of this l.d.r. are detected and compared with a reference level by a transistor amplifier (Q_1). If the light level is high, the l.d.r. R_{10} will have a low resistance, and the output from the amplifier will be low. The oscillator will then be prevented from operating and the lamp will remain off. As the ambient light level falls, the R_{10}'s resistance rises. At some point the output of the amplifier will rise positive as Q_1 switches off and the oscillator will be free to run. Pulses, at a frequency determined approximately by R_3C_1, will appear at the output of the UJT low frequency oscillator to drive the thyristor bistable circuit. The lamp will flash on and then off for every two pulses received from the oscillator. While the light level

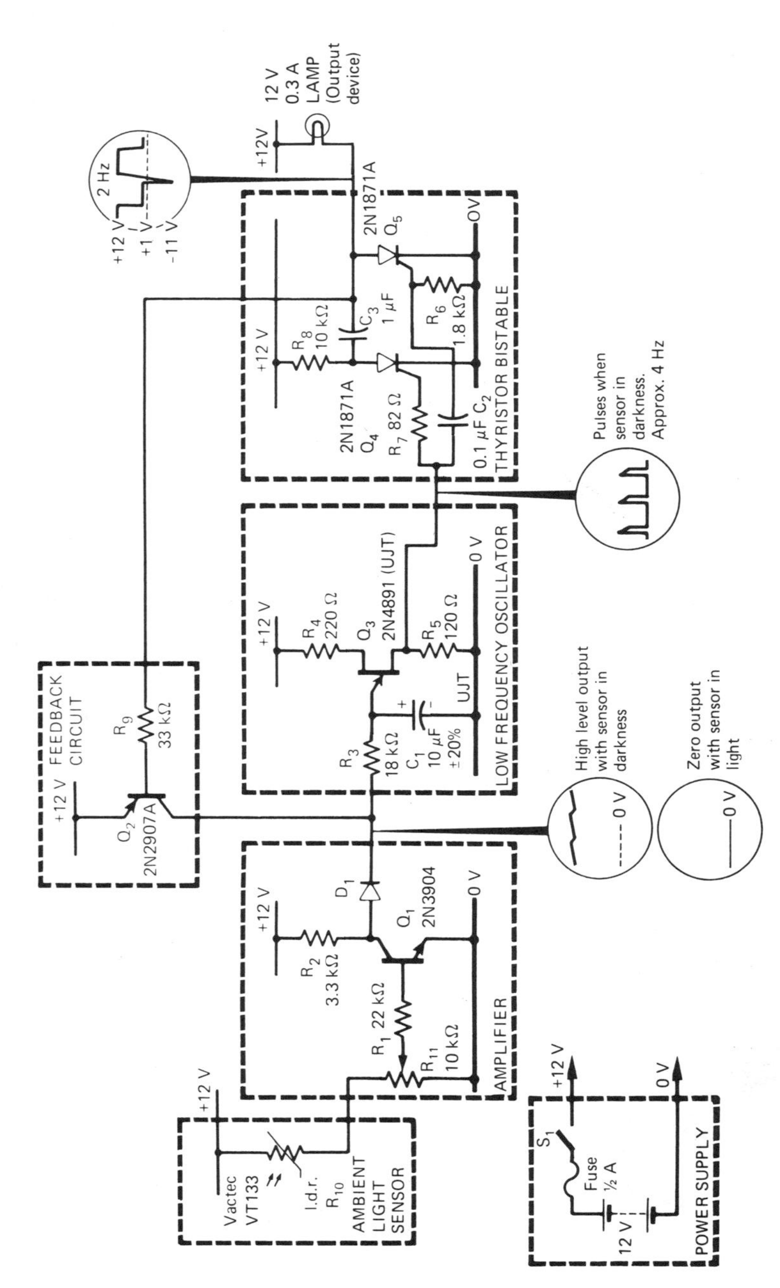

FIGURE 7.12 Light-controlled flashing unit.

sensed by R_{10} remains low, the lamp will continue to switch on and off. A feedback circuit, of the modifying type, is used to ensure that the oscillator always produces another output pulse every time the lamp switches on. In this way the possibility of the lamp remaining on, consuming and wasting power, when the ambient light has returned to a higher level is avoided. When the lamp is on, the input to the feedback circuit via R_9 goes low (1 V); this causes Q_2 to conduct and the input voltage to the oscillator is raised to almost +12 V. This will occur even if the output from the amplifier has in the meantime returned to zero volts.

The preceding paragraph contains a description of the operation of the overall system and an understanding is what is required to enable faults to be narrowed down to one block. Imagine a system failure in which the lamp does not light or flash even with the sensor in complete darkness. This could be caused by a fault in any block except in the feedback circuit. The steps to locate the faulty block would be as follows:

1. Check +12 V power line.
2. Test the operation of the lamp. This can be easily done by momentarily shorting Q_5's anode to ground. The lamp should light and then turn off when the shorting link is removed.

Having proved that the power supply is providing the +12 V d.c. to all blocks and that the lamp operates correctly, a troubleshooting method can then be used to isolate the fault. This could be:

3. Proceed with input to output tests.
   (a) Put the l.d.r. into darkness and measure the input to the amplifier. This should be low, typically less than 1 V.
   (b) Measure the output of the amplifier. This, with the l.d.r. in darkness, should be high. A typical value is +9 V.
   (c) Test that pulses are produced by the UJT oscillator. The presence of the 4 Hz pulses can be just detected by a multirange moving coil meter on the lowest d.c. voltage range. Alternatively a CRO can be used.

If these checks prove correct, then the fault must be in the thyristor bistable circuit. Since the lamp is not switching on, the suspect component will be Q_5 either open circuit anode to cathode or open circuit gate to cathode, or an open circuit connection or component to its gate.

4. Alternatively the operation of one-half the circuit could be quickly verified by a form of half-split method as follows.

Temporarily connect a 3 kΩ resistor from Q_2 base to ground. Q_2 should conduct and apply nearly +12 V to the oscillator's input. This will override any output state from the amplifier. The oscillator should then cause the lamp to switch on and off continuously, irrespective of the condition of the l.d.r., while the test resistor is left connected. If the lamp does not switch on and off, the fault lies in either the oscillator or the thyristor circuit. If the lamp does switch correctly, then the fault must lie in the l.d.r. or its amplifier. Suppose, however, that the lamp does not operate, the next check would be to measure the output of the oscillator. Note that, with the half-split method, only two tests would be required to isolate the fault to one block.

Consider a fault with the symptoms that the lamp switches on and off continuously irrespective of the ambient light level. A fault in either the oscillator or the thyristor bistable can be ruled out, since it follows that if the lamp is switching then both of them are operating correctly. The fault must lie in the l.d.r., its amplifier, or possibly in the feedback circuit. For example, a short circuit collector/emitter on Q_2 will cause the symptoms. To locate the fault the first check would be on the l.d.r., testing the input to the amplifier with the l.d.r. first in a high level and then in a low level of light. With many systems the input transducer or actuating device may be positioned remote from the rest of the blocks. The transducer and its connecting leads are therefore more prone to damage than other component parts of the system and should be checked first. If, however, the l.d.r. and its leads are proved correct, then the amplifier or the feedback circuit is at fault. The output of the amplifier would be measured. If this was high at, say, nearly +12 V, then the fault would be in the feedback circuit with Q_2 short circuit collector to emitter.

QUESTIONS

1. **(a)** Which block would be faulty if the lamp, which tests good, does not turn on when the power is applied?

   (b) State all the possible component failures within the faulty block that would give the symptoms listed in **(a)**.

2. **(a)** What would be the symptoms for an open circuit in the feedback circuit (i.e., R_9 open circuit)?

   (b) Write a troubleshooting guide for the above fault.

3. What would be the difference in troubleshooting symptoms for:

   (a) The l.d.r. open circuit?

   (b) The l.d.r. short circuit?

4. The oscillator is suspected of having failed. Describe a quick method for verifying the operation of the thyristor bistable.

5. Following tests, a fault is suspected in the amplifier. The voltage readings taken with a 20 kΩ/volt meter on Q_1 are as follows:

Q_1 Base	Q_1 Collector
0.7 V	0.7 V

   State the nature of the fault and explain how the symptoms indicate that the amplifier is faulty.

6. Briefly explain the operation of the UJT oscillator and state the probable maximum and minimum limits to the operating frequency. Which components determine these limits?

7.5 EXERCISE: OVER-TEMPERATURE ALARM SYSTEM (FIGURE 7.13)

A thermistor R_{13}, type GM473, is used in this system to sense the temperature changes in an enclosure. As the temperature varies, the changes in thermistor resistance are used to unbalance a Wheatstone bridge circuit. The output of the bridge is fed to U_1, a 741 op-amp, which is used as a comparator circuit. While the temperature is below the trip point, the thermistor has a higher resistance than R_{12} and the inverting input of the op-amp is at a more positive voltage than the noninverting input. Consequently, the op-amp output voltage is low, typically less than +2 V. This low level is well below the threshold of the CMOS gate U_2A and both the low and audio frequency oscillators are inhibited from operating. As the temperature increases, the thermistor resistance falls, until at some point the thermistor resistance becomes less than the set value of R_{12}. Then the d.c. level to the inverting input goes more negative than the noninverting input. The comparator output switches to a high output state (typically greater than +7 V), and the low frequency oscillator is free to run. The output of this CMOS oscillator switches between a high and low output state at approximately 1 Hz and is connected directly to Q_1, the gate of a VMOS power FET. VMOS FETs are capable of switching currents of several amps and have the advantage of very high input impedance, thus presenting negligible load to a CMOS gate output. The lamp, a 12 V 12 W type, will be switched on and off at 1 Hz and will take about 1.25 A peak from the supply. At the same time the output of the 1 Hz oscillator is used to gate the operation of the 2 kHz audio oscillator. A burst of 2 kHz pulses, every 0.5 sec, drives Q_2 the second VMOS FET to give an audible alarm tone from the 15 Ω speaker. A small amount of hysteresis is provided by R_3 to ensure that positive switching takes place at the op-amp output.

An a.c.-derived power supply is used to provide +12 V unregulated for the output devices and +9 V, zener regulated, for the op-amp U_1 and CMOS IC U_2. The supply to U_1 and U_2 is kept to a lower value than +12 V to ensure that the switching of the output devices does not affect the circuit operation and also to prevent damage to U_1 and U_2 if any fault condition arises. The 4011B, CMOS Quad 2I/P NAND gates, have a maximum power rating of 400 mW. If a +12 V supply is used, a fault causing the input point (A) to reside at +6 V, the upper threshold value, causes each gate to draw a current of about 10 mA, and the device dissipation is exceeded. This cannot happen if a 9 V supply is used.

Because two output devices are used, system troubleshooting is much simplified. For example, suppose that when the thermistor is heated the audio alarm sounds correctly but the warning lamp does not flash. It is obvious that the fault must lie in the lamp or Q_1 circuit, since both oscillators must be functioning correctly for the audible alarm to sound.

A further example could be a condition in which both alarm circuits fail to operate when the temperature is above the trip value. If the +12 V power indicator is on, the fault is most probably located in the thermistor bridge, the comparator or the CMOS oscillator circuits. It is unlikely, although possible, that both output circuits will have failed simultaneously. One measurement at point (A) will then indicate

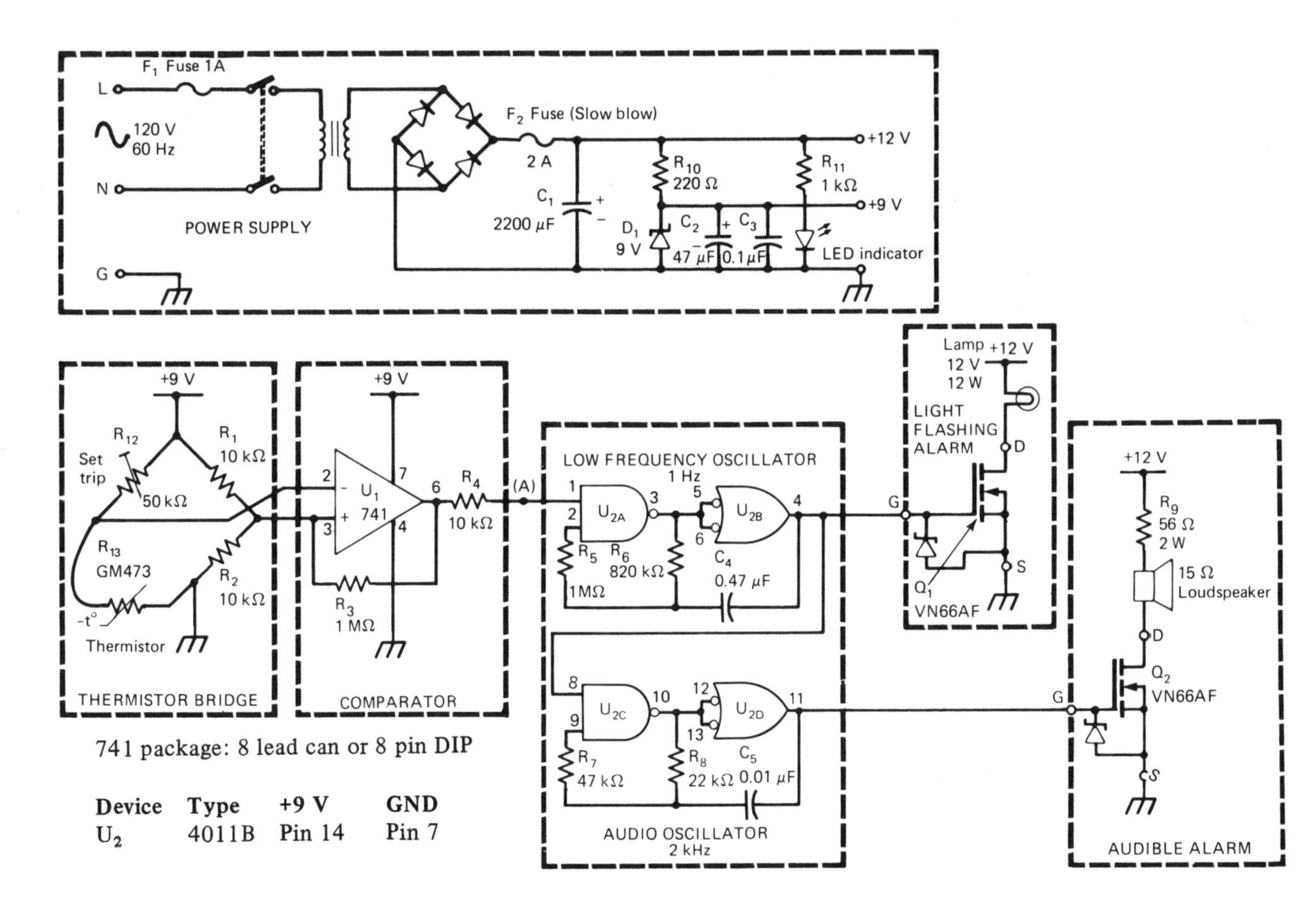

741 package: 8 lead can or 8 pin DIP

Device	Type	+9 V	GND
U_2	4011B	Pin 14	Pin 7

FIGURE 7.13 Over-temeprature alarm system.

which section of the circuit is at fault. If the voltage at (A) is high at +7 V, then the fault must be in the oscillator block.

QUESTIONS

Note that the GM473 has a resistance of approximately 47 kΩ at 25° C which falls to about 10 kΩ at 80° C.

1. The system fails with both the lamp on and the audible alarm giving a continuous 2 kHz tone irrespective of the input temperature. The lamp does not flash on and off. State the block that has failed and the likely type of failure.
2. Devise some simple test modifications to enable rapid testing of
   (a) The oscillator and output alarm circuits.
   (b) The whole system except the thermistor.
3. Write a troubleshooting guide for a failure of the +9 V regulated supply. Assume either R_{10} has gone open circuit or that C_2 is short circuit.
4. The 2 A fuse has blown and a replacement blows immediately at power on. Which component would be suspect and how could this be quickly checked?
5. Describe the symptoms for the following component failures and in each case list the most appropriate series of test that should be made to narrow down the fault to one block:
   (a) Thermistor open circuit.
   (b) C_5 short circuit.
   (c) R_4 open circuit.
   (d) R_5 open circuit.
   (e) Q_2 drain to source short circuit.

7.6 EXERCISE: PULSE GENERATOR (FIGURE 7.14)

A pulse generator is a very useful instrument for troubleshooting and testing both analog and digital systems. The essential features of such a generator are that the frequency, pulse width, and pulse delay of the output be variable over a wide range. At the same time these output pulses must be presented via a low impedance drive and have fast rise and fall times. In addition, facilities such as external gating and a synchronizing pulse output are also very useful. By gating externally it is possible to modulate the generator so that a burst of pulses is provided when the gate input is active. This has obvious uses in the testing of some sequential logic systems. A synchronizing output pulse of short duration is essential when the generator is used in conjunction with an oscilloscope in system testing. The pulse, which occurs before the pulse from the generator output, is used to externally trigger the oscilloscope timebase. At the same time the pulses transmitted through the system can be examined by the Y channel. In this way the leading edges of the pulses being

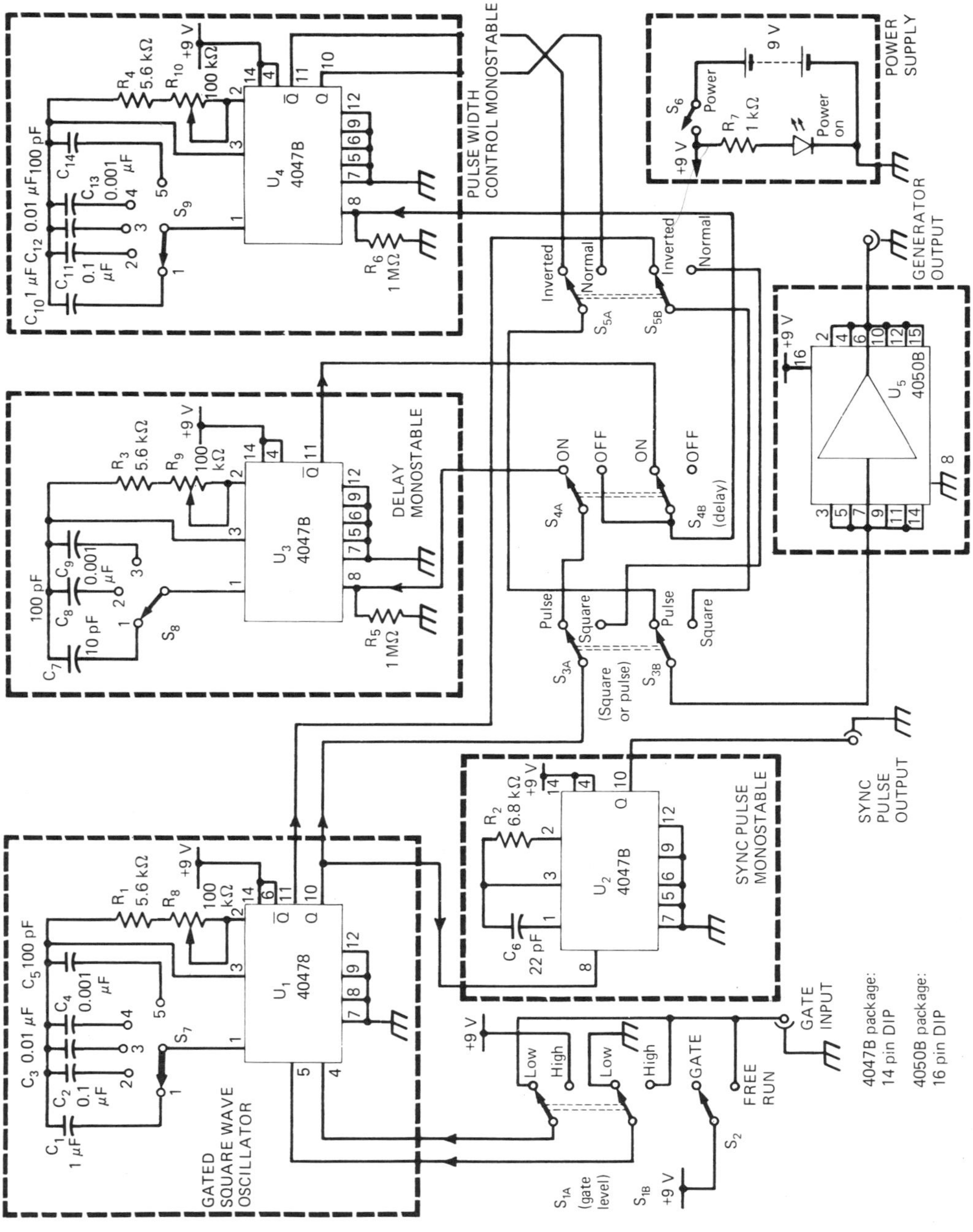

FIGURE 7.14 Pulse generator.

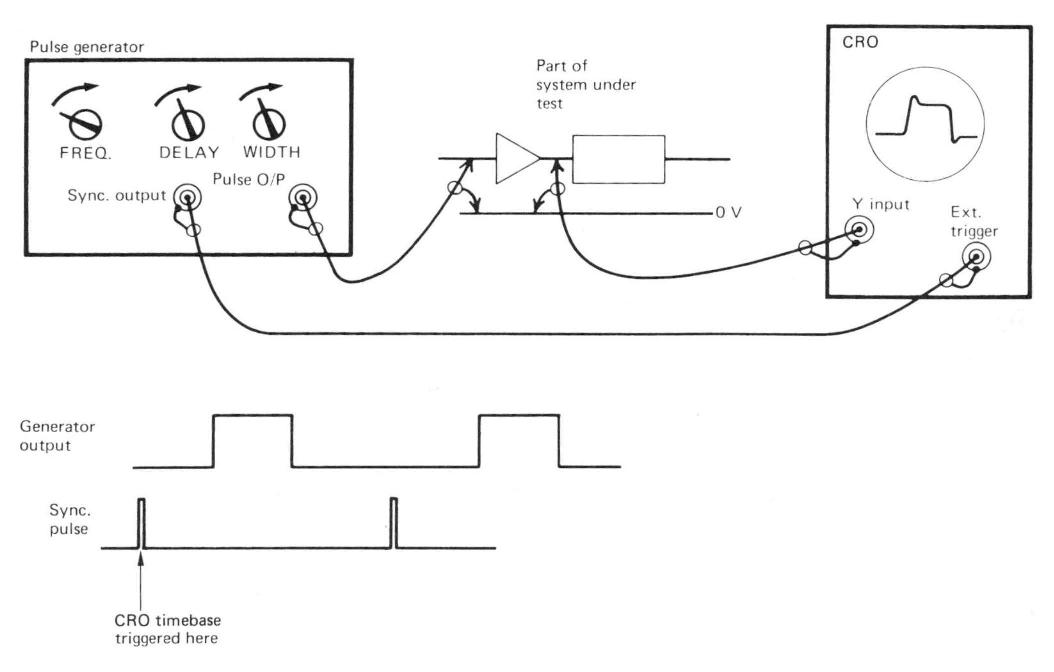

FIGURE 7.15

examined are not lost, and the full pulse is displayed on the oscilloscope. This is illustrated in Figure 7.15.

A relatively inexpensive pulse generator can be built using the 4047B CMOS monostable/astable multivibrator IC. This example provides most of the facilities described in the preceding paragraph. The most expensive items are the switches and case. The 4047B is a 14 pin DIP IC shown in Figure 7.16 in which the lettering of the pins is also explained. Basically, by suitably connecting an external capacitor and resistor, and the enabling of triggering inputs, the IC can be made to act as an astable or monostable multivibrator. To achieve the astable mode of operation, pins 7, 8, 9, and 12 of U_1 are connected to 0 V and pins 6 and 14 are connected to $+V_{DD}$, where V_{DD}, in common with other CMOS ICs, can be any voltage in the range of +3 V to +15 V. By these connections the trigger inputs and master reset are disabled. If the enabling input E_{A0} on pin 5 is made high ($+V_{DD}$) and the $\overline{E}_{A1}$ input on pin 4 made low (0 V), then the IC will oscillate at a frequency determined by the external components R_X and C_X. A bistable is included to divide the oscillator frequency (available on pin 13) by two and to give buffered (low impedance) outputs on pins 10 and 11. These will be perfect square waves with a 50% duty cycle. The output frequency on pins 10 or 11 is given approximately by the formula:

$$f \approx \frac{1}{4.4 R_x C_x}$$

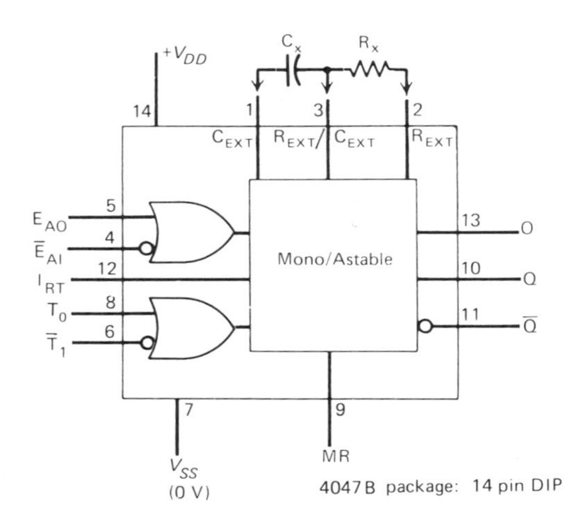

FIGURE 7.16 4047B Monostable/astable CMOS multivibrator:

C_{EXT} external capacitor connection
R_{EXT} external resistor connection
R_{EXT}/C_{EXT} common external connection
I_{RT} retrigger input
T_0 Trigger input (low to high)
$\bar{T}_1$ Trigger input (high to low)
E_{A0} enable input (active high)
$\bar{E}_{A1}$ enable input (active low)
MR Master reset
O Oscillator output (2*f*)
Q True } Buffered 50% duty cycle
$\bar{Q}$ Complement } square wave outputs

By rearranging the pin connections the 4047B can be made to act as a monostable, giving one output pulse for every trigger input. The output pulse width will be determined by the external components. In this example four 4047B ICs, U_1 through U_4, are used to give the functions as follows:

U_1 Gated square-wave oscillator (frequency range set by S_7 and fine control by R_8).
U_2 Sync pulse monostable (fixed pulse width).
U_3 Delay monostable (delay times set by S_8 with fine control adjusted by R_9).
U_4 Pulse width control monostable (Pulse width set by S_9 with fine control by R_{10}).

The fifth IC U_5, a 4050B, is used as the output amplifier with its six noninverting buffers wired in parallel to provide a low impedance output with good drive capability.

Having grasped the operation of the various ICs, the most difficult task is that of working out the purpose of the various switches. This is especially true for troubleshooting, since the switching in even a simple example such as this can appear at first sight rather difficult. The functions of the various switches are as follows:

S_1 Gate level control—in the "low" position a logic 0 (i.e., a low level) on the gate input will enable the main oscillator to run. In the "high" position a logic 1 (i.e., a high level) is required.
S_2 Gate/free-run switch. In the free-run position +9 V is connected to pin 5 (E_{A0}) of U_1 and the oscillator runs.
S_3 Square/pulse output. Signals from either U_1 or U_4 are routed via this switch to U_5.
S_4 Delay ON/OFF switch. This switches U_3 either in or out of circuit.
S_5 Invert/normal. Allows negative going pulses from logic 1, or positive going pulses from logic 0 to be selected.

In the switch positions as shown in the diagram the true output at pin 10 of U_1, when gated by a low level at pin 4, will be passed via S_{3A} and S_{4A} to the trigger input of U_3 pin 8. The pulse from U_3 pin 11 is routed via S_{4B} to the trigger input of U_4 at pin 8. Note that the complement output of U_3 ($\overline{Q}$) is used so that U_4 is triggered on by the trailing edge (i.e., the low to high transition) of the delay pulse. Finally the complement output of U_4 is passed via S_{5A} and S_{3B} to the output amplifier.

Once the operation of the IC blocks and the switches and the switch wiring is understood, troubleshooting is relatively straightforward. For example, if correct pulse outputs can be obtained but no square waves, then the fault must be in switch S_3 or its wiring, since all the ICs must be operating correctly to give pulse outputs. On the other hand, if square waves can be obtained but only nondelayed pulses with S_4 in the OFF position, then the fault may be in S_4, its wiring or in U_3, the delay pulse generator.

With this generator the outputs are suitable for testing CMOS logic. The five frequency ranges selected by S_7 are from 2.5 Hz up to about 500 kHz. The output pulses can be delayed from the sync. pulse from about 1 μsec to 200 μsec, and the output pulse widths selected by S_9 and R_{10} can be varied from 1 μsec to 200 ms.

QUESTIONS

1. What would be the most suitable service instrument for troubleshooting on the pulse generator?
2. The pulse generator fails to produce square waves or pulses in the free-run position of S_2, but can be gated into correct operation. State the location of the fault and any tests that should be made to identify the actual faulty part.
3. Describe the symptoms for the following faults:
   (a) An open circuit connection to pin 10 of U_1.
   (b) R_{10} wiper open circuit.
   (c) 9 V line to pin 14 of U_3 open circuit.
   (d) An open circuit wiper on S_{4B}.
4. A failure occurs so that neither square waves nor pulses can be obtained at the generator output. However, sync pulses are present at the correct frequency. Which areas of the system are suspect?
5. Write a troubleshooting guide for the symptom of no sync output pulses.
6. State the probable cause of failure for the following symptoms
   (a) Delayed pulses can be obtained, but when S_4 is in the OFF position no pulses can be obtained from the output.
   (b) The power indicator is off and all outputs are dead.
   (c) No variation in pulse or square wave frequency can be obtained.
   (d) No inverted output pulses can be obtained.

7.7 EXERCISE: FUNCTION GENERATOR (FIGURE 7.17)

This is an analog system using op-amps to generate sine, square, or triangle waveforms over the frequency range from 1 Hz to 10 kHz. Triangle waves are produced at the output of the integrator because of the closed feedback loop formed by the comparator, integrator, and reference switch. This means that square waves are produced simultaneously at the output of the comparator. The frequency of the waveforms can be selected by S_1 and varied by R_{27}. The amplitude of the triangle wave is set by the reference switching circuit to 5 V pk-pk.

A simple waveform shaper is used to convert the triangle wave into an approximate sine wave, and the square and triangle waves are reduced to the same amplitude as the sine wave by attenuators. Switch S_2 selects sine, triangle, or square wave output, and the selected waveform is presented to the output socket via the output amplifier. This has a basic gain of 5, but the amplitude of the output signal can be adjusted from a few millivolts to nearly 10 V pk-pk by varying R_{30}.

D.C. power is supplied from a circuit using two IC regulators U_4 and U_5, and the ±6 V lines are set by two 6.2 V voltage regulator diodes D_{13} and D_{14}. Using IC regulators makes the design of the power supply relatively simple and also the regulators have built-in overcurrent protection.

The operation of the triangle generator is as follows. Imagine that the comparator output has just switched positive to +10 V. This +10 V is fed back to both the integrator input and the reference switch. With R_{27} fully clockwise, a current of about 0.3 mA will flow through R_2 and this will charge the capacitor selected by S_1. The output of the integrator therefore moves negatively at a uniform rate. While the comparator output is at +10 V, diodes D_1 and D_4 in the reference switch will be off, while diodes D_2, D_5, and D_3 will be forward biased and on. Since D_2 is on, the voltage level at the comparator input (pin 3) will be clamped at +0.7 V by D_5. As the ramp from the integrator approaches −2.5 V, D_5 comes out of conduction, and the input on pin 3 of the comparator is determined by the potential divider formed by R_{28}, R_6, D_2, and R_5. When the ramp is at −2.5 V, the input on pin 3 is arranged by the setting of R_{28} to go through zero, and this causes the comparator output to switch to −10 V. At the integrator input, the current through R_2 now reverses, and the capacitor is charged in the opposite direction giving a positive going ramp at the emitter of Q_1 moving from −2.5 V to +2.5 V. Since the comparator output is at −10 V, the reference switch is in its opposite state with D_2 and D_3 off and D_1, D_4, and D_6 on. D_6 will hold the comparator input at −0.7 V until the ramp approaches +2.5 V, then D_6 comes out of conduction and the comparator switches to a positive output. The point at which the comparator switches is set by R_{29}. Thus at the integrator output a triangle waveform is generated with a ±2.5 V peak amplitude. The frequency can be trimmed by adjusting R_{26} by test.

This system is an example of sustaining feedback, since a connection from the comparator output to the integrator's input and the reference switch is essential for oscillations to be maintained. A break in any connection within the closed loop of these three blocks will cause oscillations to cease.

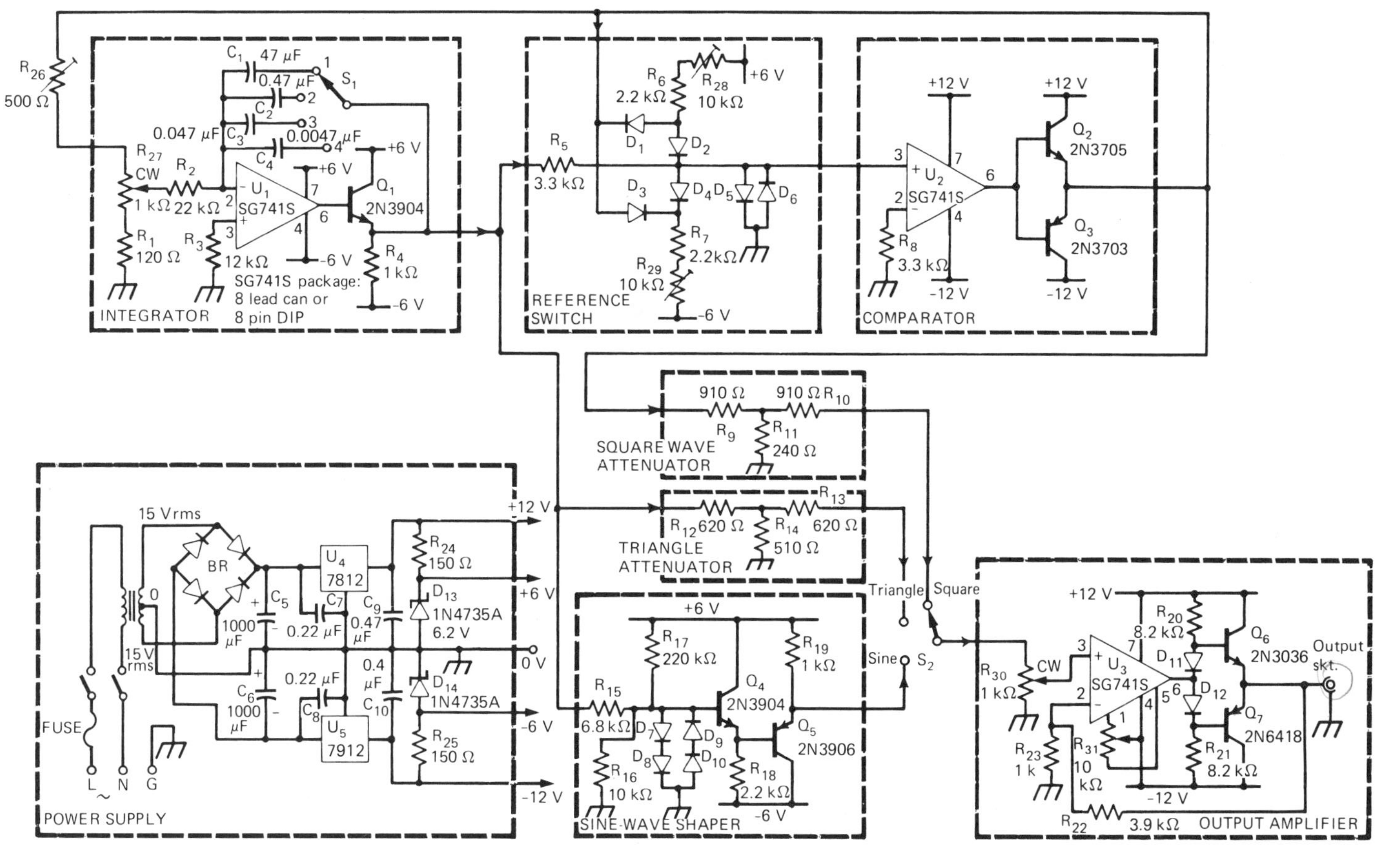

FIGURE 7.17 Function generator.

The symptoms for this will be that no output will be obtained with S_2 in square, triangle, or sine position. This could also be caused by a failure of the power supply, the output amplifier, S_2, or the three blocks within the feedback loop. If the unit fails with symptoms of no outputs, probably the first check should be to test for a triangle output at Q_1's emitter. This can be observed by a multimeter set to d.c. (with S_1 in position 1) or by using an oscilloscope. If no triangle wave is present, the fault is located in the three blocks that create the oscillation or in the power supply. Therefore, the next check would be to measure the +6 V and −6 V supply lines. Suppose though that a triangle wave is present. Then the fault must be in S_2 or the output amplifier. A suitable test would be to check for a signal at the wiper of S_2 or the top end of R_{30}. This will prove either S_2 or the amplifier to be at fault.

A failure of one block within the feedback loop can be deduced by a combination of measurement and testing. Testing may involve disconnection of the feedback loop and injecting a suitable test signal. The operation of the integrator can be checked by disconnecting the comparator output from R_{27} and injecting at this point a ±10 V amplitude square wave of suitable frequency (1 Hz with S_1 in position 1). A triangle output should appear at Q_1 emitter. Even if a square wave generator is not available, this test can be made at low frequency by alternately switching the integrator input between the +6 V and −6 V supply lines. The operation of the reference switch and the comparator can be tested in a similar manner. First the input wires to R_5 and the junction of D_1 and D_3 can be disconnected and then a square wave, or d.c. test level, can be applied to the junction of D_1 and D_3. When this test signal is positive (10 V), the voltage at D_5, D_6 should be +0.7 V. The comparator operation can be verified by using the same test. The comparator output should be +10 V when the test level from the reference switch is +0.7 V, and it should be −10 V when the test level is −0.7 V.

QUESTIONS

1. Sketch the time related waveforms that should appear at the outputs of the integrator, reference switch, and comparator if S_1 is in position 1.
2. State the probable location of a fault in the system for the following symptoms:
   - (a) No square waves obtainable.
   - (b) Sine wave output distorted.
   - (c) Reduce amplitude on all waveforms.
   - (d) All waveforms grossly distorted on positive half cycle.
3. Describe the symptoms for the following faults:
   - (a) R_{11} open circuit.
   - (b) Q_2 base/emitter open circuit.
   - (c) D_8 short circuit.
   - (d) D_7 open circuit.
   - (e) D_6 short circuit.
   - (f) R_{29} open circuit.
4. Following a failure, investigations

give the results shown below. In each case state the location of the fault and the faulty component(s).

(a) No output waveforms of any kind with S_1 in position 1.
Comparator output +10 V.
Reference switch output +0.7 V.
Integrator output zero volts.

(b) Very high frequency distorted waveforms obtained when S_1 is in position 4.

(c) Only fixed frequency waveforms can be obtained. The minimum frequency is about 12 Hz.

(d) No waveform of any kind with S_1 in any position.
Integrator output −5.5 V.
Comparator output − 10 V.
Reference switch output −0.6 V.
Pin 6 of U_1 +5.2 V.

5. Write a short troubleshooting guide for the following faults:

(a) R_{13} open circuit.

(b) U_4 open circuit.

(c) Q_5 base/emitter open.

7.8 EXERCISE: TEMPERATURE-MEASURING SYSTEM WITH DIGITAL READOUT (FIGURE 7.18)

A thermocouple is used as the input transducer to sense temperatures in the range from 0° C to +200° C. Thermocouples are devices made of two dissimilar materials, such as copper and constantan, that generate an output voltage proportional to the difference in temperature between the sense and reference junctions of the materials. The temperature range over which thermocouples can be used is very wide, above 1500° C for some pairs of materials, but the sensitivity is low, a typical output voltage being in the range of 40 μV per °C. A stable high-gain amplifier is therefore required to increase the thermocouple output signal to a level sufficient to drive a display.

In this example IC op-amp U_1 is used, type OP-07C (Precision Monolithics), which is connected as a differential amplifier. The OP-07C has a very low value of offset voltage and an extremely low offset voltage drift. Low drift in the op-amp is essential for the system to remain stable over long periods. Another requirement is that the thermocouple usually has to be positioned some distance from the amplifier and in an electrically noisy environment. This will result in large interference signals being picked up by the leads. A differential amplifier has to be used to give high rejection of common mode interference signals and to amplify the very small difference voltage between the two junctions of the thermocouple. The connecting leads are a twisted pair so that any electrical interference will be common to both wires.

The system is designed for use over the temperature range from 0° to +200° C which means that, with the resistor values shown, a full-scale voltage of about +2 V will appear at the differential amplifier output when the difference in temperature between the sense and reference junctions is 200°C. This output voltage is converted

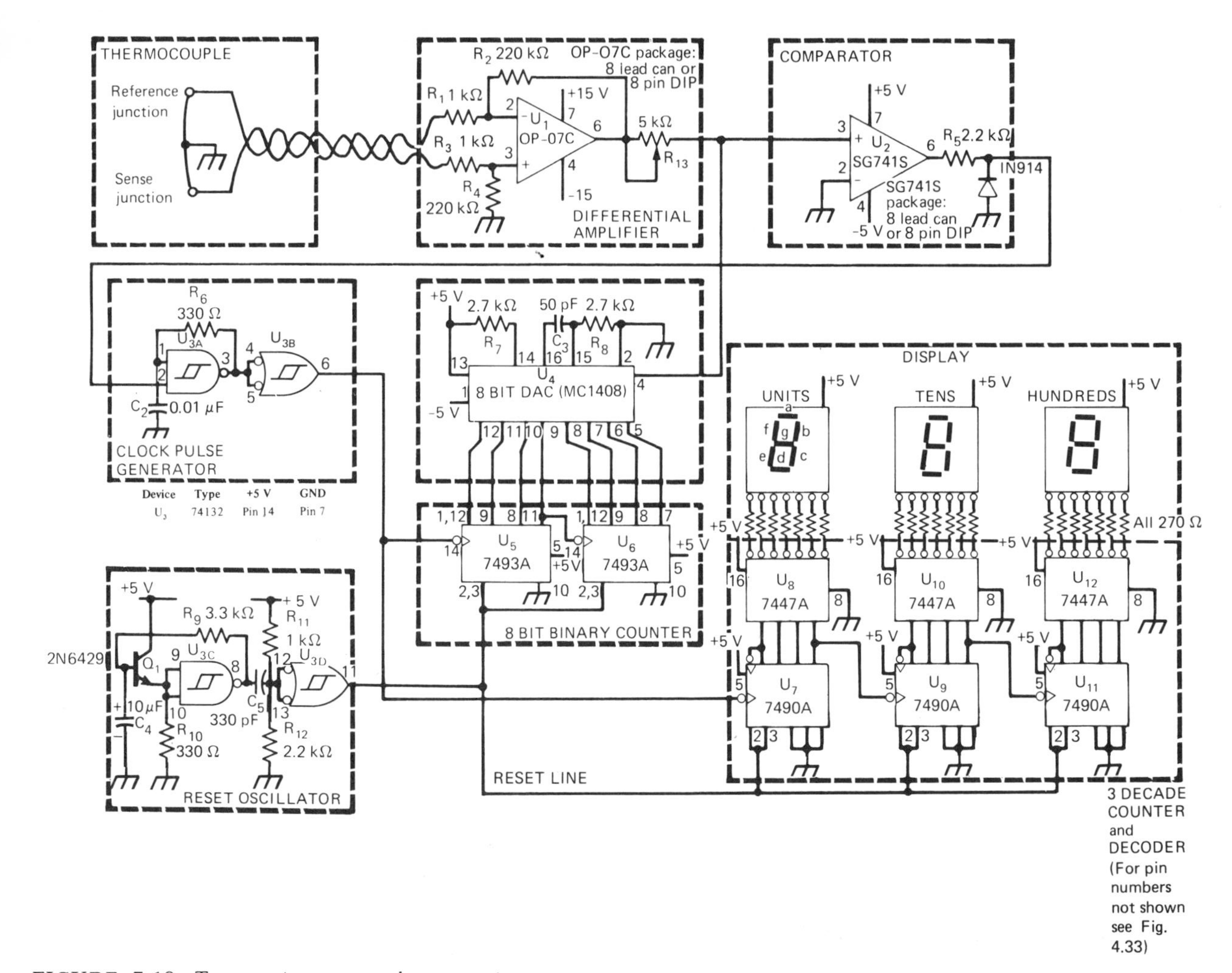

FIGURE 7.18 Temperature-measuring system with digital readout.

into a digital form by the rest of the system blocks for final display on 3 seven-segment LED indicators. When the system is fully calibrated, the input temperature will then be presented directly in digital form on the indicators. Basically this is achieved by the rest of the system blocks acting as a type of digital voltmeter.

A clock pulse generator, at a frequency of about 250 kHz, is used to drive two counters. The first counter is an 8-bit binary type using two 7493A 4-bit binary counters U_5 and U_6. This counter divides the clock by 256, and its binary outputs are presented directly to an 8-bit digital-to-analog converter (DAC). The second counter is made up of three decade counters U_7, U_8, and U_9 (7490A) each decoded by a 7447A to drive the seven-segment indicators. The output from the digital-to-analog converter is a negative going ramp, increasing in amplitude as the count in the binary counter increases. The ramp could therefore be made up of 255 steps with a full-scale value of about 2 mA set by R_7. The ramp output is compared directly with the current set up by the output from the thermocouple amplifier.

When the DAC's current ramp just exceeds the value of current set by R_{13}, the output of the comparator switches to a low value and stops the clock. In this way the number of clock pulses counted by the three decade counters and displayed by the LED indicators will be equal to the input temperature at the thermocouple sense junction. Because a relatively fast clock is used, the ramp time is typically less than 1 msec and no flicker will be observed in the display. For this reason the output from each decade counter does not need to be stored in latches as is usually the case in digital frequency meters. The input is resampled about every 500 msec by a reset oscillator. This low-speed oscillator running at about 2 Hz produces a short duration pulse to reset both counters to zero. This causes the comparator output to switch high, allowing the clock pulse generator to run, and the voltage at the output of the thermocouple amplifier is resampled. This sequence is continuous and is illustrated in Figure 7.19 where the output of the thermocouple is assumed to be about 460 mV equivalent to an input temperature of 52° C. In this case the ramp will consist of 52 steps, and the decade counters will count 52 pulses to display 052°C on the LED indicators.

For reasons of space the power supplies are not shown in the diagram. The voltages shown are required to be well regulated and free of ripple and each TTL package should be decoupled with a 0.1 μF ceramic capacitor.

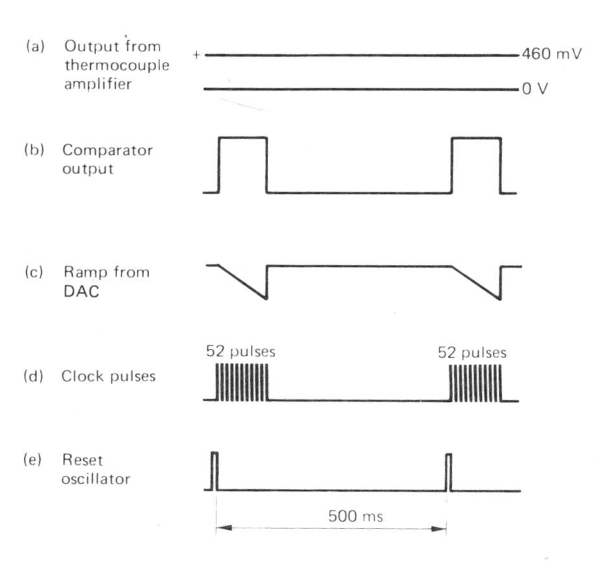

FIGURE 7.19 Waveforms in digital readout system.

The whole system can be calibrated by placing the thermocouple sense junction at a known temperature, say, steam at 100°C, and then adjusting R_{13} until the display is correct. The linearity can then be checked by varying the input temperature over the whole range. Note that the maximum temperature that could be indicated is 255° C.

Troubleshooting a system such as this is assisted because of the digital display. For example, if the maximum indication obtainable is 099° C, even though the input temperature is known to be above this, then the fault can only be located in the third decade counter U_{11}, its associated decoder U_{12} or the hundreds LED display. However, the most vulnerable component within the system is the thermocouple and its connecting leads simply because it is positioned some distance from the rest of the circuits and is therefore more prone to damage. An open circuit connection at the thermocouple would probably give the symptoms of full-scale display (i.e., 255) or an erratic display. This is because the open circuited lead, acting as an antenna, would pick up interference signals. A quick check to verify the operation of the rest of the system would be to temporarily short together the two input connections on the differential amplifier. The display should then be stable at or very near zero. Alternatively the input leads from the thermocouple could be disconnected and a d.c. reference voltage of, say, 1 mV applied to the differential amplifier input. The display should then read about 022 and should double if the d.c. reference input is increased to 2 mV. In this way a thermocouple fault can be distinguished from a fault in the rest of the system.

Suppose, however, that the thermocouple is proved to be sound. Then further checks on the rest of the system are required to locate the fault. Input to output tests can be made, measuring the output from the differential amplifier, the comparator, the clock pulse generator, and so on until the fault is located. Troubleshooting could, however, be speeded up by testing various portions of the system by applying appropriate test signals or levels. One such check would be to apply a logic 1 level (i.e., a high level) to the clock input, thus overriding the output from the comparator. The ramp should be of maximum amplitude, and the digital display should read 255. While doing this check, the system clock speed could be slowed down so that the progress of the counters may be more easily viewed. To do this a 20 μF capacitor could be temporarily connected in parallel with C_2 and a 200 μF capacitor in parallel with C_4. The speed of the clock pulse generator will then be about 100 Hz and that of the reset oscillator 0.2 Hz. The ramp will take about 2.5 sec full scale, and the system will be reset every 5 sec. Slowing down the clock in a digital system is often a useful procedure to adopt in an effort to locate some faults.

QUESTIONS

Since power supplies are not shown, assume that all power lines are correct.

1. What would be the typical resolution and short-term accuracy of the system? Give your answer in °C.
2. The system fails with all three LEDs

indicating zero. The comparator output is at +3.6 V. Which block within the system is suspect?

3. A fault develops such that the indicated temperature is always low. For example, when the input temperature is 100°C the indication is 042 and at 150°C it is 064. Which part of the system is suspect and what is the fault?

4. Explain briefly the symptoms that would result from the following faults:

   (a) R_6 open circuit.

   (b) An open circuit path to pins 2 and 3 of U_9.

   (c) R_5 open circuit.

   (d) Q_1 base/emitter short.

5. Explain the operation of

   (a) The reset oscillator.

   (b) The comparator.

6. A failure occurs such that the system only gives correct readings for temperatures up to about 128°C. What portion of the system is suspect and how could the fault be verified?

7.9 EXERCISE: SPEED CONTROLLER FOR A D.C. MOTOR (FIGURE 7.20)

The speed of permanent magnet d.c. motors can be controlled by varying the voltage applied to the motor. With a fixed supply this can be achieved by placing a variable resistor of high power rating in series with the motor, but this is inefficient, since power is wasted in the form of power loss in the resistor. A better method is to vary the amount of power applied by using a pulse width modulated switching circuit—the switch in series with the motor being on for a shorter period than it is off to give low speed, and vice versa for high speed. The switching rate, or frequency, is kept constant but the duty cycle is varied.

In this system a unijunction oscillator running at a frequency of about 400 Hz applies narrow positive pulses to the Schmitt input of the 74121 TTL monostable U_1. The width of the negative going (from logic 1 to logic 0) pulses from the $\overline{Q}$ output of the monostable is controlled by the potentiometer R_{15}. The pulse width can be varied from about 0.1 msec up to about 2 msec by rotating R_{15} clockwise. The monostable can be inhibited from operation by taking pins 3 and 4 to logic 1 by S_1, which will cause the $\overline{Q}$ output to be held in a high state. S_1 is therefore used as an on/off control for the motor. The pulses from the monostable are then fed to a power switch (Q_4) via a switch drive circuit. The purpose of the switch drive is to ensure that Q_4 is switched rapidly between the two possible states of either hard on and saturated, or fully off. This is essential so that power dissipation in the switch is kept to a low level.

While the output $\overline{Q}$ of U_1 is high, Q_1 conducts, and its collector will be low at about 0.2 V. Under these conditions Q_3 is held off, since its base current supply is diverted through D_1 into Q_1's collector. At the same time Q_2 is conducting, ensuring that the power switch Q_4 is held off with its base clamped to its emitter via Q_2. When the $\overline{Q}$ output of the monostable goes low,

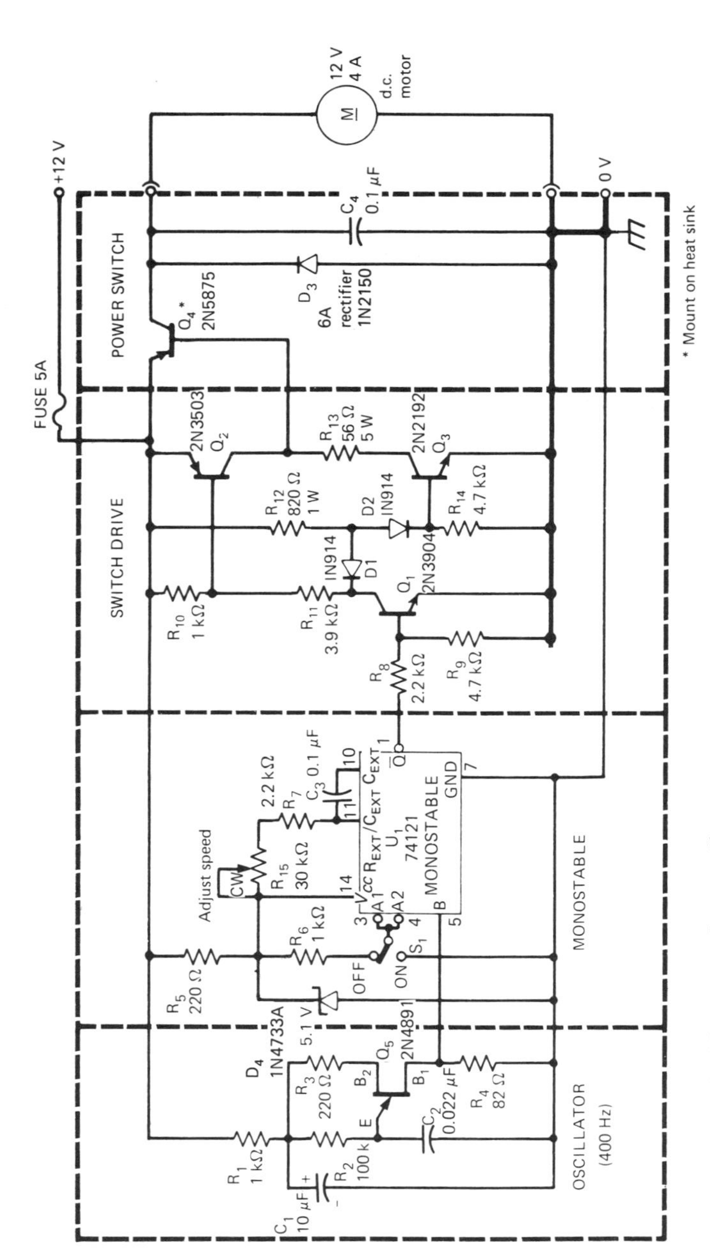

FIGURE 7.20 Motor speed controller.

Q_1 turns off thus turning off Q_2. Q_3 now switches on with base current supplied by R_{12} and this switches Q_4 hard on with its base current supplied by R_{13}. The motor receives nearly 12 V and draws current for the negative pulse period. On the trailing edge of the negative pulse Q_4 turns off, but D_3 conducts to limit any transients across the motor.

Faults in a system such as this are usually one of the following three types:

(a) Motor runs at high speed with no control.

(b) Motor does not run at all.

(c) Motor runs at low speed with no control.

Fault (a) could be caused by a failure in the power switch (for example, short circuit emitter to collector), or in the switch drive, or in the monostable. A quick test to isolate the fault would be to put S_1 into the OFF position and then to measure the $\overline{Q}$ output of the monostable. This should be a high level greater than +2 V. If this is correct but the motor still runs, the next check could be at the collector of Q_3, which with the stated conditions should be +12 V. The fault can then be rapidly isolated to one block.

If the system fails as (b) the fault could be in any of the blocks. To isolate the fault to one half of the system, Q_1's base/emitter could be temporarily shorted. This should turn off Q_1 and Q_2, force Q_3 and Q_4 to conduct, and the motor should run at maximum speed. If it does, then the fault has to be in either the oscillator or the monostable.

QUESTIONS

1. Which test instrument would be most suitable for troubleshooting the system?
2. List the possible faults within the switch drive block which would cause the motor to run at maximum uncontrolled speed.
3. Sketch the time-related waveforms you would expect to find with R_{15} fully anticlockwise and clockwise at
   (a) The emitter of Q_5.
   (b) B_1 of Q_5.
   (c) The $\overline{Q}$ output of U_1.
   (d) The collector of Q_3.
4. State the symptoms for the following faults:
   (a) R_{13} open circuit.
   (b) Q_1 collector/emitter short.
   (c) R_{15} wiper open circuit.
5. If the system fails with the motor running at maximum speed, suggest a simple temporary modification that can be made to isolate the fault to one half of the system.
6. List the faults within the monostable block that would cause the motor to run at maximum uncontrolled speed.
7. The fuse is found to have failed and a replacement blows very soon after the +12 V supply is reconnected

and switch S_1 is moved from the OFF to ON position. The motor does not run. What is the most likely cause of this fault?

8. The system fails with the symptoms that the motor will only run at low speed even with R_{15} fully clockwise. The motor can be stopped by setting S_1 to the OFF position. Describe with supporting reasons the block which is faulty and the type of fault.

9. The motor fails to turn at all and the motor itself is suspected of being open circuit. As no other motor is available, devise an alternative method to test the rest of the system.

Answers to Exercises

CHAPTER 1

1. (a) *Specification for TTL power supply*

Output voltage	5 V ± 50 mV
Output current	1 A maximum
Current limit	1.25 A
Load regulation	Better than 0.5% from zero to full load.
Line regulation	Less than 1% change for 10% a.c. power line change.
Output ripple	Less than 25 mV pk to pk at full load.
Temperature coefficient	Better than ±250 ppm per °C.
Temperature range	- 10°C to +40°C.
Overvoltage protection	Crowbar trip at 6.75 V.

(b) *RF generator*

Frequency range	10 kHz to 100 MHz
Frequency accuracy	±2.5% of dial setting
Stability	7 parts in 10^5 in 1 hour
Output amplitude	2 μV to 200 mV rms (adjustable by attenuator)
Output impedance	50 Ω
Amplitude modulating frequency	400 Hz
Degree of modulation	zero to 50%
AF output	2 V pk-pk

2. (a) *Typical unijunction data* (2N4891)

Maximum reverse emitter voltage	30 V

Maximum interbase voltage	35 V
Peak emitter current	1 A
Maximum rms emitter current	50 mA
Maximum power dissipation	360 mW
Intrinsic stand-off ratio	0.55 to 0.82.
Interbase resistance r_{BB}	4 kΩ to 9.1 kΩ
Peak point emitter current	5 μA max.
Valley point emitter current	2 mA min.

(b) *Thyristor short form data*
(Values for 2N6333 SCR)

V_{RRM} maximum repetitive reverse voltage	50 V
$I_{T(av)}$ maximum average forward current	2 A
V_T forward voltage drop	2.5 V at 3.9 A
V_{GT} maximum gate trigger voltage	0.8 V
I_{GT} maximum gate trigger current	0.2 mA
I_H maximum holding current	5 mA

4. The test specification should include:

   (a) A list of the test equipment
   Multirange meter
   5 V regulated power supply
   Digital frequency meter
   Oscilloscope (bandwidth 10 MHz)

   (b) Continuity and resistance tests.

   (c) Power consumption check—each 7490A TTL IC requires about 32 mA.

   (d) Measurement of output frequencies using digital frequency meter: 1 MHz, 100 kHz, 10 kHz, 1 kHz, all with a tolerance of ±0.05%.

   (e) Check that all outputs are within TTL specifications.

5. (a) Multirange meter.
   Digital voltmeter.
   Standard load.
   CRO bandwidth 50 MHz complete with low capacitance probes.

Digital frequency meter.

(b) Multirange meter.
Standard load (50 Ω or 600 Ω).
Digital frequency meter.
Distortion meter.
CRO.

(c) Multirange meter.
Digital voltmeter.
Waveform generator (sine, triangle, square) 1 Hz to 20 MHz.
Calibrated attenuator.
D.C. calibration source.

CHAPTER 2

1. $FR = 0.06 \times 10^{-6}$ per hour.
2. $MTTF = \frac{1}{FR} = 1.66 \times 10^7$ hours
3. Total $FR = 25 \times 10^{-6}$
   $\therefore$ MTBF = 40 000 hours
4. MTBF = 400 000 hours
   $R = 0.9753$ or 97.53% for a 10 000 hour operating period.
5. System reliability = 0.9063 or 90.63% for a 1000 hour operating period. To find the $MTBF_{system}$ first find each FR, add the FRs, and then find the reciprocal.

CHAPTER 4

Exercise 4.6 Car Seat Belt Alarm

1.

D	$\overline{B}_d$	P	$\overline{B}_p$	U_{2B} (Pin 12)	
1	0	0	0	1	Driver without belt fastened; no passenger.
1	0	1	0	1	Driver and passenger with neither belt fastened.
1	0	1	1	1	Driver without belt fastened; passenger with belt fastened.
1	1	1	0	1	Driver with belt fastened; passenger without belt fastened.

2. U_{2A} output "stuck at 0" or a short from pin 8 to pin 7 (ground) on U_2.

3. (a) The LED indicator will remain off. The output of U_{2A} at pin 8 of U_2 will be found to be correct, i.e., logic 0 if the input conditions are not satisfied.

   (b) An open circuit on a TTL input allows that input to assume a logic 1 (or high) state. Thus, pin 13 of U_{2B} will be logic 1, and the alarm will not be operated when a passenger fails to fasten his or her seat belt.

   (c) A logic 0 should only occur from the output of U_{1A} when the driver's belt is fastened. Therefore, if U_{1A} is "stuck at 0," the LED will not switch on if the driver starts the car without fastening the belt.

4. The fault is located in U_{1B} and is either:

   (a) U_{1B} output on pin 6 "stuck at 1."

   (b) A short circuit to 0 V at $\overline{B}_p$ input.

   (c) An open circuit from pin 6 to pin 9 of U_1.

   To locate fault, first check that switch $\overline{B}_p$ goes to logic 1 when the belt is fastened, then measure levels at U_1 pins 4 and 5, 6, and then 9 in turn while actuating switch.

5.

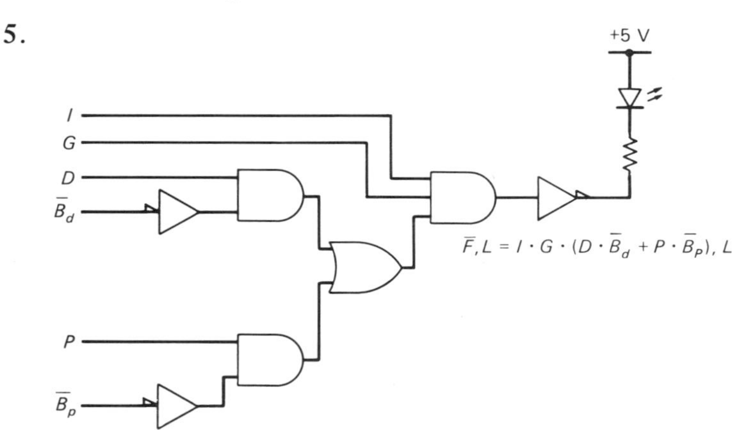

6. With no +5 V supply to U_1 the outputs of U_{1D} and U_{1C} will be in a high impedance state. This forces the inputs to U_{2B} to assume a logic 1, causing the output of U_{2B} to be logic 0. The LED will remain off under any input conditions.

Exercise 4.7 Contact Bounce Eliminator

1. +5 V supply lead to U_1 open circuit, or a failure of the clock pulse.
2. (i) Ground line (0 V) to U_1 open circuit.

   (ii) Position A of the switch open circuit or the connection to pin 1 open circuit or U_1 output at pin 3 "stuck at 0".
3.

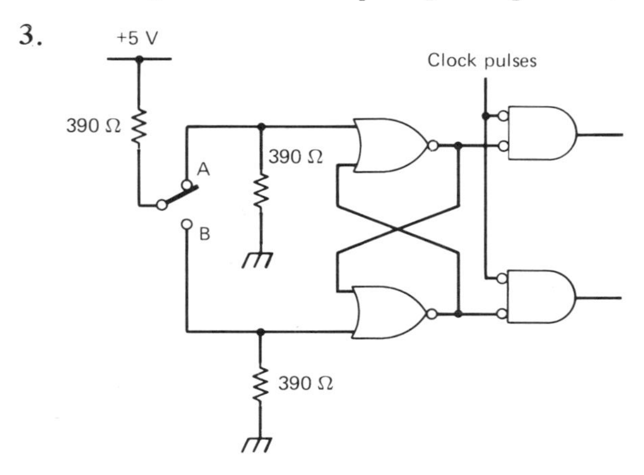

Exercise 4.8 Alarm Circuit Using CMOS Logic

1. Because NAND gates are used, a logic 0 on pin 2 of U_{1A} will hold the output on pin 3 at logic 1 irrespective of the logic state on pin 1. The output of the U_{1B}, connected as an inverter, will therefore be logic 0. This is a stable condition.
3. (i) Either R_3 open circuit or C_1 short.

   (ii) Either R_4 open or Q_1 base/emitter short circuit.
4. Temporarily connect a 10 kΩ resistor from Q_1 base to +9 V. The transistor should conduct and forward bias the LED.
5. C_1 open circuit.
6. (a) Since pin 7 of the IC is ground, a short from pin 6 to 7 will cause the output of U_{1B} to be permanently high. The LED will be on.

   (b) Without the reference of ground, the outputs of all four gates would tend to be in a "maybe" or high state, causing the LED to be on.
7. Use two gates from another 4011B IC to create a 400 Hz oscillator ($R_3{}' =$

180 kΩ, $C_1' = 0.01\ \mu F$). Remember to disable the unused gates in the IC by connecting their inputs to ground or to $+V_{DD}$. The 400 Hz oscillator can then be gated from pin 4 of the existing circuit and its output can be used to drive a small speaker via a transistor.

Exercise 4.9 Telephone Tone Generator

1. A JK bistable will act as a binary divider to the clock input when its J and K inputs are connected to logic 1.
2. (a) The 400 Hz oscillator will fail with its output "stuck at 1." Only the dial tone output will be correct.

   (b) The 25 Hz oscillator will fail with its output "stuck at 1." The dial tone will be missing.

   (c) The 1.6 Hz oscillator will fail with its output "stuck at 0." The output symptoms for the above faults are:

	Busy	Number Unobtainable	Dialling	Ringing
(a)	1.6 Hz	logic 1	correct	25 Hz highest frequency
(b)	correct	correct	logic 1	25 Hz missing
(c)	logic 1	correct	correct	logic 1 or logic 0

3. (a) A failure in the divide-by-4 circuit causing the Q output of U_{2B} to be "stuck at 0."

   (b) Since all other outputs are correct the fault can only be in U_{3A} or its connections.

   (c) The 400 Hz oscillator has failed with its output "stuck at 0." Possible causes are R_1 open circuit, Q_1 collector/base or collector/emitter short, or an open circuit connection to pin 2.

   (d) Open circuit lead from the 1.6 Hz oscillator (U_{1C} pin 8) to pin 1 of U_{3A}.

Exercise 4.10 Decade Counter with Decoder and Display

1. The input must be a TTL type signal that switches from zero (less than 0.4 V) to just greater than +2.4 V, generally with rise and fall times of less than 50 nanoseconds.
2. R_4 open circuit or pin 10 of the decoder "stuck at 1."
3. Either the reset switch is open circuit or there is an open circuit connection to pins 2 and 3.

4. The fault is in the 7490A decade counter with either an open circuit link between pins 12 and 1 or an internal open circuit to the ÷5 section. First check for a signal at pin 12 and then pin 1 which, in both cases, should halve the input frequency. Then test the ÷5 circuit of the 7490A by removing the link from 1 to 12 and injecting a test signal at pin 1.
5. (a) Segment g will not light when required.

   (b) The display will be off completely. Inputs to the decoder will be correct but all outputs from the 7447A will be high.

   (c) Pins 6 and 7 are the "reset to 9" inputs of the 7490A. With the connections open circuit, these inputs will assume a logic 1 state and the display will stick at 9.
6. An open circuit from pin 9 of the 7490A to pin 1 of the 7447A. This effectively puts a logic 1 to pin 1 of the decoder continuously.
7. The basic requirements are a light source and photocell arrangement. The output signal from the photocell must be "conditioned" into a TTL pulse by a shaping circuit such as a Schmitt trigger.

Exercise 4.11 Two-Phase Clock Pulse Generator

1. Frequency 250 kHz with 1 μsec pulse width.
2. To provide a low output impedance in both logic states.
3. Approximately +4.2 V.
4. Temporarily short the base and emitter of Q_1. Q_1 should turn off and the ϕ_1 output should switch to −22 V.
5. (a) ϕ_1 output will be correct but the ϕ_2 output frequency will increase to 500 kHz.

   (b) ϕ_1 output correct. ϕ_2 output stuck at −22 V.

   (c) No direct effect on circuit operation. Pins 5 and 13 of U_3 will be open circuit while S_1 is open, but an open circuit on a TTL input allows the input to assume logic 1, which therefore allows the circuit to operate. The only problem is that the inhibit line on pins 5 and 13 will be more susceptible to interference signals.
6. Fault A: R_5 open circuit or Q_1 base/emitter short.
   Fault B: Q_1 collector/base short.
   Fault C: Q_1 base/emitter open.
   Fault D: Q_2 base/emitter open.

Exercise 4.12 Digital Ramp Generators

1. Frequencies are approximately *a*) 160 Hz, *b*) 0.2 Hz.
2. (a) R_8 open circuit.
   (b) The 4-bit counter would be held in a reset state with all outputs at logic 0.
   (c) R_7 open circuit.
   (d) The ramp will consist of only 4 steps, but its frequency will be unchanged.
   (e) R_4 open circuit.

Exercise 4.13 Programmable Divider

1. R_2 and C_2. Pulse duration 0.5 μsec.
2. The maximum operating frequency would be limited to less than 70 kHz with $R_2 = 6.8\ k\Omega$.
3. (a) U_3 will be held in the reset to zero state. This means that a logic 0 will be present continuously on the 0 switch position of S_2. The circuit will only be capable of correct division from 1 to 9.
   (b) The inputs to U_{5A} will assume a logic 1 state, forcing U_{5A}'s output to be held at 0. U_{5B} will be closed and no trigger pulses can reach the monostable. Therefore there will be no output pulses.
   (c) The circuit will divide by the setting of S_2, i.e., 10, 20, 30, 40, etc., irrespective of S_1 setting.
   (d) The output of U_{5D} will be held at 0. There will be no output pulses.
4. (a) Open circuit link from U_3 pin 11 to U_4 pin 12.
   (b) Either the reset line to pin 2 of U_3 is open circuit or an open circuit exists from U_1 pin 11 to U_3 pin 14.
   (c) Open circuit link from U_3 pin 9 to U_4 pin 14.
5. Add another decade counter (7490A), a decoder (7442A), and switch (S_3). U_{5B} must now be a triple input NAND feeding the inverted signals from the three switches to the monostable input.

CHAPTER 5

Exercise 5.6 Audio-Frequency Amplifier Using a 741 Op-Amp

1. Voltage gain $\approx \dfrac{1}{\beta} = \dfrac{R_3 + R_2}{R_2} = 47.81.$

2. Sensitivity is the rms value of the input voltage required to give maximum undistorted output power. 110 mV rms to give 3.5 watts into 8 ohms.

3. (a) C_1 and R_1.

   (b) C_2 and R_3 and the slew-rate of the op-amp.

4. (a) 1.2 V.

   (b) -0.5.

   (c) 32 mA.

   (d) 370 mW.

5. Q_1 base/emitter open circuit or R_7 open circuit. This cuts off Q_3 and forces Q_2 to conduct taking the output to nearly -12 V. F_1 blows to protect the loudspeaker. Note that TP(A) has gone high.

6. R_2 open circuit or C_2 short circuit. Either of these faults will increase the negative feedback and reduce the gain of the amplifier to almost unity.

7. (a) Gross distortion on the positive half cycles of the output waveform.

   (b) There will be no feedback to control the gain nor the d.c. operating point. The output will be very unstable.

   (c) Q_1 and Q_3 will be cut off.
   Q_2 will conduct and the fuse should blow.
   Voltage at TP(B) - 11.3 V, at TP(A) 0 V.

   (d) Q_3 and Q_2 both overheating, since the forward bias between their base connections has been increased.

Exercise 5.7 Square Wave Generator Using a 741 Op-Amp

1. Approximately 3.3:1. [$(R_6 + R_3)/R_4$]
   To increase mark-to-space ratio, reduce values of R_3 and R_4 to 8.2 kΩ and increase R_6 to 75 kΩ.

2. (a) 16 V peak to peak. Depends on the saturation voltage at the op-amp output.
   (b) Approximately 16 μsec. This is because the 741 slew-rate is typically 1 V/μsec.
3. (a) No oscillations and output stuck at V_{sat}^{+}, i.e., +8 V.
   (b) No output with the switch set to position 2.
   (c) No oscillations and output stuck at V_{sat}^{-}, i.e., −8 V.
4. (a) Either D_1 or D_2 short circuit.
   (b) R_2 high in value.
   (c) R_6 wiper open circuit.

Exercise 5.8 Timer Unit Using 555s

1.

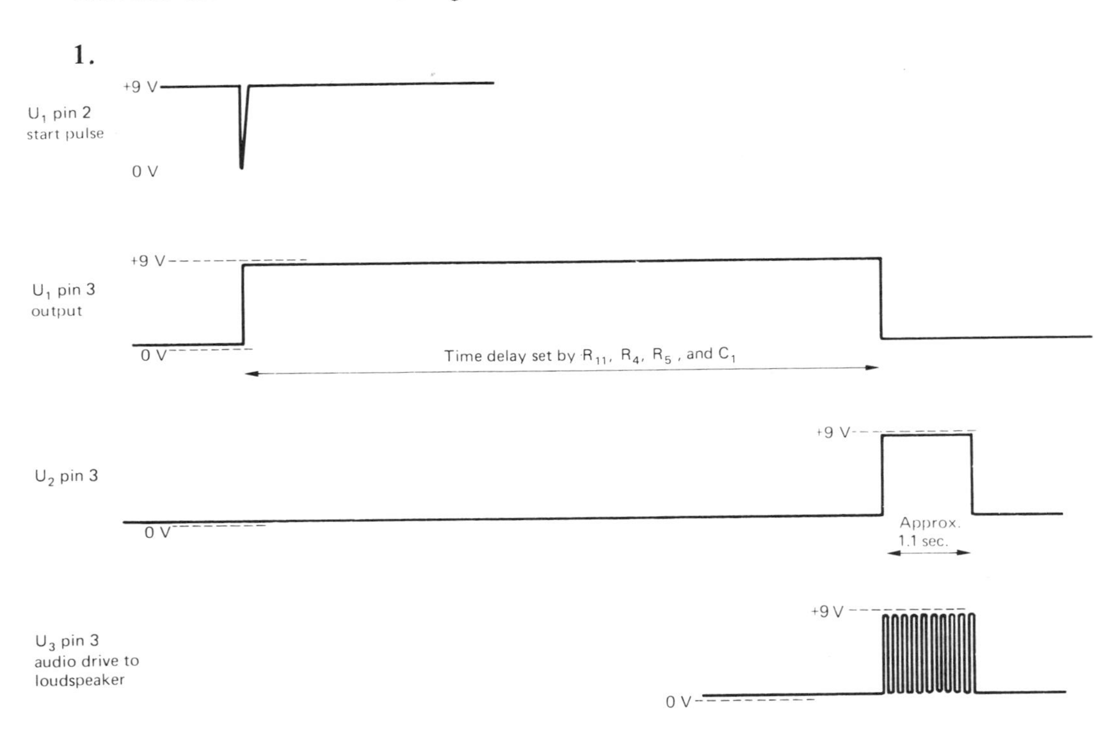

2. 600 Hz.
3. 1.27 seconds maximum; 0.94 seconds minimum.
4. (a) R_4 wiper open circuit.

(b) R_7 open circuit or C_2 short circuit.

(c) Reset line to U_2 open circuit. When the reset switch is operated, U_1 resets but U_2 instead of being held off is triggered on.

5. (a) The alarm tone will be on continuously and will not be reset.

(b) The output of U_1 will remain high after S_1 is pressed. R_4 will have no control and the alarm tone will not sound.

(c) Since U_2 cannot receive a trigger pulse when C_6 is open circuit, there will be no alarm tone at the end of the timing period. However, correct timing will be indicated by the LED.

Exercise 5.9 Wien Bridge Oscillator

1. Approximately 1.2 V peak to peak.
2. 1500 Ω with R_5 adjusted to zero.
3. Replace R_6, D_1 and D_2 by the thermistor and reduce R_3 to a value of 1.5 kΩ. As the output amplitude increases, the thermistor resistance will fall thereby increasing the negative feedback. Thus the overall gain of the circuit will stabilize at a value that just maintains oscillations.
4. (a) R_6 open circuit.

   (b) D_2 open circuit.

   (c) An open circuit from the junction of R_3 and pin 2 to the junction of R_6 and the diodes.
6. A ganged switch must be used to switch capacitors of value 0.01 μF in place of C_1 and C_2.
7. R_3 open circuit }
   D_1 or D_2 short circuit } These faults will give almost 100% negative feedback.
   C_2, R_2, or R_4 track open circuit or
   C_1 short circuit.

Exercise 5.10 $1\frac{1}{2}$ Hour Timer With Audible Output

1. Use the formula $t_p = K\ 4095\ C_tR_t$ where $K = 0.668$ when pins 11 and 12 are shorted together. $C_t = 4.7\ \mu$F, $R_t = R_1$ for minimum time.
   $\therefore t_p = 10$ minutes.
2. Measure the period of the timing waveform on pin 13 using a high impedance instrument. A chart recorder with 10 MΩ input impedance would

be most suitable. The oscillator time period should be equal to the total time delay of the circuit divided by (4095 K).

3. (a) To provide a holding current path for the thyristor when the buzzer operates.
   (b) To provide decoupling for the 5 V regulator inside the IC.
4. SCR anode to cathode short. Note that TP6 is at zero volts.
5. (a) SCR gate to cathode open circuit.
   (b) R_2 open circuit or SCR gate to cathode short.
   (c) Failure of the 5 V regulator in the IC.
   (d) R_4 open circuit or C_2 short circuit.
6. (a) The alarm buzzer will sound only once.
   (b) The timing period will be fixed at maximum.
   (c) Pin 1 is the trigger input to the IC and must be grounded to start the timing. Therefore the timer will remain in the off state. The alarm will sound continuously.

Exercise 5.11 Low Frequency Function Generator

1. (a) The frequency of the circuit will go high to about 8 Hz and no control of frequency will be possible.
   (b) The sine wave output distortion will be more pronounced.
   (c) The frequency will increase by a factory of 3 and the triangle wave amplitude will be $\frac{1}{3}$ its normal value.

   The sine wave output will be almost identical to the triangle wave.
2. Either D_3 or D_4 open circuit.
3. The frequency of the circuit will be much lower than normal. For example, even if the 100 Ω variable resistor (connected in error) is set to maximum, the frequency will only be about 0.1 Hz.
4. With D_1 short circuit the sine wave output will distort and limit on the negative half cycle.
5. R_2 open circuit.

Exercise 5.12 Pulse Generator Using a 555

1. 100 microseconds minimum.
   1.2 milliseconds maximum.

3. (a) R_4 open circuit or Q_1 emitter to B_2 open circuit.

   (b) Q_1 emitter to B_1 open circuit.

   (c) Either C_1 short circuit or S_1 wiper open circuit.

4. (a) No output pulses when S_1 is set to position 3.

   (b) The unijunction oscillator will function correctly but no pulses will be passed by Q_2. Q_2 collector will be at +9 V and its base at zero volts.

   (c) The output pulses will be of fixed narrow width as the 555 IC will follow the pulses from the unijunction oscillator.

5. Temporarily connect a 10 kΩ resistor from Q_2 base to +9 V. With a meter, check that Q_2 collector voltage falls to nearly zero volts (typically +0.1 V).

Exercise 5.13 Temperature Measuring Circuit

1. Meter protection. If a fault caused the 741 output to go into saturation, then R_7 will limit the current through the meter to a safe value.

2. If $V_Z = 5.1$ V, then
   $I_{R2} = 2.3$ mA and $I_{R1} = 5.73$ mA
   $\therefore I_Z = 3.43$ mA

3. To keep the heat generated within the thermistor to a very low value.

4. Meter reading almost zero. The 741 op-amp will behave as a voltage follower with the large amount of negative feedback supplied via R_5.

5. R_4 open circuit; R_2 open circuit; D_1 short circuit; or R_1 open circuit.

6. The voltage across R_3 will increase to about 70 mV and all the temperature readings will be 30% high.

7. Either R_5 open circuit which will open circuit the negative feedback loop, or R_3 open circuit causing a large negative input signal to be presented to the amplifier.

Exercise 5.14 Bridge Circuit with Null Indicator

1. $Q = 3.33$ kΩ, $P = 6.67$ kΩ.
2. The feedback line of the differential amplifier would be open circuit. The very high open loop gain of the op-amp would make null adjustment almost impossible.
3. (a) Q_2 collector/base short.

   (b) R_{10} open circuit.

   (c) D_1 or LED_3 open circuit.

   (d) R_{11} open circuit.

   (e) R_{12} open circuit.

   (f) Either Q_1 base/emitter open or short circuit or R_7 or LED open circuit.
4. Either an open circuit to the p-section of R_{13} or an open circuit connection to the bottom end of R_X. This will cause the amplifier to go into negative saturation.
5. (a) A null will be indicated irrespective of the setting of R_{13}.

   (b) A null will be indicated when R_{13} is in mid-position irrespective of the value of R_X.

Exercise 5.15 Active Filter Circuits

1. Low pass 1.6 kHz. High pass 339 Hz.
2. (a) The circuit changes to an amplifier with high pass characteristics. This is because no negative feedback path exists for frequencies higher than 1.2 kHz.

   (b) Oscillations at the center frequency.

   (c) The amplifier now has low pass characteristics but the gain is only 0.5.
3. 995 Hz.
4. Bottom end of R_6 open circuit.
5. (a) Filter will have low Q. No adjustment possible with R_6.

   (b) Low pass section of twin-tee will be inoperative causing the response of the whole circuit to be that of a high pass filter.

(c) There will be no positive feedback and the output from pin 10 will be unreferenced. The overall response will be flat with almost unity gain.

CHAPTER 6

Exercise 6.7 Simple D.C. Power Supply

1. Approximately 2 V pk-pk at 120 Hz.
2. 25 mA. Since the current through R_2 is 75 mA.
3. (a) Power dissipated by R_2 = 563 mW.
   Power dissipated by D_5 = 188 mW.

   (b) Zero load. Power dissipated by R_2 = 563 mW.
   Power dissipated by D_5 = 563 mW.
4. (a) D_5 open circuit. Note that, because a load of 150 Ω is connected across the −7.5 V line, the voltage at TP2 will only increase to about −9 V.

   (b) The transformer resistances appear normal, and 250 Ω is correct for the resistance from TP1 to 0 V. The fault is either D_1 or D_4 short.

   (c) Open circuit of the primary or the secondary transformer windings.
5. (a)

TP	1	2
MR	−14.9	0 V

   Additional symptom: R_2 running hot.

   (b) Circuit will have only half wave recitification. The voltage at TP1 will fall, the ripple frequency will be 60 Hz, and the ripple amplitude will be high.

   (c)

TP	1	2	
MR	−15.2	0	R_2 cold

Exercise 6.8 Discrete Linear Regulator

1. (A) R_5 open circuit.

   (B) Zener short circuit.

   (C) Zener open circuit.

(D) Either C_2 or C_3 short circuit (a short across the load). This has caused the current limit to operate with TP4 at 1.3 V.

(E) R_2 open circuit. The current limit operates with only a very small current flowing through Q_1. This causes all test voltages to be low.

(F) Q_1 base/emitter open circuit.

2. (a) 544 mW; $P_{CE} \approx I_C (V_{in} - (V_{out} + (5.6)(I_C)))$, b) 1.79 W; $I_C = \dfrac{0.7}{R_2}$

3. V_O max approximately 12.6 V.
   V_O min approximately 6.3 V.

4. (a)

TP	1	2	3	4	
MR	5.7	5.7	11	12.4	No control of output voltage with R_5.

(b)

TP	1	2	3	4	
MR	5.7	6.3	6.4	7.3	No control with R_5.

(c)

TP	1	2	3	4	
MR	5.7	6.3	13.7	15 V	Q_3 overheating.

(d) With C_1 short there will be zero volts on all four test points. R_1 will be warm.

Exercise 6.9 Op-Amp Power Supply

1. See Chapter 1.

2. (a) Since the unit is not loaded all the current flowing through R_3 must flow into Q_1

$$\text{Current through } R_3 = \frac{16-9}{R_3} = \frac{7}{56} = 125 \text{ mA}$$

(b) $I_B = \dfrac{I_C}{h_{FE}} = \dfrac{125}{40} = 3.125$ mA

(c) I_Z = current through R_2 + base current of Q_1 = $\dfrac{16-8.3}{820} + 3.125 \times 10^{-3} = 12.52$ mA

3. (A) D_6 short or Q_2 base/collector short circuit.

(B) D_5 open circuit.

(C) C_1 open circuit.

(D) Q_2 collector/emitter short or C_4 short.

(E) R_2 open.

(F) Q_2 base/emitter open circuit.

Exercise 6.10 An Inverter

2. (A) Either C_1 or C_2 open circuit. Both transistors are conducting.

   (B) Either Q_1 collector/emitter short circuit or Q_4 base/emitter short circuit.

   (C) R_4 open circuit.

   (D) Either R_2 open circuit or Q_1 base/emitter short. Note that Q_4 is conducting and will be overheating.

3. (a) The astable multivibrator will still oscillate but the amplitude of the waveform at TP4 will increase to 5 V instead of the normal 750 mV. This will cause the oscillator frequency to fall and will give an asymmetrical output. Q_4 will still switch on and off but the a.c. output will be reduced to about 25 V rms.

   (c) The oscillator will switch off with TP4 going high, Q_3 will conduct and be overheating.

TP	1	2	3	4	5	6
MR	0.15	0.7	5.8	0.75	4.8	6

Exercise 6.11 A General Purpose Laboratory Power Supply Using a μA723 Regulator

1. (A) Either R_8 wiper or R_5 open circuit.

   (B) R_2 open circuit.

   (C) D_2 short circuit or an internal short in the μA723 IC.

   (D) Q_1 collector/emitter short.

   (E) Q_1 or Q_2 base/emitter open circuit.

2. (a) 12.64 W.
   $P_{CE} = V_{CE} I_C$
   where $V_{CE} = 36 - (V_O + I_L R_7)$
   and $I_C = I_L$
   Note: base dissipation of $(V_{BE} I_B)$ has been ignored.

   (b) 39 W.

(c) 42.64 W.

3. (a) The noninverting input of the error amplifier inside the IC will be set to V_{ref}. Therefore, the minimum obtainable output voltage will be 7.2 V, not 3 V.

   (b) The output voltage will be fixed at 3 V and R_8 will have no control.

   (c) The current limiting transistor inside the IC will be conducting, and the output voltage will be held low even with R_8 at maximum.

Exercise 6.12 An HV Power Supply

2. (a) R_1 is a forward-biasing component to the bases of Q_3 and Q_2 to ensure that oscillations are maintained. C_2 bypasses R_1 so that, when power is first applied, the switching transistors are forced into oscillation.

   (b) R_2 and C_5 form a smoothing filter for the HV output. Note that R_2 is within the feedback loop and does not therefore degrade the output impedance.

3. (a) R_3 open circuit. This open circuits the feedback loop, and the control transistor is switched hard on, forcing the HV output to be maximum.

   (b) Q_1 base/emitter open circuit, *not* C_1 short as this should blow the fuse.

   (c) R_2 open circuit.

4. (a) The oscillator will be off and the HV output will be zero. However, Q_1 will be forward biased and the voltage across C_1 will be -9.5 V. Q_2 will be conducting and overheating.

   (b) The oscillator will be on but there will be no HV across C_5 or C_4. However, an a.c. voltage of about 600 V rms will appear across the secondary winding.

   (c) The HV output will be at maximum, in excess of 2 kV, and no control will be possible. The d.c. voltage at Q_1 emitter will be the same as its collector, i.e., -12 V.

Exercise 6.13 Remotely Controlled 5 V Logic Supply with Overvoltage Protection

1. The advantages are:

(i) Power dissipation in Q_5 and U_1 is kept to a low level, thereby giving higher efficiency.

(ii) By using switch drive to Q_5, a fault such as collector/emitter or short in Q_5 can be detected, so that the possibility of the 5 V logic supply being on when the control signal is off will be avoided.

2. A relatively high audio frequency is preferred because

   (i) it creates minimum disturbance and

   (ii) the values of smoothing filter components L_1 and C_6 can be kept small.

3. Wire an LED with 1 kΩ series resistor across the fuse F_1. When the crowbar operates and blows F_1 the indicator will light because of the current path provided by the thyristor.

4. (i) Protection for the base/emitter junctions of Q_1 and Q_2. During the switching cycle of an astable multivibrator the base voltage falls to nearly $-V_{CC}$. Most small signal transistors have maximum reverse base/emitter voltage ratings of about 6 V. Thus, a voltage higher than 5.6 V as the supply to the astable might cause the transistors to be damaged.

   (ii) R_{10} and D_4 act as an additional filter on the control input line ensuring that any ripple or interference signals at TP2 are kept to a low level.

5. Since $V_{GT} = 0.8$ V and $V_D = 0.7$ V (for a silicon diode), then the crowbar will operate when the voltage across R_{12} is 1.5 V. Let this voltage be labeled V_G. Now

$$V_G = \frac{(V_T - V_Z)}{R_{11} + R_{12}} R_{12}$$

where V_T is the voltage at TP5 that will trip the crowbar. Rearranging gives

$$V_T = V_G \frac{(R_{11} + R_{12})}{R_{12}} + V_z$$

$$\therefore V_T = 1.5 \frac{250}{150} + V_Z = 2.5 + 11 = 13.5 \text{ V}$$

6. (A) Either C_4 short circuit or Q_2 collector/base short.

   (B) U_1 open circuit.

   (C) Either L_1 open circuit or an open circuit junction on Q_5.

(D) R_9 open circuit.

7. Replace F_1 with a suitable indicator such as a 12 V bulb and series resistor. Ensure that the +12 V control input is switched off. Disconnect the top end of D_6 from TP6 and apply a positive d.c. voltage to the cathode of D_6. Gradually increase this d.c. test voltage and note its value that causes the indicator bulb to light. Switch off and reconnect D_6.

   Repeat method to determine trip voltage for D_5 sense circuit by disconnecting D_5 from TP5.

8. (a) It is unlikely that both sensing circuits will have failed; therefore, the fault must be confined to those components that are common to both sensing circuits. These are: R_1 open circuit, C_2 short circuit, R_2 open circuit, Q_6 gate cathode open or short, or Q_6 anode open circuit.

   (b) Either D_5, R_{11}, or D_1 open circuit or C_5 short circuit.

9. The fault can only be in the output overvoltage sensing circuit since the crowbar does not operate without U_1 in circuit. The fault is D_6 short circuit which will cause Q_6 to trigger the instant TP6 rises positive.

Exercise 6.14 Lamp Brilliance Control Using a VMOS Power FET

1. Under maximum brilliance condition the VMOS FET is on for 90% of the switching cycle. Therefore,

   Mean dissipation of VMOS $= V_{DS(on)} \times I_D \times 0.9$
   $= 2 \times 1.25 \times 0.9$
   $= 2.25$ W

2. To protect the CMOS gate input when point (x) goes positive.

3. An oscilloscope.

4. C_1 open circuit.

5. (a) The control of the lamp brilliance will be limited. As R_4 is varied the frequency, not the duty cycle, of the oscillator will change. The frequency change will be from about 250 Hz to 1 kHz.

   (b) The bulb will light and be at maximum brilliance as soon as the +12 V supply is switched on irrespective of the setting of S_1. The drain to source voltage of the VMOS FET will measure zero.

   (c) The bulb will be off and the output of U_{1B} of the oscillator (gate of VMOS FET) will be at zero volts under all conditions.

(d) The bulb will be off when S_1 is in position A but will be fully on when S_1 is placed in position B. No oscillation will occur from U_1B output on pin 4.

Exercise 6.15 Full-Wave A.C. Power Controller

1. These components form a filter circuit to prevent transients or "spikes" on the a.c. power line from triggering on the thyristors.
2. One thyristor is not switching on. Possible faults are:
   (i) Q_2 (or Q_3) gate/cathode or anode/cathode open circuit.
   (ii) R_4 (or R_5) open circuit.
   (iii) Either secondary winding to T_2 open circuit.
   (iv) One diode in the bridge rectifier open circuit.
3. (a) F_2 blown. Zero power in load.
   (b) Zero voltage across D_5. Zero power in load.
   (c) The unijunction Q_1 will trigger on very early in each half cycle. Therefore maximum power will be dissipated in the load and R_7 will have no control.
   (d) The control of power in the load by R_7 will become too coarse, and as R_7 is increased in value a point will be reached when there is zero power in the load.
   (e) Zero voltage across D_5 and zero power dissipation in load. R_1 will be overheating.
4. A CRO for the control circuit up to B_1 of the unijunction. Replace the load with a 100 W bulb to check full operation.
5. Either R_7 track or wiper open circuit.

CHAPTER 7

Exercise 7.4 Light Controlled Flashing Unit

1. (a) The fault can only be in the thyristor bistable circuit provided the lamp is good.

(b) Q_5 gate/cathode open circuit or anode/cathode open circuit.
R_6 short circuit
C_2 open circuit

2. (a) The operation of the system would appear normal, apart from the fault condition that the lamp could remain on when the l.d.r. is taken from a low to high light level. This would not occur in every case.

(b) Symptom

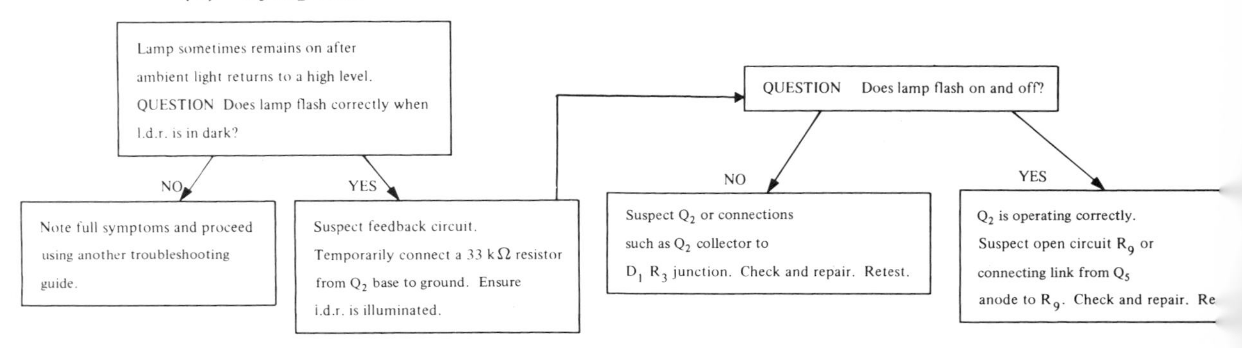

3. (a) With l.d.r. *open circuit.* The input to the amplifier will be zero volts. Q_1 will be off, and the lamp will flash continuously irrespective of ambient light conditions.

(b) With l.d.r. *short circuit.* The input to the amplifier will be highly positive. Q_1 will be on and the lamp off, irrespective of ambient light conditions.

4. Momentarily connect a 2.2kΩ resistor from Q_5 gate to +12 V. Q_5 should conduct and the lamp should light. Then momentarily connect the test resistor from Q_4 gate to +12 V; this should force Q_4 to conduct and the lamp should go out.

5. Q_1 has failed collector/base short circuit. The lamp will remain off irrespective of the light level on the l.d.r. However by temporarily connecting a 33 kΩ resistor from Q_2's base to ground, the oscillator will operate and the lamp will flash.

6. With the l.d.r. in low light conditions the output from the amplifier is high and C_1 charges via R_2, D_1, and R_3 toward +12 V. As the voltage on the UJT emitter rises, a time t is reached when this voltage just exceeds the peak point. The UJT triggers on and C_1 is rapidly discharged through R_5. This causes a short duration positive pulse to appear at B_1, and this is

used to trigger the thyristor circuit. The frequency of the UJT oscillator, if the effect of D_1 is neglected, is given by:

$$f \approx \frac{1}{CR \log_e \left(\frac{1}{1-\eta}\right)}$$

where $C = C_1$ (10 μF ± 20%)
$R = R_2 + R_3$ (21.3 kΩ ± 5%)
η = intrinsic standoff ratio for the UJT = 0.55 to 0.82.

The minimum frequency will occur when C, R, and η are all at their maximum values, i.e., C_1 = 12 μF, R = 22.4 kΩ, η = 0.82.

Then f_{min} = 2.17 Hz.

The maximum frequency will occur when C_1, R and η are at their minimum values, i.e. C_1 = 8 μF, R = 20.2 kΩ and η = 0.55.

Then f_{max} = 7.75 Hz.

Exercise 7.5 Over-Temperature Alarm System

1. The low frequency (1 Hz) oscillator has failed with its output on pin 4 "stuck at 1."

2. **(a)** Connect a push-to-make switch between point (A) and +9 V. When this switch is made, both oscillators will be gated on and the alarms should operate irrespective of input temperature.

   (b) Connect a medium value resistor (4.7 kΩ) in series with a push-to-make switch across the thermistor connection in the bridge. When the switch is operated the test resistor will be connected in parallel with the thermistor, the bridge will be unbalanced, and the alarms should operate.

3. Symptoms: both alarms fail to operate when input temperature is high.

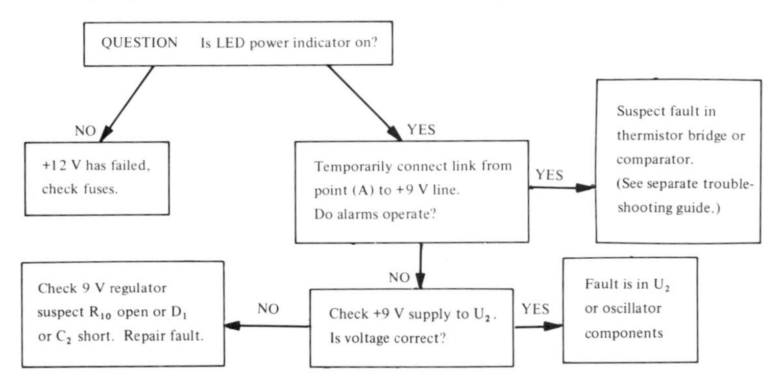

4. The only component that is connected directly across the +12 V line and might fail short circuit is C_1. This could be quickly verified by using an ohmmeter to measure the resistance (power off) across the +12 V line to 0 V with the 12 W bulb and loudspeaker removed.

5. (a) The inverting input to the op-amp (741) will be held positive causing the comparator output to be low irrespective of input temperature. The alarms will not operate when the temperature is above the trip value. However, taking point (A) to +9 V or connecting a 4.7 kΩ resistor across the thermistor connections will cause the alarms to come on.

   (b) With C_5 short the output of the audio oscillator U_{2D} pin 11 will be permanently "stuck at 0." Therefore, the lamp alarm will function correctly but no pulsed audio tone will be given.

   (c) Point (A) will drift positive taking the CMOS gates into their linear regions. The oscillators will operate erratically causing the lamp to switch on and off and an audio tone to be given irrespective of input temperature. To isolate the fault first ground point (A) and both oscillators should switch off. Then take point (A) to +9 V, and oscillators and outputs should be normal. The next test will be to check the output of the op-amp, and this will pinpoint the fault.

   (d) While the temperature is held low the alarms will be off. However, when the trip point is exceeded, the high level at point (A) will cause the output of the 1 Hz oscillator to switch to logic 1 and to remain at this level. The lamp will come on and will not flash, and the audio tone will sound continuously. The alarms will switch off when the temperature is reduced. It will be seen that the 1 Hz oscillator has failed and that the two gates are simply acting as inverters.

   (e) A d.c. current will flow through the speaker and R_9 continuously and R_9 will be running hot. No pulse audio tone will be available under any conditions, but the lamp alarm will function correctly. A quick voltage check across Q_2 from drain to source will show that this is zero volts, and with R_9 hot this will pinpoint the fault.

Exercise 7.6 Pulse Generator

1. An oscilloscope with a bandwidth of at least 10 MHz and good external triggering facilities.

2. Either **(i)** An open circuit +9 V connection to the wiper of S_2 or **(ii)** S_2 has failed open circuit in the FREE RUN position.
   Check by measuring the d.c. level (Y-channel of CRO switched to d.c.) on the wiper of S_2.

3. **(a)** Only square wave output obtainable with S_5 in the inverted position. No sync. pulse output to generator output obtainable.

   (b) No fine control of output pulse width available. The output pulse widths will be fixed at 200 msec, 20 msec, 2 msec, 200 μsec, and 20 μsec.

   (c) The generator will only produce output pulses with S_4 in the OFF position. No delayed output pulses will be obtainable.

   (d) Very similar symptoms as for **(c)** except that measurement will show no pulses from U_3 at S_{4B} wiper when S_4 is in the ON position.

4. Since sync. pulses are present then both U_1 and U_2 must be working correctly. The areas common to both square and pulse outputs are S_3 and U_5. If pulses and square waves can be obtained at S_{3B} wiper then U_5 or its connections is suspect.

5. Symptom: no sync. pulse output.

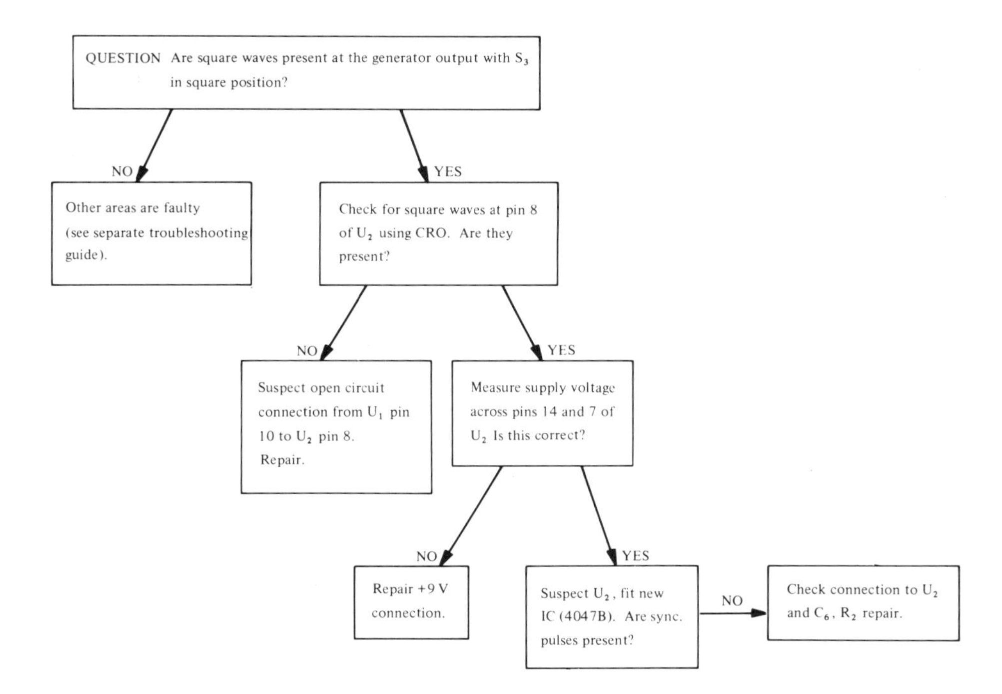

6. (a) Open circuit link from S_{4A} (OFF) to S_{4B} wiper or S_{4A} open circuit.
   (b) S_6 open circuit or more likely open circuit battery connections.
   (c) R_8 wiper open circuit.
   (d) Either an open circuit from U_1 pin 11 to S_{5B} or S_{5B} itself is open circuit.

Exercise 7.7 Function Generator

1.

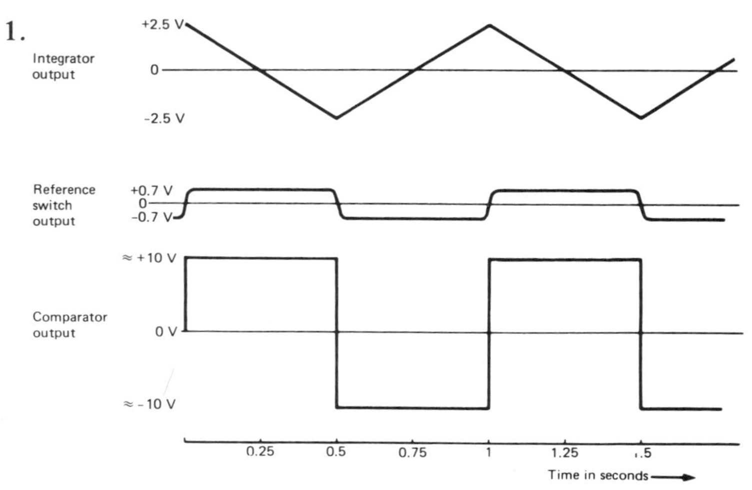

2. (a) Square wave attenuator or its connections.
   (b) Sine-wave shaper. (c) Output amplifier. (d) Output amplifier.
3. (a) The square wave output will be high because only a small attenuation will take place in the square wave attenuator.
   (b) With Q_2 open circuit the impedance from the comparator to the integrator will be very high on positive half cycles. This will cause the output from the integrator to become a sawtooth wave shape and the output from the comparator to become very asymmetrical, the wave being a series of negative pulses at low frequency.
   (c) The positive half cycles of the sine wave output will be reduced in amplitude and will be distorted.
   (d) The positive half cycles of the sine wave output will increase in amplitude and will resemble the triangle wave.
   (e) No output waveforms obtainable. Comparator output at about +10 V and integrator output stuck at about −4.5 V.

(f) The triangle wave will be reduced in amplitude and will be negative going from zero volts. This will result in a distorted sine wave output.

4. (a) C_1 short circuit. (b) C_4 open circuit.

(c) Either R_1 open circuit or an open circuit connection to the bottom end of R_{27}.

(d) U_1 base/emitter open circuit.

5. (a) Symptom: no triangle wave output.

(b) Symptom: no output waveforms of any kind obtainable.

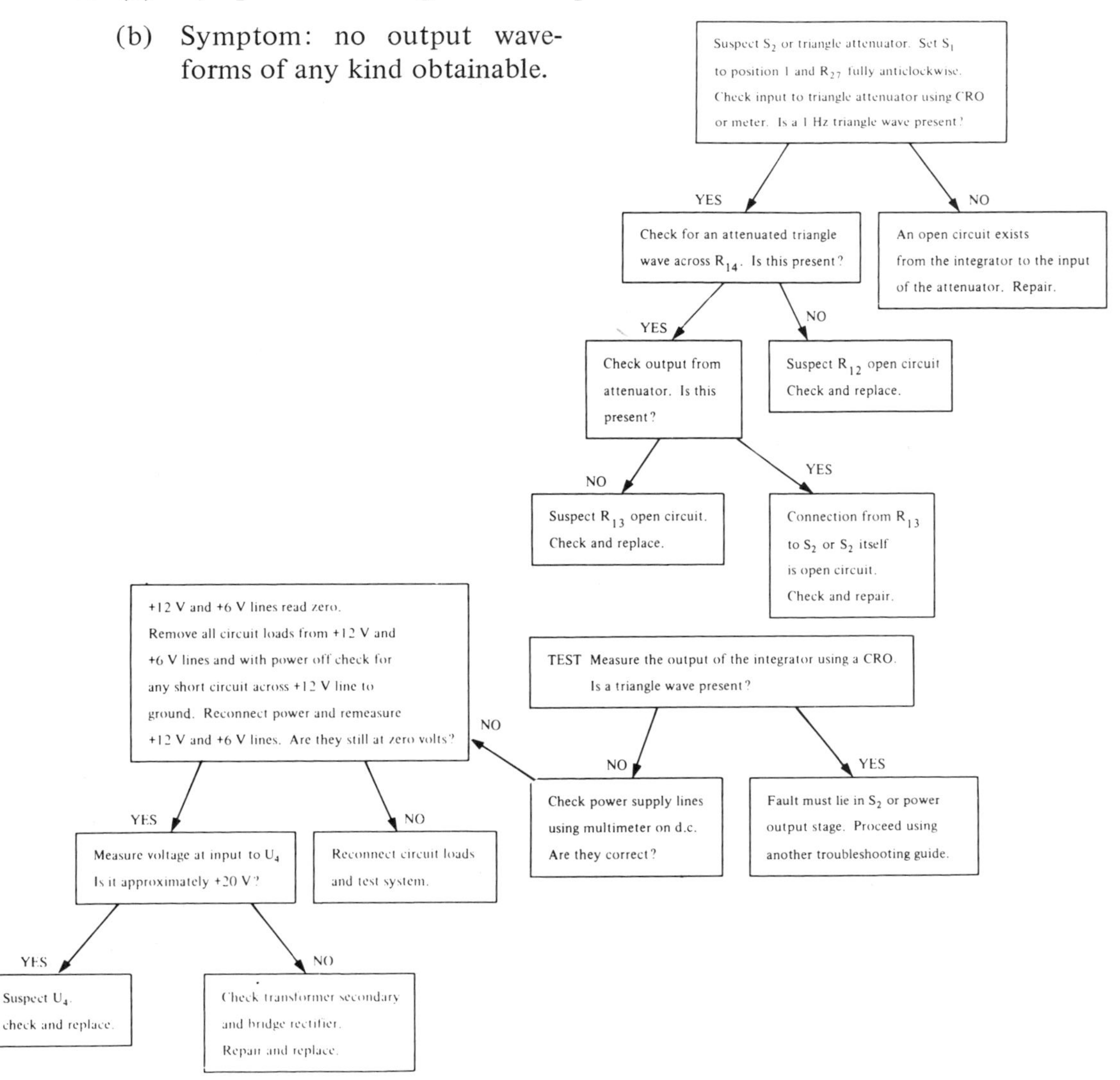

(c) Symptom: no sine wave output.

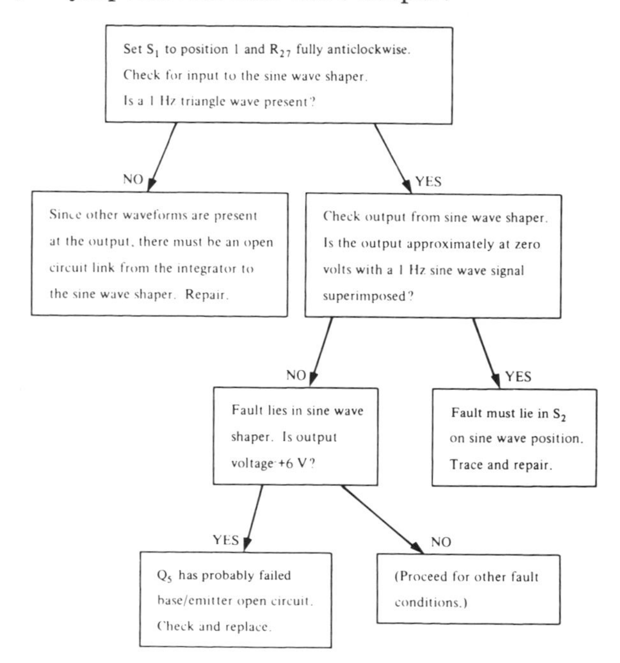

Exercise 7.8 Temperature Measuring System with Digital Readout

1. Resolution ±1°C since the display reads in 10 mV steps. Accuracy depends naturally on the thermocouple itself, but could be better than ±2.5% of display.
2. Since the comparator output is high the clock pulse generator must have failed.
3. R_{13} wiper open circuit. The rest of the system is functioning but an open circuit of R_{13} wiper will cause all readings to be low.
4. (a) No output from the clock generator and its output will be "stuck at 1." All LED displays will be indicating zero.

   (b) If the reset line to U_9 goes open circuit then pins 2 and 3 of that IC will assume a logic 1 state. The tens and hundreds displays will stick at zero.

   (c) The input to the block will assume a logic 1 state and cause the clock to run continuously. The display will be erratic and may indicate any number up to 999.

(d) No reset pulses will be present and again the display will be erratic.

6. The fault is in the digital to analog converter, the 8th bit is missing.

Exercise 7.9 Speed Controller for a D.C. Motor

1. An oscilloscope.
2. Q_1 base/emitter or collector/base open circuit or D_1 open circuit or Q_3 collector/emitter or collector/base short circuit or R_8 open circuit.
3.

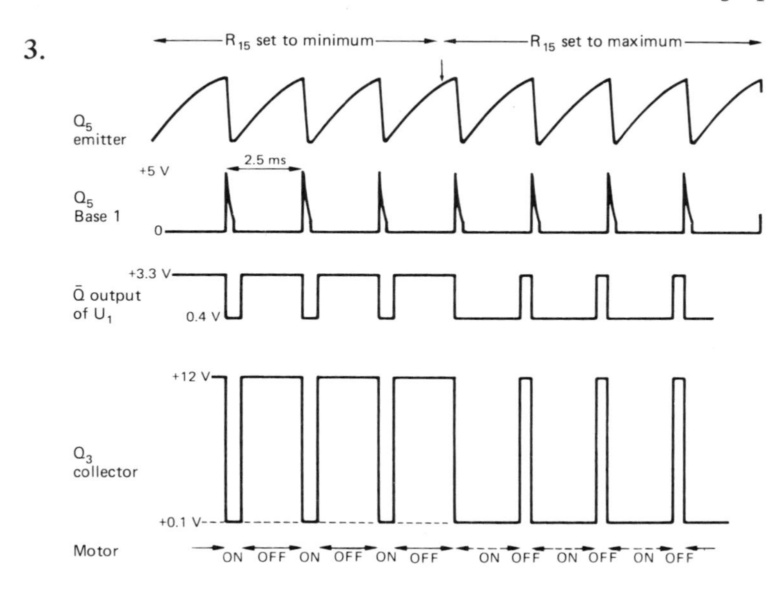

4. (a) Motor off. No base current drive to Q_4.

   (b) Motor off. Voltage at the junction of D_1D_2 will be 0.7 V.

   (c) The motor will run at full speed. The on/off switch will function correctly but R_{15} will have no control.

   The pulse width output from the monostable will be at maximum value.

5. Temporarily connect a 4.7 kΩ resistor from Q_1 base to the +12 V line. This should turn on Q_1 switch off Q_3 and Q_4, and the motor should stop. If this is true, then the fault can only be in the oscillator and monostable circuits, but if the motor continues to run then the switch drive or the power switch is at fault.

6. R_{15} wiper or track open circuit or R_5 open circuit or D_4 short circuit or R_7 open circuit or $\overline{Q}$ output of U_1 "stuck at 0."

7. An overload current is taken only when pulses are applied to the motor via the switch. The most likely cause is that the motor has stalled or seized and is therefore taking a large current from the supply. Other possibilities are D_3 or C_4 short circuit.

8. Since the motor can be stopped by S_1 the fault must lie in the monostable or oscillator. To give the symptoms of fixed low speed, either the monostable output pulse is limited to a narrow width or the oscillator frequency has changed to a much lower value. The latter is the more probable and could be caused by R_2 going high in value.

9. Replace the motor by a 12 V, 1 A lamp. The brilliance of the lamp should be varied when R_{15} is rotated over its range.

Index